T0136430

# THUNDER BELOW!

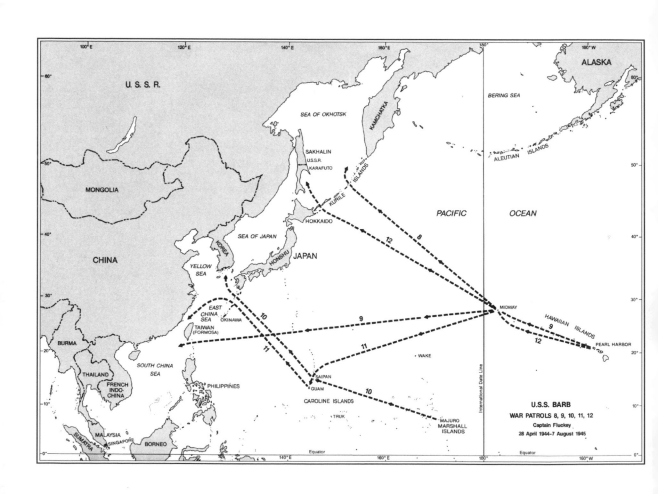

U.S.S. BARB
WAR PATROLS 8, 9, 10, 11, 12
Captain Fluckey
28 April 1944-7 August 1945

# Thunder Below! The USS *Barb*

## Revolutionizes Submarine Warfare in

## World War II

**Eugene B. Fluckey**
**Rear Admiral, USN (Ret.)**

**University of Illinois Press**
**Urbana and Chicago**

Frontispiece (hardcover edition only): "The 'BARB' Strikes," courtesy of General Dynamics' Electric Boat Division

©1992 by the Board of Trustees of the University of Illinois
Manufactured in the United States of America
16  17  18  19  C  P  19  18  17  16
∞ This book is printed on acid-free paper.

Library of Congress Cataloging-in-Publication Data
Fluckey, Eugene B., 1913–
Thunder below! : the USS Barb revolutionizes submarine
warfare in World War II / Eugene B. Fluckey.
p.  cm.
Includes index.
ISBN 0-252-01925-3 (cloth : alk paper)—ISBN 0-252-06670-7 (pbk.)
ISBN 978-0-252-01925-8 (cloth : alk paper)—ISBN 978-0-252-06670-2 (pbk.)
1. World War, 1939–1945—Naval operations—Submarine. 2. Barb (Ship).
3. World war, 1939–1945—Naval operations, American. 4. World War,
1939–1945—Personal narratives, American. 5. Fluckey, Eugene B., 1913–   .
I. Title.
D783.5.B36F58      1992
940.54'5974—dc20      91-45582
CIP

To Margaret, my beloved Scot, who inspired me to write this story of the submarine *Barb* and who assisted me in my research in Tokyo, China, Taiwan, and beyond. Critiquing every word, her suggestions have been invaluable.

To Marjorie, my beloved, diabetic wife, who, half a world away during World War II, absorbed my entire bucket of fear, so that I could grapple intelligently with a concrete enemy to shorten the war. God rest her brave soul.

To Barbara, my beloved daughter, who as a six-year-old stood my watch for me and twice saved her mother from comas.

To each of my shipmates in the *Barb* for the extraordinary heroism above and beyond the call of duty that you displayed.

To the wives, families, and sweethearts who boosted our morale by using that long arm of love, correspondence.

# Contents

# Illustrations

# Preface

This is the history of the U.S. submarine *Barb* while she was under my command during World War II. It is written in the sense of the lively nonfiction that it was.

Reconstructed conversations abound to breathe life into the story. These have a factual basis from the written log, reports, messages, letters, notes, oral histories made during and after World War II, stories that the Navy Department's public relations office asked me to write for their publications, reunions, interviews, and an illegal diary kept by a shipmate. One crew member who reviewed the manuscript remarked, "You must have had a tape recorder."

Regarding the events depicted on other submarines, their stories are taken from reports—both Japanese and their own—personal conversations with the captains involved, and consummate knowledge of how they operated.

There are two divisions for the Japanese commands. The naval warships made voluminous "after action" reports that depicted the gallant and courageous fights to save their ships. The actual names of the captains and the convoy commanders are used where made available. The individual merchant ships' actions were derived from my personal observation during the battles, interrogation of Japanese prisoners we picked up, and reports made to the shipping companies that were obtained by the Japanese Department of War History.

Our outstanding wolfpack commander, Admiral Elliott Loughlin, verified the *Barb*'s actions on three patrols.

Conversations between Fleet Admiral Chester Nimitz and Admiral Charles Lockwood, the Submarine Force Commander, were related to me while I was Nimitz's personal aide when he was Chief of Naval Operations after the war.

All messages, letters, and the diary are verbatim.

My extensive research occupied over ten years. Japanese officer friends assisted me while I was in Tokyo, and afterward through correspondence. This enabled me to unfold the factual Japanese side of events, where known. In June 1991 I conducted my research in China.

My objective in *Thunder Below!* has been to provide the reader with the best and most complete account of every attack, whether against land or sea targets, as reported by those concerned. There are still mysteries to unravel—due to the disappearance of certain records—but I have not fabricated this history. I was there.

When I joined the *Barb* in January 1944, war with Japan was in the limited offensive stage, Japan having reached the maximum limits of its empire. U.S. submarine warfare was to be intensified to strangle Japan. The Central Pacific drive was ready to begin, as was the movement of the Southwest Pacific forces into New Guinea. The South Pacific forces had taken Guadalcanal; the North Pacific forces were to take the Aleutians. Now we would begin the tightening of the steel belt around Japan. The Chinese Army could not be deployed for it was engaged in a major Japanese offensive in China. The overall Japanese and Allied forces were relatively equal. Unscathed, U.S. production was steamrollering along at flank speed.

# Acknowledgments

I am deeply indebted to the following people and organizations who assisted me in World War II and in over ten years of research for this book:

Publishers Robert Hale Ltd. of London for permission to quote from *Bamboo and Bushido* by Alfred Allbury, a British soldier and prisoner of war.

General Dynamics' Electric Boat Division, Groton, Connecticut, for their excellent construction of the submarine *Barb* in 1942. Their outstanding workmanship enabled her to withstand the pounding punishment meted out by enemy warships and aircraft. Though her test depth was designed for only 312 feet, I took her much deeper, with confidence, to evade the enemy. I also greatly appreciate their courtesy in permitting me to use their famous lithograph "The 'BARB' Strikes" as the frontispiece to this book.

Leading Torpedoman Charles Tomczyk of New Jersey for permitting me to use sections of his illegal diary. They portray a variation on events I too had witnessed.

Neville Thams and Jack Flynn, Australian Army prisoners of war and close friends, for permitting me to quote from their speeches.

*Barb* officer Everett P. "Tuck" Weaver and Rita Weaver for the innumerable hours they spent researching, investigating, writing letters, obtaining translations and stories, editing, and tracking down evidence through the maze of debatable history. *Barb* officers Max Duncan, Dick Gibson, and Bob McNitt (the executive officer during war patrols 8 and 9) for their research and assistance in reproducing conversations accurately. Paul

Saunders, chief of the boat, for his help in verifying actions and conversations that took place in the *Barb*. Jack Dittmeyer for his assemblage of shipmates' addresses, and my other shipmates for photos and anecdotes. Professors Henry Sage, Karel Montor, and Frank Gambacorta, Admiral Robert Long, and Linda O'Doughda for their astute editing. Top World War II combat artist Al Murray for the use of one of his many portraits. The Naval Imaging Command's John Lewin for resurrecting the *Barb's* World War II film. William Crispin for his astute assistance with patrol area charts.

Dr. Dean Allard and his capable team for their tremendous assistance throughout my years of research in the Naval Historical Center's Operational Archives and Contemporary History Division.

My Japanese friends, Admirals Morinaga and Inouye, for arranging my month-long research in the Tokyo Naval History Archives. I give my utmost appreciation to Captain Shin Itonaga, Commander Koike, and Lieutenant Yoshimatsu for digging out records that correct a number of errors made by the Joint Army-Navy Assessment Committee. Although difficult to find, authentic documentation was available in many cases.

Fleet Admiral Nimitz and Vice Admiral Lockwood, both lifelong friends, related to me their connection with the *Barb's* activities during patrol. Both urged me to write this book.

Robert Sinks, a naval China coastwatcher during the *Barb's* attack in Namkwan Harbor, and his son Murray, who made my extraordinary revisit for verification possible in 1991.

The USS *Barb* on arrival home at New London, Connecticut, 21 September 1945.

## USS *Barb* (SS-220)

| | |
|---|---|
| Keel laid | 7 June 1941 |
| Launched | 2 April 1942 |
| Commissioned | 8 July 1942 |
| Decommissioned | 13 December 1952 |
| Loaned to Italy | December 1954 |
| Renamed the *Enrico Tazzoli* (S-511) | |
| Scrapped | 1975 |

The *Barb* received the Presidential Unit Citation, the Navy Unit Commendation, and eight battle stars for her World War II service.

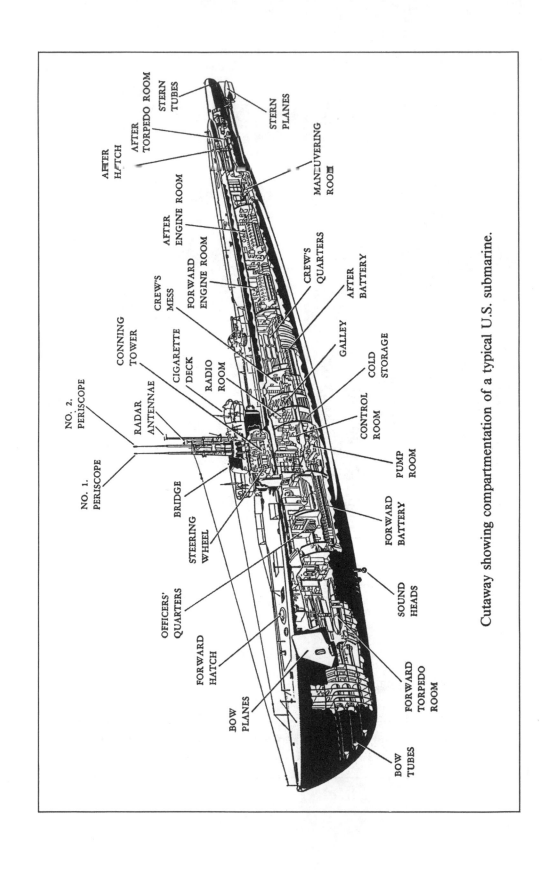

NO. 1.
PERISCOPE

NO. 2.
PERISCOPE

RADAR
ANTENNAE

CONNING
TOWER

CIGARETTE
DECK

RADIO
ROOM

CREW'S
MESS

AFTER
ENGINE ROOM

FORWARD
ENGINE ROOM

AFTER
HATCH

AFTER
TORPEDO ROOM

STERN
TUBES

STERN
PLANES

MANEUVERING
ROOM

CREW'S
QUARTERS

AFTER
BATTERY

GALLEY

COLD
STORAGE

CONTROL
ROOM

PUMP
ROOM

FORWARD
BATTERY

SOUND
HEADS

FORWARD
TORPEDO
ROOM

BOW
PLANES

BOW
TUBES

FORWARD
HATCH

OFFICERS'
QUARTERS

STEERING
WHEEL

BRIDGE

Cutaway showing compartmentation of a typical U.S. submarine.

# Part I The Eighth War Patrol of the USS *Barb* in the Sea of Okhotsk, North of Hokkaido, 21 May–9 July 1944

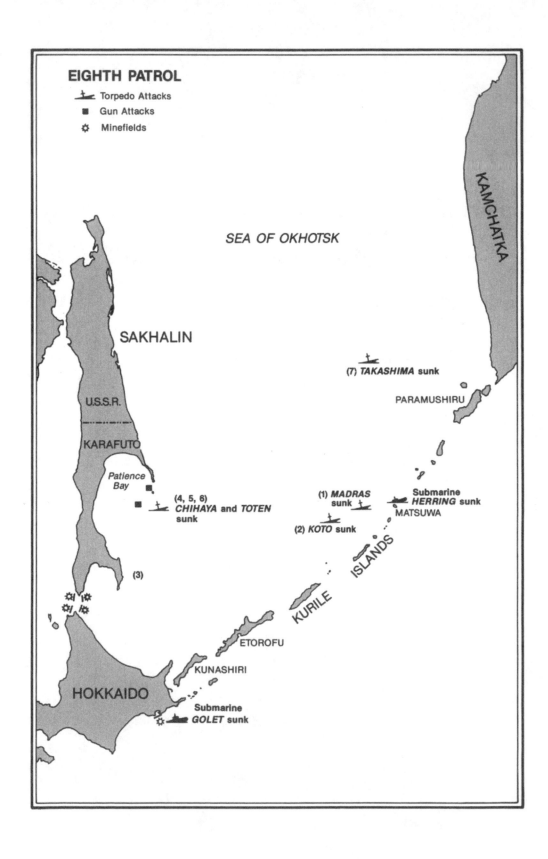

EIGHTH PATROL

⊢ Torpedo Attacks
■ Gun Attacks
✿ Minefields

KAMCHATKA

SEA OF OKHOTSK

SAKHALIN

U.S.S.R.

KARAFUTO

Patience
Bay

(4, 5, 6)
*CHIHAYA* and *TOTEN*
sunk

(3)

(1) *MADRAS*
sunk

(2) *KOTO* sunk

Submarine
*HERRING* sunk

MATSUWA

(7) *TAKASHIMA* sunk

PARAMUSHIRU

KURILE ISLANDS

ETOROFU

KUNASHIRI

HOKKAIDO

Submarine
*GOLET* sunk

# Chapter 1 Polar Circuit

"*Barb* Ho-o-o-o!" The call echoed from the bridge of the submarine and dissipated, smothered in the icy fog.

I smiled at Tuck Weaver, the officer of the deck. "Sure as hell, the *Herring* has arrived." The bellowing was that of her captain, Dave Zabriskie. No wonder he had been such a great football player at the Naval Academy. "Get me the electronic bullhorn, Tuck. I can't outblast him with a megaphone."

"Aye, captain." Tuck pulled his parka behind his ears and doubled his lanky frame over the conning tower hatch. "Below! Helmsman, tell the electricians to break out the bullhorn."

As we waited, my mind drifted back to my assignments after graduating from the Naval Academy in 1935. All had been on antiquated warships built before or just after World War I: the battleship *Nevada,* the destroyer *McCormick,* then duty in Panama on the submarines *S-42* and *Bonita.*

Tuck burst upon my reverie as he slid the bullhorn into my gloved grasp. I caressed the bridge rail of my *Barb:* my first command, and she was only two years old.

Leaning the horn on the rail, I squeezed the trigger and yelled, "Ho-o-o *Herring!*" In that instant we heard the clap of main ballast tank vent valves opening.

Tuck muttered, "She's diving. Why?" From the bridge speaker came "Bridge! Sonar reports submarine diving close aboard."

"What in the devil is Dave doing, Tuck? Does he think we're Japanese?"

"Captain, I hope he doesn't let us have a torpedo. Lying to here, we're a sitting duck."

"He wouldn't do that, Tuck." I had radioed the rest of my wolfpack—the subs *Herring* and *Golet* (whose skipper, Jim Clark, was a classmate)—to assemble so I could explain the search plan that I wanted them to follow. We had received an ULTRA radio from our naval intelligence codebreakers telling us that a convoy of three ships and one destroyer was leaving sometime that day from Matsuwa Naval Base, halfway up the Kurile Islands chain. It was probably their first convoy this year inside the frozen Okhotsk Sea, now that the ice had broken. "Strange that Jim Clark hasn't replied to my signal. I need all three subs to cover their possible tracks."

We waited, fingers crossed, for the *Herring* to surface. And waited. I thought of all my war patrols, including five "no soap" (no enemy) patrols in the *Bonita,* out of Panama, protecting the canal against an attack that never came.

I had known the former skipper of the *Barb* when he was skipper of the *S-45* in Panama. On his *Barb's* first six war patrols—five in Europe and one in the Pacific—she was not credited with sinking anything. At Pearl Harbor, before leaving on her seventh patrol, the *Barb's* skipper had grown distraught. One morning at 0200 he came to my cabin on the submarine tender. Waking me, he explained that he had become a fatalist. His *Barb* was scheduled to depart on her seventh patrol in a few days. This would be his last patrol before being sent to noncombat duty. He was sure the *Barb* would be sunk this time, yet his career would be ruined if he asked to be relieved of command. So?

Knowing that I had top ratings from my former skippers for developing the means of winning the coveted "E" for excellence in engineering and torpedoes, he wanted me to join the *Barb* as his prospective relief and to take command every other night, for he was exhausted. Rather backhanded, I thought! Still, the offer was attractive. I had arrived in Pearl from the Command School in New London, Connecticut, the day before. My division commander, Captain Karl Hensel, had assigned me to two months' duty repairing the new fleet subs coming in from patrol, since I had served only in old subs and had no combat experience. Then he planned to send me on a war patrol with an experienced skipper. Wanting combat duty above anything else, I had been disappointed again.

Seeing me hesitate, the *Barb's* captain assured me that if we came back alive, the boat would be mine. This was my chance to get a command. I accepted!

Captain Hensel was most annoyed with the request, but once I had convinced him that I knew the job of every single member in the *Barb,* he gave me his blessing. "Gene, if you've got that many ants in your pants, get going. You'll get orders."

Thirty minutes later I was in the *Barb.*

Through my pushing a little and changing some tactics, the *Barb*'s seventh patrol was successful. We sank one ship and made our first-ever double bombardment with our sister submarine the *Steelhead.* In the middle of the bombardment our 4-inch gun misfired. The captain wanted to scram, but I convinced him that I could drop down on deck from the bridge and order the gun captain to eject the shell through the breech. I would then catch it and throw it overboard. The captain agreed. I dropped down on deck, caught the shell, and got rid of it. The bombardment was a success, and the crew was elated to actually see sections of Japanese factories being blown up.

We returned alive. I looked forward to the captain's promise being fulfilled.

When the *Barb* arrived at Midway Island for post-patrol repairs, the senior officers congratulated the captain and presented him with his orders as a submarine division commander based ashore there. He in turn showed them his letter recommending that Lieutenant Commander Eugene Fluckey relieve him as the captain of the *Barb.*

This surprised them, for they did not know that I was already on board as the prospective commanding officer. Alas, they already had an officer who was one year junior to me, Jake Fyfe, standing by as the relief. My heart plummeted! Luckily, Admiral Lockwood, then submarine force commander, had ridden the *Barb* from Pearl Harbor to Midway and knew I was on board. The squadron commander agreed to send a message immediately, urging that Jake's orders be changed to another submarine and that the *Barb* command be given to me, an unknown.

A day of frantic waiting ended with a heartening approval. To avoid any reversal of the decision, the *Barb*'s departing skipper quickly arranged to have the change of command ceremony held two days later.

Now I had the *Barb!* 28 April 1944 became a red-letter day in my life.

But the fates were still gambling with me. Admiral Lockwood sent me a message stating that he would visit Midway and wanted to talk with me on 20 May, the day before we were scheduled to depart for the *Barb*'s eighth patrol in an area some 600 miles long and 400 miles wide in the Okhotsk Sea north of Japan. Was he having second thoughts about giving me command of the *Barb* after having had only one combat patrol?

There was nothing to do but go ahead and prepare for the patrol. We completed our repairs and conducted four days of sea trials with the division commander kibitzing.

First Class Torpedoman Charles Tomczyk, in charge of the forward torpedo room, broke out his illegal, secret diary (something I did not know about until much later) after the final test—an attack on a destroyer.

> Fired three bow tube exercise torpedoes for three hits and two from the stern tubes for two more hits. Captain received nickname "Dead-Eye Fluckey." It's a cinch we're ready for patrol. Planes dropped miniature bombs and the escort dropped two depth charges for the benefit of new men and to see if boat is tight. Depth charges shattered all center lights but ship is in good shape.

Three days of loading followed, then Admiral Lockwood arrived. Leaving his plane, accompanied by Squadron Commander Commodore "Shorty" Edmonds, he headed straight for the *Barb*. We were standing on deck as he walked down the dock. I could see a deep frown on his face. Saluting the colors as he crossed the brow, he then returned my salute. Looking me straight in the eye, he responded abruptly to my cheery "Good morning, admiral."

"Good morning, skipper. How do you feel about taking the *Barb* out on patrol?"

This was it. I had to take the offensive or lose the *Barb*. "We're ready in every respect, sir! How many ships do you want us to sink?"

The frown faded. "How many ships do you think you can sink?"

"Will five be enough, admiral?"

A smile appeared. "Yes, five will be enough."

"What types? Tankers, freighters . . . ?"

"Five of any type will be more than enough." He thought for a moment. "Captain, the *Barb, Herring,* and *Golet* will be covering the whole of the Okhotsk Sea jointly. For all three skippers, it's their first command. Whenever anyone makes a contact with a convoy, you will be the Wolfpack Commander. I know you have no experience in a wolfpack. Can you handle it?"

I looked him in the eye. "No problem, sir. You will have your ships."

He returned my look, then nodded. "Fine, captain. I'm confident you'll do well. Good luck and good hunting." With a hearty slap on my back and a strong handshake, he departed. I grinned and sighed happily in relief.

Tuck brought me back again from my musing. "Port Lookout. Hey, Miller, can you spot the *Herring?*"

"Yes, sir. She's got her scope up and is just rounding our stern close aboard."

Tuck and I watched the *Herring* plane up alongside. Their conning tower hatch sprang open and Zabriskie appeared.

"Gene, had to check you out in this on-again, off-again fog. What's the dope?" he hollered across the water.

"Dave, I called the wolfpack together, yet haven't been able to contact the *Golet*. A four-ship convoy leaves Matsuwa sometime today. My guestimate is that they'll head for La Perouse Strait and into the Sea of Japan. Here's my search plan. Draw a line between Matsuwa and La Perouse. Perpendicular to it, you take the 10-mile square to the north and east of our present position. I'll take the same to the south and west. Whoever makes contact sends a contact report before diving. This is about the maximum we can cover without Jim Clark. If you hear from him, let me know. Good luck!"

"Aye, Gene. I concur. See you later."

Our two subs separated at full speed on the surface, headed for our stations. The time was 1120; air temperature was 36°, the sea 34° and calm.

At noon Lieutenant Al Easton poked his nose up through the hatch. "Permission to come on the bridge? Lunch is soup and sandwiches. Ready to relieve the watch."

Tuck lowered his binoculars—"Come up"—then resumed the relentless search of sea, horizon, and sky to avoid a deadly surprise.

I went to the loudspeaker and informed the crew of our situation. Then I turned to Al. "Keep a taut watch. In this splotchy fog we must not be sighted. Still, we must stay on the surface at high speed to search the greatest area. I'm going below."

Tuck and Al replied in unison, "Aye, captain," then Al went on with the business of relieving the watch and instructing the two new lookouts who stood on the railed-in platform welded to the periscope shears above.

Down below I took a turn through the boat to see how the men were doing. Their obvious anticipation of probable action filled me with pride. In the after torpedo room, the men were greasing and caressing the four stern tubes.

One stopped his work to ask, "Captain, are we going to get them this time?" To my reply, "Absolutely," he turned back to his work for a moment, then said, "Captain, may I ask a favor?" I nodded.

"Fire your stern tubes, so we can turn the empty skids into bunks and ease our hot-bunking. Ever since I've been in the *Barb*, we've had nothing less than three men to a bunk in eight-hour shifts. We've never fired any stern torpedoes." It was a reasonable request.

"Don't worry, I won't forget. Besides, I promised the admiral that the *Barb* would sink five ships, and five ships he'll get come hell or high water."

They grinned, knowing that was as many as any sub had ever sunk in one patrol.

As I hopped over the coaming of the watertight door to the maneuvering room, two electricians were at the controllers of the main propulsion motors. "How are the batteries?"

"Fully charged, captain. We just finished topping off with one engine, so if you want to increase to flank speed, all four engines are available. Are we going to catch them?"

"You bet we are. Just keep squeezing that rabbit's foot you have hanging there."

In the after engine room it was too noisy to talk. The three machinists at the throttles gave me thumbs up and victory signs. I got the same response in the forward engine room from those who were monitoring the two forward 1600-horsepower diesels.

Passing through the crew's quarters, I noticed lights on and no men asleep. They were just too excited, and everyone was double-checking his battle responsibilities. As I entered the crew's mess, a loud cheer erupted. The men at lunch were eager to get into action.

"Sink 'em all, captain!"

"The *Barb* will do it!"

"Give them back more than they gave us at Pearl Harbor!"

The watch squad in the control room, led by Gunner's Mate First Class "Swish" Saunders, was straining at the leash with expectation. Everyone was talking at once, so I gave them the clenched fist and moved on to lunch.

Seated at the head of the table, I inquired of the six officers there whether anyone had any problems. My super executive officer, Bob McNitt, sitting on the bench to my right, spoke. "Captain, I've just checked all departments. No problems—we're ready. Al has the watch topside. Dave Teeters is searching for the convoy on radar. Radioman Jack Dittmeyer sent a message to the *Golet* again: no response."

"Bob, before we left Midway, Jim Clark told me that he would start his patrol on the east coast of Hokkaido, just south of the Kurile Islands. He should have no difficulty in receiving messages unless his radio is on the blink. Still, the *Herring* and *Barb* can handle this convoy, if our plan . . ."

"Captain! Radar contact 21,000 yards. Time 1222."

"Bridge," I shouted, "head for him. Sound BATTLE STATIONS TORPEDOES!

I'm coming up." I gulped my soup and stuffed a sandwich into my parka pocket.

The battle stations' gongs resounded through the boat. Men scurried to their stations. I let them pass and crossed the passageway to my cabin, picked up my binoculars and gloves, and headed aft. Passing to the conning tower ladder in the control room, I spotted Paul Monroe, the engineer and diving officer. "Flood negative tank, Paul; visibility is dubious, so we may have to take her down fast."

"Aye, aye, captain."

I climbed up into the conning tower. As the assistant approach officer, Bob was scrutinizing the radar scope with Dave. Louis Brendle, the chief of the boat and battle station radar operator, relayed the radar info to Al and Dick Gibson at the torpedo data computer. Quartermaster Ellis Shankles had the helm hard over, coming to the radar bearing of the enemy. I hurried up the ladder to the bridge. "Tuck, anything in sight?"

"Nothing, sir."

I called down the hatch to Bob. "Coach us ahead of the convoy until plot has their zigzag plan. Keep closing at full speed on the surface as long as the fog hides us. I'll be on the bridge until we dive."

"Aye, captain. Range closing fast."

While I was munching on my sandwich, Tuck, the battle station officer of the deck, apprised me of the situation.

"Captain, we're working ahead. The fog is thinning, sky clearing, and their smoke is visible at times. Must be a convoy of coal burners."

"Aye, Tuck. Have the battle lookouts concentrate on the sky. We'll take the surface to the horizon."

Tuck shouted, "Bluth, Bentley, confine your search to the sky. These clouds are likely plane lockers, hiding one."

"Aye, aye, sir."

*Time 1242.* The haze lifted suddenly, leaving us with a stripped feeling. "Take 'er down, Tuck!"

"CLEAR THE BRIDGE! DIVE! DIVE!" The klaxons rang out: A-oo-ga! A-oo-ga! I took two steps down the hatch, grabbed the side rails of the ladder, and jumped the remaining six feet to the conning tower deck in order to avoid having the two lookouts riding my shoulders as they came down in this customary fashion. Tuck, always last so that no one would be left topside, grabbed the lanyard of the hatch, swung it closed, and spun the hatch wheel to secure the locking lugs. The *Barb* now had a 15° down angle and was passing 80 feet on the depth gauge. "Left full rudder."

*Time 1244.* A sharp BANG! echoed through the boat—a bomb, but not close.

"Level off at 90 feet, Paul. Must be a single air cover spotting our wake. We'll just stay here for a minute until we get rid of the plane, then bring her up to 60 feet so I can take a look."

*Time 1246.* "Rudder amidships. Up periscope." A quick swing around. "Aircraft bearing—mark." Bob called out the bearing for plot in the control room. "Distance about five miles. I can only see the top of the mast of one ship. Let's start the approach on her. Bearing—mark. Down scope."

*Time 1252.* Distant depth charges. The *Herring* must be attacking.

Japanese records relate what happened earlier that day:

As the light of day penetrated the dense fog, Captain Nozu Kajiro, the commanding officer of the frigate *Ishigaki* and convoy commander, signaled, "Weigh anchor. Follow me." The convoy already had sea watches set, and the sound of groaning capstans and clanking chains echoed off the nearby cliffs above the small harbor. These were topped by a hastily constructed airbase. The largest ship—the army transport *Madras Maru*\*—had a civilian skipper, Kikuji Okamota, and carried 109 troops. She took her position astern of the *Ishigaki*. Following in single file came the naval transport *Koto Maru,* under Navy Commander Imano Haruo, and the *Hokuyo Maru*.

As the convoy cleared the harbor, the fog lifted. Kajiro hoisted signal flags. "Course west. Speed 10 knots. Use Zigzag Plan Four. Radio silence." The convoy lumbered on with the ships weaving back and forth together. On board the *Madras Maru* the troops were rejoicing. After a long winter during which the Okhotsk had completely iced over, they had been relieved by an army company from the east coast of Japan. Ordered back to their home base on the shore of the Inland Sea, they would have a month with their families before leaving for the southern battlegrounds.

At 1100 Kajiro received a dispatch that the fog at Matsuwa had lifted sufficiently. A scout bomber would be on station shortly after noon to cover his convoy. His command plan was to head the convoy across the Okhotsk to Karafuto (the Japanese-occupied half of Sakhalin Island). By so doing he would be able to use the ice floes—which would hinder submarine operations—then follow the coast south to Soya Kaikyo (i.e., La Perouse Strait).

*Time 1252.* The frigate *Ishigaki* exploded and sank, hit by the *Herring's* torpedoes. The three convoy ships scattered: *Madras Maru* headed south, the *Koto Maru* west, and tail-end Charlie, the *Hokuyo Maru,* east, returning to Matsuwa.

---

\* *Maru* means "merchant ship."

*Time 1300.* With my left arm hooked over the handle of the periscope, swinging it around in a complete circle, I summed up the situation.

"Down scope. The plane is not in sight. The convoy must have dispersed. We have only one large ship heading this way, so we'll get her. Make ready tubes 4, 5, and 6 for a small gyro, starboard track, bow shot. Should she zig across us at the last minute, expect a quick shift to a stern tube shot. Pass this by phones to all compartments. Now let's get a setup.

"Up scope. Range 3 divisions. Bearing—mark. Down scope. Identification, straight bow, foremast, bridge with passenger superstructure, funnel, mainmast, spoon stern. Look her up in Naval Intelligence Manual 208J. Big guns fore and aft."*

Al called from the computer: "Checks. She's steady on course one nine zero, speed 9 and ¾ knots, range 4300 yards. Tubes are ready."

"This should be a cold turkey, unless our scope is sighted. Bob, just as we have practiced—I don't want the scope exposed for over 4 seconds. Keep it just clear of the well with the calculated bearing of the ship set. I'll follow it up from the deck. When I say 'Down,' run it down. On firing, have the JK-QB sonar stay on the ship and the JP sonar on the torpedoes."**

"Time for a setup, captain. Sonar checks solution."

Crouching on the deck, eye glued to the eyepiece, I shouted "Up—Down!" From the deck Bob read the bearing. I looked at the computer. "Angle on the bow, range, bearing—everything checks. Lot of uniformed men looking over the side." So sorry, I thought, but we did not start this war. "Spread torpedoes from aft, amidships, and near bow. Plot, have you identified the ship?"

John Post called up from below. "Sir, similar to transport *Kasima Maru.* Length 495 feet."

"Aye. Dick, set spread at 50 yards between torpedoes."

Bob piped, "Coming on to a 90° starboard track. Time for a final bearing, then shoot."

*Time 1315.* "Up—Down."

"Bearing checks."

"FIRE 4!"

Pharmacist's Mate William Donnelly pushed the firing plunger near the

---

* Periscope ranging is accomplished using reticles, a network of fine horizontal lines placed in the focus of the objective lens. By estimating the height of an object, such as the masthead height to the waterline of a ship, the number of lines included is converted, using a table, to a ship's range.

** The JK-QC sonar head is bottom mounted and can either listen or echo range, as does the QB head. The JP sonar is mounted on deck and can listen only. Each type signals the direction of the noise source.

helm. The *Barb* shuddered as the torpedo was ejected by high-pressure air and roared off. The tube then vented inboard to avoid a large air bubble rising to the surface.

"FIRE 5!" Again the jolt. "FIRE 6!"

"Conning tower, JP sonar reports all torpedoes hot, straight, and normal."

"Stay with them. Let me know if you detect any deviation. Bob, what's torpedo run time?"

"For 1400 yards, any second now..."

BOOM! Cheers everywhere! "Up scope and leave her up. Camera quick! The water column aft is still in the air."

BOOM! "Directly under the stacks!" I shouted. BOOM! "Waterspout near the bow. WOW! A secondary explosion from ammunition within is curling the plates back! The ship is slowing! She's taking a down angle by the bow! All hands in the conning tower, come take a quick look!"

Bob wisely took a complete circle sweep before turning it over. "All clear, sir; she'll sink. Suggest we reload while we're waiting."

"Good idea. Do it."

After those in the conning tower had a quick peek at the sinking ship, I returned to the raised periscope. Landing barges crowded with soldiers floated off the forward part of the ship as the deck went under. *Madras* was now dead in the water. Lifeboats were being lowered and automatically unhooked as they touched the water, engines running. Quickly, they all pulled away to avoid the suction as the ship upended and headed for the bottom. Only one rowing lifeboat was engulfed in the swirl of the whirlpool. A few seconds later the occupants came swimming back up. A barge picked them up.

*Time 1320.* Official report: Ship sank. Latitude 48-21.1N. Longitude 151-19.6E.

Ten minutes later, awaiting completion of reload, I sighted a medium-size merchant ship with her hull down, too distant to close submerged. Though the *Herring* hadn't sent us a contact report before diving, she must have hit the convoy. With the diverse courses of the two ships, either she sank the escort, or the escort was holding her down like a good shepherd while his sheep were getting clear.

With our reload completed, we planned to assist the lifeboats, should some need emergency medical attention.

I said, "Bob, the escort plane's been gone for over an hour. If the ship got a Mayday message off before she sank, he could be close by in the haze. As we surface, keep the lookouts here until we know it's safe. Tuck, follow me. Let's go!"

"SURFACE—SURFACE—SURFACE" went out over the speakers, followed by the klaxons—A-oo-ga! A-oo-ga! A-oo-ga! Opening the hatch with Tuck close behind, I said, "Take air search to starboard, I'll take the port side."

We swung our binoculars skyward, ignoring the lifeboats and landing barges 50 to 80 yards on our port beam. I heard a Japanese officer scream a command.

As I spun around, bullets whined over my head with the rat-tat-tat of a machine gun. "FLATTEN, TUCK! ALL AHEAD FLANK SPEED! Tuck, get below tail or head first. Someone will catch you."

I crawled on my stomach to the hatch and went down head first into waiting arms. "Bob," I said with grudging admiration, "the gall of those men machine gunning us when they can see our big 4-inch gun. What guts! I thanked God the lousy gunners didn't realize the recoil was pitching the gun up so that the bullets passed over our heads. Only a couple hit the shears." We could have blown them all out of the water—and they deserved such—but there was no time. Our new target was now hull down. To end around her, unseen, we couldn't dally.* Bob quipped, "Well, captain, we now know we can't trust lifeboats. Secure from battle stations?"

"Sure, but keep the tracking party." Then I chirped, "One ship down, four to go."

*Time 1410.* Our target (unbeknown to us, the *Koto Maru*) was on a southeasterly course, heading for the Kurile chain. She might make a strait and pass between two islands or get air cover before we attacked. Bob thought an unseen end around on the surface was not the answer. I agreed, and we moved in from 12,000 to 8000 yards on the port quarter, permitting the enemy to get a good look at us. Maybe he would see the danger in his plan.

The ship's captain seemed to agree, for the *Koto Maru* changed course to head southwest for La Perouse Strait. The fog closed in. We let our foe think he was secure, and we ended around. Unbeknown to us, he had radioed "HELP! SUBMARINE CHASING ME!"

*Time 1700.* Visibility was 2000 yards in patchy fog, with light haze. We were nearly ahead of the target for the approach position. "Bridge, radar contact closing fast—3000 yards! DIVE! DIVE!"

---

* Due to our maximum 9-knot submerged speed, we would seek an attack position *ahead* of the target on the surface (using our maximum 19-knot speed) and thus ended around, or circled.

We pulled the plug.* No bombs were dropped. "BATTLE STATIONS TORPEDOES!"

*Time 1710.* Radar depth, 45 feet. Setup looked good. I remembered what Rufus Roark had said about the *Barb* never having fired torpedoes from the after torpedo room. "Make ready all tubes aft, 7, 8, 9, 10. Bob, Al, I'm going to cross ahead of him and fire a stern salvo to ease the hot-bunking aft." Everyone smiled. "Well, to the after torpedo room crew it's not funny."

*Time 1730.* "Up scope." A quick swing revealed no planes. A clear sky. "Put the scope on the sonar bearing of the ship." Nothing. A bit too foggy. "Bring me up to 45 feet—radar depth."

Dave reported, "We have the target. Range, 4000 yards."

"Solution checks. Ease her down to 63 feet."

*Time 1742.* "Up scope. Down. She's coming out of the fog. Sonar, get me a single-ping range! I don't want her to spot the scope."

"Sonar reports 1520 yards."

"Checks."

"FIRE 7!" Whoosh—jolt! "FIRE 8!" Whoosh—jolt! "FIRE 9!" Again, ears popped as the air pressure in the boat increased. "JP SONAR, STAY ON THOSE TORPEDOES! Dick, what's our torpedo run?"

"Sixty-five seconds, sir."

"Sonar reports all torpedoes running hot, straight, and normal."

"Bob, they haven't got time to avoid our fish. Up scope. They're scurrying around pointing at the wakes."

WHAM! "Beautiful hit three-quarters of the way from the funnel to the stern."

WHAM! "Right smack under the funnel. Big secondary explosion venting and ripping her deck like tearing a piece of cardboard."

WHAM! "Tower of water two-thirds of the way from the funnel to the bow. Looks like the bow is breaking off. Give me the camera quick! Her tail is upending to the vertical! Lifeboats are hanging straight down from their davits, spilling men into the sea. I've got the photo. She's gone!"

*Time 1745.* Official report: *Koto Maru* sank. Latitude 47-52N. Longitude 151-02E.

"Stand by to surface on the batteries. Tuck, after this morning's hail of bullets, we'll check the water area first."

*Time 1746.* Click—Bang! Click—Bang! Two bombs close, like a giant hammer slamming against the hull. "DOWN SCOPE! RIG SHIP FOR DEPTH

---

* We dived as fast as possible, flooding our negative tank as well as our main ballast tanks.

CHARGE! Take her deep—300 feet. Left full rudder." Stupid me to have left that periscope up. Watertight doors were being hastily closed throughout the *Barb*. "All stations report damage."

*Time 1749.* Two more bombs. Not close.

Paul called up from the control room. "Captain, the only damage is leaks around the bow plane shafts and one electric cable pushed in, causing a firehose effect. Some of us down here got wet. The gland nuts have been tightened. Nothing leaks now."

"Good!" That was a relief. "Bob, we'll stay down for a while to reload torpedoes and have some chow while the plane shoves off. Secure from battle stations and depth charge. Have the watch let me know when we've reloaded."

While we were eating, Roark came forward from the after torpedo room. "Captain, reload completed. My men want to thank you for using the stern tubes. It's the first time on a *Barb* patrol. We've already turned the three empty torpedo skids into nine bunks. Please shoot one more from aft, then everyone will have a bunk of his own."

"Rufus, we'll do just that, providing you'll guarantee a hit."

"You've got a deal, sir. Shake on it."

Our hands met. "Two sunk, three to go!" I couldn't conceal my pride.

*Time 1900.* Surfacing in a heavy fog, we returned to the wreckage. I needed a prisoner who knew the area. In half an hour we found flotsam and a few survivors on top of the floating wooden hatch bails. Sadly, most of them were suffering from hypothermia. As we slowly moved through the human wreckage, I noticed that none of the officers had survived. A gunner's mate in a near-collapsed condition beckoned to us. Swish, Traville Houston, and Fred Campbell hauled him aboard.

"Pass the word—Doc Donnelly to the bridge," I said. "This prisoner needs immediate help to survive." Jim Lanier, the officer of the deck (OOD), called the order down in his typical Alabama drawl and added, "with a blanket."

Our admirable pharmacist's mate arrived and took charge.

"Doc, we need to know what happened to the rest of the convoy to plan our search. As soon as he has revived and is in a satisfactory condition, bring him to the wardroom for questioning."

"Aye, aye, sir."

Swish and Houston passed the man up to the 20-mm gun platform and Doc wrapped him in the blanket. He was then passed through the access door to the bridge and down the hatch. Since the after torpedo room now had more space and a continuous alert watch, this became his cell.

*Time 2200.* Bob had laid a chart of the Okhotsk Sea out on the wardroom table. No one in the *Barb* had interrogated a prisoner before. None had any knowledge of Japanese. Fortunately, being a French interpreter, I had an interest in languages and knew the value of even a few words of an enemy's tongue. So I had with me a one-page Japanese phonetic vocabulary that I had torn out of a General Bureau of Personnel bulletin in 1942. There was nothing else on board.

The prisoner entered. I greeted him, *"Komban wa."* Smiling, he bowed and responded. Motioning him to sit on the bench to my right, I looked at my prepared list of questions, then at my vocabulary page. "Your name, please?" The prisoner smiled understandingly. Then, looking me straight in the eye, he placed his index finger vertically in front of his lips and waved it slowly back and forth. This was an obvious indication that he was not permitted to divulge any information and must remain silent.

Anticipating this (since our own code required we give only name, rate, and service number), I reached under my left thigh and pulled out a .45, caressed it, and slowly laid it out of his reach on the table. "YOUR NAME, please?"

A bit shaken, he blurted out, "KITOJIMA SANJI!"

Using a bit of English, Japanese, sign language, and facial expression, I indicated, "That's better. Now that we fully understand each other, we will get along happily. We call you Kito, you understand?"

For some reason it seemed necessary to speak louder when working against a language barrier. He nodded, understanding. I went back to my impoverished Japanese. "Kito, what you rate?" His answer was incomprehensible.

Seeing this, he shifted to sign language. Cocking his hand at me like a pistol, he uttered "BOOM." Then, with his index finger, he drew two chevrons on his sleeve. Gunner's mate second class—good. Now we could get going on the information we desperately needed. "What name your ship?" He hesitated. I patted the .45.

*"Koto Maru."*

"Where convoy start?"

"Matsuwa."

"When?"

"Today."

"Number ships?"

"Four."

"Name ships?" Grimaces—slight hesitation—slight pat on the .45.

*"Ishigaki, Madras Maru, Koto Maru, Hakuyo Maru."*

The vocabulary defaulted. We could get not a word regarding sinkings.

Bob and I held a conference. He would act out being a convoy and I would be the submarine. As Bob pantomimed the four ships steaming along, I said, "Whoosh," then ran my finger like a torpedo toward the center ships.

Kito's eyes brightened. Joining the game, he gently took my finger and shifted it toward the lead ship, striking Bob's hand. "BOOM! You (sinking motion) *Ishigaki*. Ships go all directions. Aircraft hunt for you. Other ships okay."

Great, now we knew that the *Herring* had been at work for sure and sank the destroyer, but nothing else. Since we had sunk two ships, the *Herring* was probably chasing the surviving ship. As we hadn't picked up any interference from *Herring*'s radar on our screen since early afternoon, she must be chasing toward Matsuwa.

"Navigator Bob, set us on a course for La Perouse now at full speed, just in case the *Hakuyo Maru* evaded the *Herring*. I've got just a few more questions for Kito.

"Aye, captain."

Twisting the chart around, I pointed to the La Perouse area where we knew mine lines existed. "Kito, where mines?" He shrugged his shoulders. "Kito, you lookout?" I held my hands up like binoculars.

"Yes."

"Where?"

"On bridge, starboard, me see your torpedoes."

"You see charts on bridge?"

Proud of his position, he said, "Yes, me old friend Captain Haruo. He show me charts."

"Kito, you charts show mines. WHERE?" My hand barely flicked the .45 while I handed him a pencil.

He gasped "Here!" and drew a diagonal line across the area in front of the strait.

Studying it a minute, I thought it looked reasonable. "More?" He shook his head. Time for a pantomime. I indicated that if there were more, we would be sunk—and he, too, would go down with us to a horrible death.

Pondering this he gritted his teeth. "Aww, aww, more!" Then he took the pencil, erased the first line, moved it farther away from the strait, and drew two additional lines.

Now we were really getting somewhere. "Good, Kito. When we are safe, you are safe. Just one more question and you can turn in. What depth, mines?"

This perked him up. "Fifteen meters," he replied, indicating with his fingers and wobbling hand that he meant "more or less."

Fifty feet was reasonable. I signaled to Doc, who had been watching. "Okay, Kito, that's all for tonight. We'll talk more tomorrow. *Sayonara.*"

With a bow and a relieved smile he said, "*Sayonara,* captain," or the equivalent—perhaps a dirty word. Doc took him in tow, and he shuffled off to his torpedo cradle—certainly happy that it wasn't the cradle of the deep.

The off-watch officers and Bob gathered around for a briefing on our search plan and the info garnered from Kito. In spite of the lengthy, busy day, practically everyone on board was awake, excitedly talking over the day's events. This included being at battle stations from 0100 to 0300 as we made an approach on a large tanker picked up by radar in a fog. Just as we reached a beautiful, close-in firing position on the surface, she came out of the fog showing Soviet lights. This was the dangerous trouble with the area. Due to the blasted neutral Soviet ships passing through, everything had to be identified before shooting.

Bob announced, "All present except Jim Lanier, OOD."

I started with our discussion with Kito and the consequent change in plan. They were fascinated that we had been able to obtain so much info with the stonewall language block. Tuck, who looks and acts much like Will Rogers, couldn't resist. "You two are miscast in your profession. You'd outdo Laurel and Hardy in pantomime."

Laughing, I looked at dark-haired, ever-smiling Al. "Al, you are in charge of Kito. He's your baby. He's intelligent. You'll be responsible for teaching him English. *Daijobe* means Okay."

"*Daijobe,* captain, but why me?"

"See how astute you are at picking up Japanese? Besides, as torpedo and gunnery officer, he's in your domain. Don't let him leave the after torpedo room without an escort. Your men on watch aft can help enormously by talking to him and showing him how to do things. Though he's an enemy, be kind. I'm sure he will cooperate. Keep him busy. Whenever you have a dirty job, put him on it with one of your men. Misery loves company. He knows he's lucky to be alive."

"*Daijobe,* captain. His lessons start tomorrow when he arises and has to go to the . . ."

"*Benjo.* But have someone flush it for him so it doesn't blow back in his face." Smirks.

"On a different subject, the *Barb*'s performance today was flawless. The numerous failures of torpedoes to explode have been solved. Admiral Lockwood fired torpedoes against cliffs and dropped some with inert loads onto concrete pads without their exploding. These proved the detonator firing pins were being deformed on contact. The fix-stronger springs were provided and the inertia of the firing pin was reduced by substituting aluminum for steel. For *Barb's* last patrol, Bob obtained the first 24 of these pins and springs made by the torpedo shop at Pearl and had them installed at Midway. *Barb* is the first ship to test this modification. Today the pins proved successful. Well done." Bob nodded in acknowledgment.

I had Bob write up the night orders while I passed the word to the crew regarding the *Herring* attack, our present objective, and the need to get some rest. Exhaustion breeds error. I added, "We never know what the night will bring."

The control room sounded like a hen party with all of the cackling. When I related the *Herring* attack and the sinking of the destroyer *Ishigaki,* cheers echoed throughout the boat. I followed with a short explanation of the need to keep our own personal batteries fully charged at all times— the *Barb* had to sink at least three more ships. More cheers! "Tomorrow at 1000 the cooks will have our traditional cake set up in the control room to celebrate today's double sinking. Then, all hands will SPLICE THE MAIN BRACE! (a naval saying from antiquity meaning to celebrate by having a drink together). Now off to your bunks and be quiet." Cheers resounded through the boat.

Like the fog creeping in on little cat feet, the men crept off to get some sleep. All except one. As lights went out, Tomczyk crawled into his upper bunk in the forward torpedo room. Flashlight in hand, he broke out his diary and recorded the day's happenings, ending with: "P.S. 6 torpedoes fired—6 hits—2 ships sunk, as much as in all previous 7 patrols. NO ERRORS. Captain is sure living up to his name and still hunting for more."

# Chapter 2    Submarines Down?

*31 May. Time 1500.* After the *Ishigaki* sank, Japanese messages were flying and the pot commenced to boil at Matsuwa Island. The naval command received the messages of her sinking and that from the *Koto Maru* pleading, "Help! Submarine chasing me!"

The command then dispatched the *Shinkotsu Maru* to pick up survivors and the frigate *Namikaze* to take over escort for the convoy. Both ships were taken from a convoy proceeding from Paramishuru to Hokkaido. This left that convoy without any protection for the *Hiburi Maru* and the *Iwaki Maru.* The scouting plane claimed he sank a submarine at midday and reported he had just found a group of survivors in landing barges. The command had been unable to make radio contact with the *Madras Maru.*

Later, the *Hokuyo Maru* reported she was returning to Matsuwa. As it wasn't possible for the *Namikaze* to catch up with the *Koto Maru,* she was ordered to pick up the landing barge group of survivors and the *Shinkotsu Maru* to pick up the *Ishigaki* survivors. Orders then went to the unprotected *Hiburi Maru* and the *Iwaki Maru* to anchor at Matsuwa for safe haven. Two additional scout bombers were sent to escort *Koto Maru.*

(On 1 June the Japanese reported that the *Koto Maru* was sunk at 1745 the previous day and that the submarine that sank her was sunk by her air cover two minutes thereafter. The *Iwaki Maru* and *Hiburi Maru* anchored in the open, narrow harbor at Matsuwa. The *Hokuyo Maru* failed to arrive.)

20

***Top:*** Periscope photographs of the *Madras Maru* sinking, 31 May 1944. The split image is due to the range-finding lens in the periscope, which brings the masthead height of the target to the waterline. ***Bottom:*** Periscope photograph of the *Koto Maru* sinking, 31 May 1944.

*1 June. Time 1000.* "Now hear this: the cooks have completed a veritable masterpiece to celebrate the *Barb's* sinkings yesterday of the Imperial Japanese Navy's *Madras Maru* and *Koto Maru*. Said culinary work of art is exhibited in the control room from now until 1045, when it will be devoured. Permission is granted to photograph from a selected angle only, to avoid revealing classified equipment.

"Further, the captain has been entrusted, by the most considerate U.S. Navy, with sufficient bourbon whiskey to provide two ounces to each man as a depth-charge ration. Our captain believes that laws are made by men and can be changed by men. According to the General Prudential Rule, a captain at sea may render a departure from such rules necessary to avoid immediate danger. Thus, the captain has decreed the following. If we get depth charged, we kick each other in the tail. Whenever we sink a ship, we splice the main brace! Hip-Hip-Hooray! The bar is now open."

The cake, three feet square and six inches high, rested in front of the diving station with the cooks, James Vogelei and Charles Dougherty, sitting behind it. The multicolored frosting depicted a submarine firing torpedoes at two merchant ships flying the Japanese flag. One of the ships was broken in the middle like a "V" for victory as she sank. The other was sinking bow first. I was amazed that they could concoct such a cake. Bob suggested that Kito, our prisoner, be brought forward to see the cake before it was cut.

Meanwhile, the men who were not abstainers lined up and received their ration of treasured sipping whiskey. Of course, each one had to check his name off on a muster list—submariners are, by nature, sneaky.

The officers received their tot of whiskey and headed for the wardroom for a few sea stories. Normally at this hour Bob held officers' school for two hours. This day was special, however. The *Barb,* as usual, stayed on the surface, sweeping back and forth near La Perouse in our never-ending hunt for the enemy.

Actually, the whiskey was of poor wartime manufacture, appropriately nicknamed "Black Death." But we felt obliged to sip it because it was forbidden fruit. Prohibited onboard in peacetime, it was now locked up for "medicinal purposes" or to be issued as a depth-charge ration.

I noticed Tuck frowning after the first sip. "No like?"

"Coming from Ottawa, Illinois, I can now understand why the Ottawa Indians called this stuff 'fire water.' Must be that the good alcohol is needed as fuel for torpedoes."

"Tuck, did you receive a ration when you were sunk up here in the Kuriles in your *S*-boat?"

"No, the bottles broke."

"Next to Dick Gibson, you're the most junior officer in the *Barb*. How did it happen you were patrolling the Okhotsk Sea a year ago?"

"Well, captain, nearly two years ago the Bureau of Personnel conducted a one-time experiment by sending a few newly commissioned reserve officers to the Aleutians for on-the-job training in *S*-boats. This was an alternative to submarine school, torpedo school, electronic school, and sound school. Consequently, the first time I ever saw either an ocean or a submarine was when I reported to the *S-30* in Dutch Harbor, Alaska, wearing my two-week-old ensign stripes. The wardroom was composed of Naval Academy graduates with years of submarine experience. There was no bunk for a trainee, and I was received like a skunk at a garden party.

"During that 16-month tour there was a change in command. Captain, you wouldn't believe that those two skippers had been trained by the same Navy. The departing commanding officer was ultra-cautious and conservative. His replacement was William Stevenson. His personal courage sure exceeded the capability of the boat, the competence of the crew, and probably his own judgment about what and when to attack.

"Nine miles east of the Paramushiru Naval Base, a routine gun attack on a trawler turned into target practice when the foot firing mechanism failed and the sights got cloudy after the first shot. Suddenly, a frigate from the base emerged through the light haze, all her guns blazing. Captain Stevenson cleared the deck, dove, and commenced his approach. The sea was glassy as our World War I relic, armed with steam torpedoes that trailed bubbles and belched geysers of water when fired, headed straight at her.

"The diving officer was having severe depth control problems. He had forgotten to add sea water to replace the weight of the ammunition fired. The planesmen couldn't figure out why the boat kept trying to rise unless the bow and stern planes were depressed. At the critical firing bearing— range 1000 yards—the periscope was dunked. The skipper bellowed, 'I can't see. Bring me up!' The planesmen overreacted, putting the planes on full rise. The boat surfaced. We quickly flooded the auxiliary tank to bring the boat back down. Meantime, the frigate passed alongside, dropping a string of depth charges. Our depth gauges flew off the bulkhead just before the lights, pumps, and motors were knocked kaput. We sank to the bottom.

"The shallow shelf on which we landed was 250 feet below the surface. It happened to be the only place in the patrol area where the ocean wasn't deep enough to crush our hull. After about four hours we completed our electrical and other repairs. During the short arctic night we surfaced in the haze amidst four patrol vessels and cleared the area.

"A week later Captain Stevenson worked his way into Kakamabetsuwan Harbor and sank a freighter. Later, he planned a gun bombardment on the airfield hangar at Matsuwa. It didn't come off because the deck gun wouldn't function."

Tuck shook his head and smiled grimly.

"Tuck, the *S-30* may be around today because the gun wouldn't fire. There are supposed to be shore batteries on Matsuwa. No surfaced submarine would last long in there."

Tuck's story was an epic. I remembered how we fried in the *S-42* in the Canal Zone without air conditioning. "I'm sure you froze," I said. "Your *S-30* has my unbounded admiration for pulling herself to the surface by her own bootstraps to fight on. Those are great lessons, Tuck. We all need to pay attention to them."

Chief Commissary Steward Stephen Malan appeared for Brendle. "Captain, Kito is present, ready for you to cut the cake."

The officers piled out and into the jammed control room. From the seat saved for me on the diving bench, facing the cake, Kito's five-foot frame wasn't visible. "Where's Kito?" The taller sailors moved aside as Kito was pushed forward into the limelight.

"G'morning, captain."

"Good morning, Kito. You're speaking English already." Pointing, to help him comprehend, I asked, "Do you like the cake?"

He smiled. "Ahh—me like great cake. Me lookout *Koto Maru.*" He put his finger on the ship, broken in two and sinking. Then quickly, simulating binoculars, he raised his cupped hands to his eyes, looking at me. "Sighted sub, glub, glub."

"Now, who taught him that?" Everyone chuckled as Swish ducked and tried to hide his small frame behind a couple of cohorts.

At this, Kito surprised us all. The ham in him burst out, and turning to the crew he gave an encore of his successful performance, throwing us into gales of laughter. All agreed that Kito should have the first slice of cake. He was delighted. One-third was set aside for those on watch; the remainder was dispatched in twenty minutes.

***Time about 1300.*** Off Point Tagan—the southeast corner of Matsuwa Island—the wakes of the torpedoes that sank the *Iwaki Maru* and the *Hiburi Maru* were sighted by shore gun batteries. At the same time the *Herring* broached with her bow, bridge, and conning tower breaking through the ocean surface. The shore batteries opened fire and scored at least two direct hits on the conning tower. The submarine sank stern first two kilometers south of Point Tagan.

As she sank, the Japanese reported that a large volume of air bubbled the surface over an estimated five-meter wide area. Another shell may have hit one of the *Herring*'s center main ballast tanks carrying diesel fuel, for the report said that a heavy gushing oil slick erupted just seaward of the air bubbles. (Submarines departing on patrol usually carried diesel oil in center main ballast tanks to extend their time on station. When these became empty they were then reconverted to normal main ballast tanks.) The reported position of the sinking was on a ledge about 330 feet deep.

As no other submarines were in the Kuriles, this had to be the *Herring*.

*2 June. Time 0101.* I was fast asleep when the buzzer above my head jarred me out of a beautiful dream. "Captain, radar contact at 10,500 yards."

"Sound BATTLE STATIONS TORPEDOES!"

The words boomed out over the loudspeaker followed by the Gong-Gong-Gong-Gong that drove everybody from their bunks. In 30 seconds I was in the conning tower, buttoning my hooded parka as I bent over the radar scope.

Bob joined me as people rushed to their battle stations. "Bob, the target's on our port quarter. Looks like we're both on westerly courses, closing the minefields. Take a peep at your charts to see how close we can safely come to his track submerged. I'll be on the bridge with Tuck for a night surface attack."

It was cold on the bridge, and I had Tuck hold my gloves while I tightened the strings of my parka hood.

"Just a bit chilly at first, when you've come from a warm bunk, sir. After a few minutes you'll forget all about it. You'll be numb all over."

"At least it's not rough, with the bridge filled like a bathtub. What's the visibility? My eyes aren't adjusted yet."

"About 6000 yards. The moon has set. No clouds. Zodiacal light fair, sea calm. The phosphorescence may cause us some difficulty."

"Bridge. Captain, Al has a setup. Target is high speed, making 16 knots, course 265°. Distance to his track is 5100 yards. I recommend that we keep about 3000 yards from him. Plot says he is barely skirting the mine line. We could have a navigational error of two miles in our position."

"Aye, Bob. We've quite a bit of phosphorescence in our wake, so we'll hold our speed down to 8 knots. I hope our camouflage paint will keep us from being detected. What's the target bearing?"

Dean McCloud, the port lookout, called down from his platform on the periscope shears. "Captain, I've got her!"

"Where away?"

"A bit forward of the port quarter, but abaft the port beam, sir."

"Got her. Good work." I asked Tuck if he could see her with the target-bearing transmitter (TBT) telescope.

"Bridge, radar bearing 150°, range 6100 yards."

"Captain, I've picked him up on the TBT."

"Stay on him, Tuck. I'll watch everything else. Bob, Tuck has the target on the TBT. Have the fire control party use his bearings. They're more accurate than radar. Target appears to be a *Chidori*-class frigate. I don't like long-range firing, but we have to stay clear of the minefields. Set us on the closing course for a bow shot. Make ready tubes 4, 5, and 6. Set torpedo depth six feet." I wished we had magnetic exploders for this one in place of our contact ones.

*Time 0205.* "Captain, we're steady on the firing course. The setup checks perfectly. No zigs, no zags, no change in target course and speed. Range 5500 yards, tubes ready."

"Great. That means we haven't been detected. This should be a sure shot. As soon as we fire, make sure the JK sonar follows the torpedoes and reports back. Good visibility—nothing on the surface around us."

"Lookouts! Lego, Wells, on your toes now. Don't bother about the target. We have it. Keep your search sweeping. We don't want any surprises. That includes mines that may have broken loose from their moorings."

*Time 0219.* "Bridge! Range 3100 yards. Recommend we go no closer and start shooting."

"Aye. Final bearing, Tuck.

"Mark."

"FIRE 4! FIRE 5! FIRE 6! Stay on the target, Tuck! Let me know if he changes course or speeds up."

"Bridge. All tubes fired. Sonar reports all torpedoes running hot, straight, and normal."

"Aye, all stop. Watch for any movement." I could follow the phosphorescent wake of the torpedoes. "What's the time of torpedo run?"

"Two minutes and nine seconds, sir."

"Let me know when they should be hitting."

*Time 0221.* "Bridge. Sonar reports torpedoes still running. The bearings have melded into the target."

"Great! They'll be hitting any second now! I can see the wakes right at the ship!" Each second became a minute.

"Captain, time is up for hitting."

"Target is turning away! He definitely is, captain, almost 90°!"

"My God, Tuck, the torpedoes passed right under him! Damn! We won't have another chance to sink him."

"And all hell is breaking loose!"

The Chidori chased down the luminous torpedo tracks with all guns blazing at ghosts. She put her depth-charge racks and sidethrowers into action, dropping 12 depth charges. They exploded some mines, which blew columns of water into the air. Following that came the automatic explosions of our torpedoes at the end of their run. Thus, the Chidori could now claim the sinking of an American submarine. We secured from battle stations and sadly returned to our bunks.

*Time 0305.* A-oo-ga! A-oo-ga! "CLEAR THE BRIDGE! DIVE! DIVE! 200 feet! Radar contact close, closing fast."

Leaping from my bunk, barefooted and clothed only in my skivvies, I dashed into the control room. Glancing quickly at the Christmas Tree (the hull opening indicator board has a red light on any opening or a green light for anything closed) as the *Barb* nosed downward, I was reassured. Post, who had the bridge watch, now acted as diving officer. "Green board, pressure in the boat, captain."

"What was it, John?"

"Don't know, sir. All from radar. We've run into fog."

Click—Wham! Depth bomb, not too close. "RIG FOR DEPTH CHARGE! Take her to 260 feet. Guess our Chidori sent some Zoomie friends to call on us. It's only been 40 minutes since we attacked her. Shows us that the defense net in this area is on its toes—quick response."

After ten minutes it was apparent that the pilot must have been trigger-happy and thought he saw a submarine in the predawn surface fog below. No more bombs fell.

I called up to the conning tower. "Bob, John's bringing her up to periscope depth. I'm chilly wearing nothing but shorts and I know it's colder up there. Secure from depth charge. Take a look with the scope. If visibility is poor, stay submerged and reload torpedoes forward. Set course for Cape Shiretoko Misaki. Let's all turn in and get a few hours' sleep until breakfast."

"Aye, captain. Sonar reports a few noises of light screws. Probably spitkits venturing out for dawn fishing. Visibility is zero."

"Okay. Make your speed three knots. Turn everything over to the regular watch. Have them call us if sonar picks up a ship."

After breakfast we surfaced. On the bridge the damp cold seemed to penetrate every pore, making us shudder. The sun helped by slowly burning off the fog. My *Barb,* with constant helming, weaving back and forth searching for prey, reminded me of a beagle wagging its tail while quail hunting.

It seemed as if we were loafing, but submariners on the bridge have an instinct for relaxed tautness. With binoculars sweeping sky, horizon, and water surface, they know the lives of their shipmates depend on their spotting the enemy first. It takes only one hole to sink a sub. Survivors are a rarity.

It was time to start interrogating Kito again. This time when he entered the wardroom, the .45 was absent. Flexing my cold fingers, I prepared for our bout in sign language, aided by my one-page Japanese vocabulary. We spread out charts, but after greetings, Kito led off. Roark had taken him to the forward torpedo room for the reload earlier. Tomczyk provided the on-the-job training. He liked the work and wanted to become a torpedoman. I chuckled. This was a tact I hadn't anticipated. I dashed off a paper making him an official member of the torpedo reload party. *"Arrigato!"* he said, obviously pleased. I glanced at my vocabulary list—it said "thanks." Ah-so.

Bob and I found that Kito had more naval experience than his rate as a second class gunner's mate would warrant. He had twelve years of schooling. He had joined the Navy in 1935, and his dislike for service arose two years later. One day he had gone on shore liberty from his ship in Amoy, China. An Army sergeant in charge of the checkpoint post handed him a machete and ordered him to slice open the head of a surly Chinaman who had just passed through. Kito objected. The sergeant threatened him with immediate imprisonment if he disobeyed orders.

Knowing the Army had complete control of everything in China, Kito reluctantly carried out the order—and immediately vomited. He then returned to his ship and told his captain, who replied with a shrug. Kito never went ashore again in China. In 1939, having served his hitch, he left the Navy and became a shopkeeper. Recalled in 1941, he was assigned to the gun crew of the *Koto Maru* and had sailed to many ports.

We excused Kito, mused for a bit over the absolute domination the Japanese Army had over the country and the Emperor, then turned back to our never-ending study of the charts.

The little-known Okhotsk Sea is phenomenal. From depths of over 18,000 feet, it jumps up vertically to the Siberian shelf in the north. All the Kurile Islands seem to have active or dormant snow-covered volcanoes encircling the deep basin from Japan northward.

With the Pacific Ocean tides filling or emptying this Japanese teacup, one encounters fantastic whirlpools, which we avoided. Between the islands we ran into currents that caused us to steer 30° off course as the *Barb* was

Tuck Weaver, Gunner's Mate 2d Class Kito (prisoner from the *Koto Maru*), Gene Fluckey, and John Post on the afterdeck of the *Barb*'s bridge, 4 June 1944.

swept along. Our area was a kaleidoscope of microclimates. To top it all off, our freshly printed naval charts were copied from Swedish data taken in 1894. Why, I wondered, did we have to depend on such shaky information in such dangerous settings?

While we were eating supper, the watch officer sighted a large icefield and changed course to circle around it. Tuck was the only one in the *Barb* who had worked around ice. The rest of us left our half-empty plates to rush up to the bridge or the conning tower for a look. The *Barb* became a tourist bus as men climbed up and down. There was more to our interest, however, than idle curiosity.

Normally, flat ice can freeze no thicker than 15 feet. Knowing that the entire sea freezes over in winter, we were not anticipating icebergs and weathered columns of ice among the flat field. Using our periscope ranging device with our radar, we calculated the highest peak at 105 feet above sea surface. That meant some 900 feet of frozen mass below. These bergs could only have broken off from glaciers. Some may have been circling around the area for years before defrosting completely. The sun glistened off the ice as we watched the eerie shapes glide by.

I figured shore observers to the westward on Karafuto would be blinded to our presence. On the other hand, we had discovered that our submarine risked sinking or severe damage should we attempt to submerge and hide from the enemy beneath the field. We would be depth-charge proof, but our test depth was only 315 feet and those ice floes reached far below that depth.

Two hours later we dived to avoid a plane sighted at a range of five miles. When we surfaced ten minutes later, he had gone by. We were happy he hadn't ventured our way while we were sightseeing. I secured our air search radar because Japanese planes could detect it. To avoid any ice at night, we set course south to Hokkaido. After the wee-hours' fracas with the Chidori earlier, most were ready to hit the hay.

*Time 2211.* "CLEAR THE BRIDGE! DIVE! DIVE!" A-oo-ga! A-oo-ga! "Plane passing close aboard down port side." Bolting out of my bunk, I jumped through the watertight door opening to the control room. My bare feet skidded on a wet spot. As the British would say, I went ass over teakettle. Ending up on my rump, I bowled over Dick, the diving officer. Grinning at me as we untangled ourselves, he said, "Green board, pressure in the boat, captain. Gad, I thought I was being depth charged!"

No bombs. We never saw the plane. Visibility was zero, but all topside had heard it. With only four hours of night left, we couldn't afford to stay submerged unless it was absolutely necessary. We surfaced and continued our course.

In five minutes I was back in the arms of Morpheus.

During the late morning we submerged quickly as the fog suddenly lifted, leaving us nude, five miles off Abashiri. With a dearth of ship traffic, we considered a bombardment of the factories. Working into Abashiri Bay on the north coast of Hokkaido, we reconnoitered the area by periscope. Bob had the bright idea of asking Kito his opinion of the operation. Kito dampened our plans. Upon hearing that there were seven major air bases within a 30-mile radius that would give us the works, we chose discretion and retired from the field undefeated.

Walking back to the after torpedo room with Kito, I was surprised to find Gaines Smith, the leading torpedoman, lolling in a hammock made of cord. Laughing, because I had never seen a hammock in a submarine, I was informed that Kito was the manufacturer. Then Kito showed me how he did it, for he was working on a cord belt for me. I was fascinated watching him deftly tie knots with both his fingers and his toes at the same time. He was proud of his dexterity.

On my way forward, Radioman Edward Hinson told me that he was receiving Radio Tokyo clearly. He would pipe it into the crew's mess on their speaker so they could hear the infamous Tokyo Rose. Her nightly program had first gone on the air in late December 1941. I remembered reading about one of her first programs during the dark days after Pearl

Harbor. In her soft-spoken, impeccable English she had whispered in her seductive voice, "Where is the United States Fleet? I'll tell you where it is, boys. It's lying at the bottom of Pearl Harbor." Thereafter, she would play a sentimental song of the Thirties. In the earlier days of the war, however, submarines on patrol couldn't listen to Tokyo Rose because our broadcast receivers put out a superheterodyne beat—a radio emanation that could be picked up by enemy direction finders. Now we had different radio receivers with no restrictions.

*3 June. Time 2000.* The men jammed the crew's mess, listening to the singsong of Japanese music. Then came the soft, jeering voice. "Good evening all you American submariners. Do you really enjoy your illegal attacks on our defenseless merchant ships? You are only denying medicines and food to the women and children of our Greater Japan. Do such abominable acts make you bullies feel brave? When you return to your families, will they be ashamed of what you have done? Or will you ever see them again? Will you be one of the increasing number that will never go home? A message has just been handed me from a brave destroyer captain protecting La Perouse Strait after a fierce battle today. I can now report that another of your submarines has been lost. I will play an appropriate recording."

To the strains of "I've Got My Love to Keep Me Warm," the men burst out laughing. "Hey, captain, did you hear that? We got sunk!"

"Jernigan, I just wish Tokyo Rose had mentioned the *Herring* or *Golet* areas. That would give us an idea what they're doing."

Carroll Stowe, an electrician, added, "Find us some ships, sir, and we'll give her something more to talk about."

The record ended. Tokyo Rose concluded, "Well, fellows, guess who's keeping your love warm tonight?"

We spent the next few days along the convoy lanes slowly working north along Japanese Karafuto Island to the Siberian half, Sakhalin, where I thought we might encounter a floating Japanese fish cannery. Drift ice became a problem: 5 to 10 feet above the surface, leaving 45 to 90 feet below. According to our arctic ice experts, a ship could tell when it was approaching an iceberg by having a man watch the temperature of the water injection to the engines. After dark, Machinist Ralph Leier had the watch. On the bridge a short time later I heard, "Bridge. Injection temperature is dropping!"

BAM! We collided before I could order "ALL STOP!" Fortunately, the *Barb* was at slow speed. Thereafter, at night or in near-zero visibility in the vicinity of ice, we were forced to keep our speed at two knots to avoid damage. This hindered our search.

*6 June. Time 0600.* My seventh wedding anniversary. My loving wife was waiting patiently in Annapolis, Maryland. Wide awake in my bunk, all my thoughts dwelled on her. We had had a long courtship from December 1935 to June 1937, only because naval regulations required dismissal of any Naval Academy graduate who married within two years of leaving the academy. The philosophy was that an ensign's pay of $143 per month was insufficient for two people to live on. Furthermore, marriage would interfere with an officer's ability to absorb all phases of his duties.

The day I met my blonde Marjorie, I wrote home to my family telling them to remember her name, for I would some day marry this young lady. Then she refused to marry because she was a diabetic, saved from a slow death only by the discovery of insulin. Her doctor had decreed that she could never have children, so she didn't think it was fair to marry. Acknowledging that there would be no offspring, I convinced her in spite of her reservations. Exactly two years after my graduation, we were married—and broke. She was in charge of contraception. Nine months and ten days after our wedding, our golden-haired Barbara arrived.

By Marjorie's time it was 1500, 5 June. She would believe the *Barb* was still at Midway. In order to keep her from worrying, which affected her diabetes, I had written 20 love letters before we left. A friend would mail one every three days. I blew her picture a kiss; it was time to get moving.

With no ship contacts, my primary interest turned again to minefields. Kito came forward for more interrogation.

Minefields? Yes, Kito knew. Having been on the bridge a lot with his captain, he knew where some of them were. Bob and I traced the mine lines in La Perouse Strait in more detail on the charts. Following along the north coast of Hokkaido, all seemed clear. Then, as I inched my finger down its east coast, Kito suddenly stopped me at Akkeshi Wan (i.e., Atsukeshi Bay). A look of horror crept into my face.

"What's the matter, captain?"

"That's exactly where Jim Clark told me he was going to start his patrol with the *Golet*. Kito, what's there?"

"New minefield. Many mines. Thirteen meters. A hundred laid about two months ago."

I sent for Tuck. When he appeared I said, "The *Golet* planned to start her patrol at Akkeshi Wan. Kito tells us a new minefield was laid there in April. Have your radiomen try to communicate with her and the *Herring* every hour until midnight. We may be picked up by the direction finder net, but our risk is small compared to theirs."

Kito knew of no other minefields. He was excused. What he had told us was believable considering the great depths and fast currents in the Kurile straits.

Late that evening Dittmeyer sadly announced that we had been unable to contact either the *Golet* or *Herring*. Pearl Harbor also had tried in vain. There was an uncanny silence as the crew was informed. With a dry eye but a heavy heart, each of us knew that we had lost both our wolfpack mates. Weeks later the message would come announcing that they were overdue and presumed to be lost. The *Barb* was alone in this cold, cruel sea.

The officers of the deck had good practice at broken field running while dodging drift ice, until we ran into a startling wonderland. The *Barb* was surrounded by slender columns and pinnacles 50 to 60 feet high that looked as if they might topple over. Beautiful white seals basked on flats in sparkling sunlight; a lost world smothered in diamonds. We stopped and let the men come up on the bridge to gawk.

Clearing that area into the open sea, we started our sweeping search once more. Fleeting starch-white clouds painted the horizon. This prevented the watch officer from sighting a vast, solid pack of ice, 10 feet high, until it was only three miles away. We had been running along its side for 20 minutes when all hands on the bridge sighted the masts and funnels of four trawlers icebound by a quick freeze. Smoke was coming out of one of the funnels. I decided to lob a few shells into them.

"MAN BATTLE STATIONS GUNS!" The gongs resounded. Men came pouring out to their stations at the 4-inch gun, the 40-mm cannon, and the twin 20-mm mounts. No aircraft were on radar, so Tuck turned the *Barb* toward the edge of the floe to shorten the range. Swish, the gun captain, reported, "Sir, the pointer and trainer have the targets sighted in their telescopes."

"Okay, Swish. We'll just use the 4-inch. Remaining guns stand easy and be prepared to shoot at aircraft if we get caught on the surface."

As we drew closer, with all binoculars on the trawlers, I was just about ready to open fire—but something went wrong with my eyes. Lowering my binoculars, I noted Tuck looking at me with a startled expression. "Captain, the ice pack is moving away. May I speed up?"

It wasn't my eyes; the pack was moving. "All ahead standard!" The *Barb* lurched forward. The ice pack disappeared completely! "All stop!" We were alone on a flat, calm sea with not one single, solitary thing in sight!

The men were shaking their heads in wonderment. "Captain, may I secure from battle stations?"

"Go ahead, Tuck. I want to think."

Dick, who had had the watch when this started, relieved Tuck. But then he turned to me and said, "Captain, I hope we never run into that again. It's not only unnerving, it's downright spooky. I'd like to have someone take my watch, sir. I'm unfit."

One look at him and I agreed. "Dick, I'll take over. Go below and send up Jim Lanier."

A few minutes later Jim appeared from the control room donning his parka. "Captain, what's all this about? We're all set to start shooting. Everyone sees the targets, even through the periscope in the conning tower. We secure from battle stations. The targets disappear. The watch officer has to go below because he's too nervous. Nobody wants to talk about it. Has everyone gone nuts?"

"Jim, I just don't know. Take charge and be quiet. I need to think."

Somewhere I remembered reading, when I was a Boy Scout, of General Adolphus W. Greely's polar expeditions. Team members thought they were hallucinating due to some of the things they saw. Experience taught them that these were arctic mirages. Under certain conditions an atmospheric lens could portray exact images of scenes many, many miles distant. As I recalled, the most famous phenomenon came from an artist in northern China who painted the skyline of Constantinople long before Marco Polo's travels.

This disappearing island of ice was only a mirage, for sure. Yet at what distance would we find this floe? Twenty, 50, 100 miles? Of course, the trawlers might be Soviet.

I was almost sane again when Jim reported an ice field on the starboard bow. Now I felt like a grizzled arctic veteran who knew all about atmospheric lenses and mirages. Taking a good look through my binoculars, I decided it was another mirage and told Jim as much.

"Come on, captain, you've got to be kidding."

"Jim, you really believe it's an ice field?"

"Unless I'm touched in the head, sir."

"Well, let's settle it. I'll bet you a quart of whiskey when we reach port that it isn't there."

"You're on. What a lead-pipe cinch."

We changed course toward the purported ice field. It was there! Silence. I left the bridge quietly and retired to the wardroom. A few minutes later a messenger approached with his regular noon report. "Twelve o'clock and all chronometers are round"—instead of wound. I decided to turn in and take stock of myself.

Tomczyk's illegal diary for that day and the next one reads:

> *8 June.* Zigzagging on surface. Made radar contact 12,000 yards. Came in till we sighted the ship and started attack. Dove at 1820 and manned battle stations. Waiting for the ship to come to papa to get its farewell before visiting Davy Jones. Made ready the bow tubes at 1930 and standing by to fire. Secured the bow tubes and battle stations. Ship turned out to be a Russian 10,000 tonner with guns fore and aft and 20-mm anti-aircraft mounts scattered all over his topside. Boy was he lucky! Surfaced and pulled all six fish for a check.

> *9 June.* Making preparations to shell Etorofu Jima. Which lives up to its name, "Island in Shallow Water." Kept going in on the surface under cover of a fog. Came into 2500 yds. Fog suddenly lifted exposing us. Had to dive in a hurry. Water here very shallow—20 fathoms. Soundings not the same as charts. Captain Gene pulled out, surfaced and started in again from a different angle. Target is Cable Station which is in the middle of a 'U' shape cove. Kept going in until only 2 fathoms of water. Had to back down. Fog lifts and settles down like a yo-yo. Land can be seen on three sides again. Tried again with the same results—4 fathoms. Backed out and tried again. Captain Gene very determined. Heard him say "DAMN CHARTS!" Never heard him swear before. And he better not catch any of us using obscene words. Says it shows one is an uneducated stupido. Plenty of words in the English language to express oneself without sinking into gutter talk. Same trouble, shallow water. Sighted three small patrol boats on fourth attempt. At present we are laying dead in the water. Target in fog. Can't see to shoot. Backed out and tried a few more approaches, same results. Able to see three volcanoes busy at work throwing out smoke—close?—you said it. Challenged by one of the patrol boats. The jerk—if he only knew. Ignored him. Captain still wants to eliminate this cable station, so all Japanese ship movements will have to be passed by radio and be intercepted by good old USA. Backed out and headed north along the Kuriles. Am sure we'll return to polish this station off when he can see it. The night is cold as hell but for some reason, unknown (ha!), everyone is sweating. No wonder I'm getting bald."

Lying in my bunk, I prayed for the *Golet* and *Herring*. I knew they had gone down on eternal patrol. Though we had made only one sight contact with the *Herring* and none with the *Golet,* the feeling they were close by was comforting. Now alone, the *Barb* could only avenge. Farewell. God bless you gallant comrades.

# Chapter 3 Up in the Clouds

*10 June. Time 1327.* "Captain, smoke is sighted by the high periscope watch!"

"Dave, crank up four main engines and head for it at flank speed. Station the tracking party." Having a quartermaster on watch in the conning tower with periscope raised, while on the surface, was paying off. It made us more visible to the enemy, yet doubled the search area that we could cover visually.

The bearing change showed our target to be a small, high-speed ship. On a nearly parallel closing course, we saw the tops of the gray masts of a *Chidori*-class frigate. In unlimited visibility, we commenced an end around. The race was on with the *Barb*—unseen on the surface—abaft the enemy's beam at 17 knots against his 15.

*Time 1546.* Our struggle to get ahead for a submerged attack was punctuated by the appearance of a patrol plane at eight miles. From his maneuvers I judged that we hadn't been sighted; we kept going. After he circled the frigate, he became suspicious and came over to investigate. We dived. Satisfied that there was nothing there, our foe returned to his patrol.

We surfaced with the plane still visible. The 20 minutes we lost became a real handicap as we entered the homestretch to our barrier, the minefield. I wished we had a fog that would slow our opponent down. I wished we had submarines that could travel at speeds higher than nine knots submerged. What a different ball game it would be if we had submerged speeds competitive with those of the surface escorts.

*Time 1830.* Puffing and panting, we lost the race as the Chidori rounded the cape to cross the minefield. This time I had believed we could sink her with the torpedo running-depth set at five feet. Nothing ventured nothing gained. Yet it was a heartbreak to use up our precious diesel fuel in a five-hour chase without furthering the war effort. Perhaps tomorrow.

*11 June. Time 0230.* "Captain, we're 10 minutes away from Sakayehama Harbor. The night orders say that you want to take a look inside. Sea is flat calm. Air temperature a cold 26°. Visibility good. Solid white overcast, highlighted by the moon. Moonset behind mountains in about 30 minutes. Our speed 10 knots. No phosphorescence. Battery charge completed."

"Coming up, Tuck. Don't wake Bob McNitt. With dawn in another hour, he'll have to be up navigating."

We slowed to a crawl as we came into the harbor so that no one would see our wake. Tuck and I glued our eyes to our binoculars hopefully. Tuck's eyes were more night adapted than mine since he had been on the bridge for the last few hours. It takes about 20 minutes for one's eyes to adjust. "Tuck, see anything?"

"Quite clear. Perfect blackout. Lighthouse is off as are the channel buoys. . . . Just a minute! There's something alongside the wharf."

"Radar, give me a range to the shoreline."

"Twenty-one hundred yards, captain."

"Okay. Left full rudder. We'll coast in quietly on the battery. All engines stop and secure. I don't want them to hear us. Tell me when the range is 1650 yards." Gradually, we snuggled in until we were dead in the water. The silence was frozen and so were we, whispering to each other. Not a light was seen from any building.

"Sixteen hundred seventy yards. Range is steady." The command rolled out over the still water. I bent over the hatch. "For God's sake turn down your volume! Tuck, anything yet?"

"Sir, the harbor's empty. A few factories, warehouses. On the hill overlooking the wharves are a couple of cannons on wheels, much like a saluting battery. Pray they don't salute us."

"Don't worry. They'd have to identify us first. We could be one of their patrol boats. What's at the wharf?"

"One, two, three, four fishing sampans—and a tug." Tuck kept peering into the semi-darkness. "Hey, there are a few men moving about the sampans now. Perhaps getting ready for dawn fishing. Unless you want to bombard, we better move out."

"Concur. No bombardment. It might cause a diversion of ships from this lane. All motors back two-thirds. When you're clear, head north across the bay and get back on your engines." I bid Tuck good morning and went below.

Toward noon we were paralleling one of the ice floes. I was admiring the beautiful white seals basking on the ice when Tony Romaszewski, the port lookout, stooped down and tugged on the hood of my parka. "Captain, I think I saw a trawler hiding from us behind an iceberg. Hold your binoculars on that berg over there. The skipper's standing on his bow, peeping around the corner now and then."

Sure enough, Tony had picked up one that both Paul and I had missed. Our orders were to sweep the seas clean. Some fishing boats were armed, others not. Most had radios and served as pickets; they also supplied Japan with much-needed protein. Starve the enemy until they sue for peace? This one was probably hunting the friendly seals who didn't seem to have enough sense to defend themselves by even sliding into the water. Between orders and sympathy for the seals what must I do?

"MAN BATTLE STATIONS! Four-inch gun only." The trawler was 3000 yards away in ice that we dared not enter. The *Barb* now stood 10 yards off the floe. The inability to see the target provided our main problem. Gunnery Officer Al shrugged his shoulders.

Bob came up with an idea. "Maybe we can flush him out by putting a shell into the water to the right of the berg behind where his stern must be."

No sooner said than done. BOOM! Our jack rabbit quarry jumped out at full speed, dashing in and out among the bergs. We fired 30 rounds, all near misses that ruined many bergs. The 31st shot hammered into a larger one, tearing out a huge block of ice. It ricocheted off another and straight through the hull of the trawler, finishing her off. It was probably the only time an enemy has been sunk by gunfire without being hit.

Halfway through lunch another trawler appeared, heading for the ice field. We exchanged fire and finished her with the 4-inch. Lunch resumed. We'd saved the seals.

*Time 1332.* "Captain, high periscope reports two smoke streaks."

The targets were zigzagging on a southerly course to our east. Their base course to the south was meaningless: it went nowhere. With perfect visibility our end around would take four hours. They had to move westward.

*Time 1604.* Targets obliged and changed course. The *Barb* was now 20,000 yards dead ahead. We dove. "BATTLE STATIONS TORPEDOES!" We commenced the attack in a calm sea. Through the periscope I saw a freighter and a transport in echelon. Then I discovered heavy black clouds racing up from the south. "Looks like a tornado! Angle on the bow still zero, heading right at us. Range—mark. Down scope."

I hoped that we could shoot before anything happened.

"Up scope! Damn, can't see anything. Paul, bring her up to 58 feet. Mind your depth carefully. It's frothing on the surface. That's good; hold your present depth if you can without broaching. Waves are over five feet. We're in a cloudburst, visibility is about 5000 yards. Nothing in sight. Bob, see if sonar has contact."

"Nothing, captain. Too much noise from the squall."

"All right, let's try radar depth. Up scope."

*Time 1630.* The targets had vanished. They must have turned south again. The squall was moving off, the sea flattening. "These queer microclimates are baffling," I said. "Bob, surface and bend on four engines. They're not going to get away." I felt like a racehorse that had been left at the starting gate.

In a quarter of an hour, radar had them again going south. Soon, the lookout Paul Bluth spotted something that appeared to be ghost ships in the white haze. "Good work. That's them, right on the radar bearing."

*Time 1652.* We dived as visibility became unlimited. Then, as we commenced our approach, the ships changed their minds again and started a new series of zigs to the southwest. With wide visibility and the sea glassy, we couldn't surface. A submerged race was our answer. For three hours, at seven knots, we kept almost abreast without being able to close enough to let loose our fish.

*Time 2000.* They finally headed west again. Now we had them. "Make ready all six forward torpedo tubes. Slow to one-third speed."

"Captain, from your observations, plot shows their zig plan to be 30° to the right and left of base course on 10-minute legs. We'll be able to shoot on their next right zig with small gyro angles for the torpedoes and perfect track angles of 90° to 105°. It's in the bag."

"Good, Bob. Open the outer doors. Up scope! Wait a minute! The ships have jumped up, sailing along in the clouds!"

"Impossible!"

"Bob, it may be impossible, but that's the way it is. Now what depth do you set on the torpedoes to hit a skyship? Down scope."

Al chimed in. "We know he's actually on the water. We have the generated range because you can't take an accurate range with the periscope on a ship in the clouds. Captain, just give us continuous bearings and we can shoot."

Bob shook his head. "We must be in the mirage business again. If we're getting a vertical distortion in the periscope line of sight, we can just as well have a distortion in the horizontal bearing."

"Close the outer doors. It's almost sunset. I don't want to waste torpedoes on phantom ships. They're both still unalerted, and they have a long way to go. As soon as they are out of sight, we'll surface and go after them for a night surface attack. That way we won't have a mirage handicap. Secure from battle stations. Have the tracking party stand easy until we regain contact. Bob, you take a break now and tell our good steward, Ragland, to bring me a mug of tea. While you're in the wardroom, take a gander at the local charts. We could have a coastal problem tonight."

*Time 2050.* We surfaced with the targets in sight at 10 miles. Visibility was decreasing. In an hour radar picked them up. The sun set and the blackest night we had ever seen descended. With our six-knot speed advantage, we were closing at 200 yards per minute.

*Time 2300.* "MAN BATTLE STATIONS TORPEDOES!" The crew hurried to their stations, happy and excited about imminent action after the tedious 10-hour chase. Tuck joined me on the dark bridge. He took off his red goggles and stuffed them into the pocket of his parka. All those on night bridge duty wore them for the half hour prior to going on watch to assist in adapting their vision to the dark. As a test I raised my gloved hand, extending three fingers. "Tuck, how many?"

"Four."

"Nope." His eyes weren't ready yet, even with all the oatmeal cookies, raw carrots, and vitamins he ate daily like the rest of us. But night vision was one of our secret weapons, and tonight we were going to need it.

"Captain," Tuck said, "I've never seen a more jubilant crew going into action. What you can't understand is that all of us have been conditioned to think that a submarine should stay submerged all day and wait for targets to come by, then surface at night for fresh air and to charge batteries. On *Barb*'s last patrol we heard a lot of your conversations with the former skipper. Frankly, we thought that your ideas were a bit far out. Your concept of using a submarine like a motor torpedo boat that went after the enemy instead of the enemy coming to her just didn't gel with most of our experience. You sure didn't convince our skipper that *Barb* shouldn't stay submerged unless absolutely necessary. Now we're all convinced you were right. We've done more in 10 days than *Barb* did in all of her past seven patrols."

I nodded. "Tuck, you've had more combat experience than I have. Let me know when the men feel I'm overreaching and I'll talk to them. How many fingers?"

"One."

"Okay. Here's my plan."

At Paramushiru Island, the largest naval base in the Kuriles—

The *Chihaya Maru* (2738 gross tons, 1161 cargo tons) loaded out for departure at 1100 on 8 June. One hundred seventy-three soldiers and 260 tons of military cargo were on board. Other cargo consisted of salted fish and miscellany.

The crab factory ship *Toten Maru* (3823 gross tons) was fully loaded with 1500 tons of canned crabmeat and dried fish and 200 troops with military cargo.

Both ships were armed with cannons. Due to the absence of an escort, their two-ship convoy was routed to avoid the area where a wolfpack had been sinking ships off Matsuwa. Their orders were to proceed westward to Soviet waters off Sakhalin Island, thence down the coast of Karafuto and through La Perouse Strait on 12 June. They departed on schedule.

"Tuck, we'll make a surface attack using radar ranges and your bearings on the target bearing transmitter. The ships are in echelon, with the largest ship 250 yards to port and astern of the lead ship. I'll come in on the starboard beam, slowing to dim the phosphorescence. Shoot the leader first. Ready?"

"I'm on. Bearing—mark."

"Okay. Let's go!"

"Bob, how are we doing?"

"Perfect setup: plot and the torpedo data computer check exactly. Bearings check. Radar range 2500 yards."

"Great. All engines ahead one-third. Make ready all tubes fore and aft. Phosphorescence is very heavy, but I believe we can mosey in another 1000 yards without being seen. Use tubes 1, 2, and 3 on the lead ship and 4, 5, and 6 on the trailer. Tuck will mark midship bearings. Spread the torpedoes from aft forward with 50 yards between torpedoes. Tell the crew what we're doing. JK sonar, stay on the torpedoes. All hands on your toes. We're heading in!"

*Time 2330.* "Captain, range 1900 yards. All torpedo tubes ready."

"Aye. Tuck, a few minutes to final bearings. How do the ships look through your TBT telescope?"

"Like clear blob silhouettes. The one astern appears closer."

"That's only because she's larger. Open the tube outer doors. Broocks, keep a watch aft. Let me know if the phosphorescence increases. I'll watch forward. Final bearing."

"Captain, range 1590 yards; we're there."

*Time 2334.* "Bearing—mark."

"Set."

"FIRE 1! FIRE 2! FIRE 3! Helmsman, come left 10°. Sonar report. I saw first torpedo broach. Final bearing trailer. Give it to them, Tuck."

"Bearing—mark."

"Set. Range 1800 yards."

"FIRE 4! FIRE 5! FIRE 6!"

"Bridge. Sonar reports torpedoes 2 and 3 hot, straight, and normal. Torpedo 1 is hot and straight and appears to have left the surface now for its set depth of six feet. This will slow it down a bit. Torpedoes 4, 5, and 6 of second salvo are all hot, straight, and normal."

"Bridge, time for . . ."

WHAM! "Bob, torpedo hit in stern of leader. Two misses."

WHAM! Pause. WHAM! "Bob, one hit in stern of trailing ship and a second hit about two-thirds of the distance between her stern and the funnel. Another miss."

"Captain, what's that noise?"

"Second ship is blowing her whistle continuously. She's settling by the stern. Hope it's a signal to abandon ship. Left full rudder. All engines ahead full. We'll pull out and see what gives."

"Bob, trailer is sinking and blinking furiously at us or firing an automatic weapon in our general direction. Tuck says the lead ship is turning away. I'll keep our rudder on and make a complete circle to go after her. ALL AHEAD FLANK SPEED. She's not going to get away! We haven't got time for a reload, so we'll use the stern tubes. The situation will be changing rapidly. Tuck will stay on her constantly."

*Time 2339.* "Captain, fire control has no setup. We're confused and have lost the picture completely, as has plot."

"Bob, trailer has sunk. Lead ship is definitely crippled and slowing. She has now reversed course. We'll pass close aboard on opposite course, then position for a stern shot."

The *Chihaya Maru* fired her forward cannon.

BOOM! I ducked behind the waist-high, thin steel shielding around the bridge. "ALL TOPSIDE, TAKE COVER! GO BELOW!" Splash! The lookouts and quartermaster dropped down the hatch into the conning tower. "Bob, they're shooting at our wake. All ahead standard. I'm slowing to reduce the phosphorescence. We'll be too close! RIGHT FULL RUDDER!"

BOOM! Squatting, I looked around. "Tuck, what are you doing up here? I ordered everybody to take cover below."

"Captain, you're going to need me to take bearings if you fire a stern shot."

"True, Tuck. Thanks. Three sunk—two to go!"

BOOM! I couldn't keep from laughing.

"What's funny, captain?"

"Tuck, it's so idiotic to cower behind this thin sheeting. Any shell would pierce it."

Chuckling, we both stood up and rested our arms on the coping, looking at the target as the *Barb* sped by close aboard.

"Bridge, range opening 200 yards."

BOOM! Breathtaking, earsplitting, like a giant firecracker going off in front of your face. The radar watch swallowed his chewing gum. Our target was turning right rapidly, coasting. We crossed her stern.

"Helmsman, steady as you go. All ahead one-third. Bob, I know you haven't time to get a fire control solution. We're astern of the target. Her propellers have stopped; She's still turning right. I'll position for a stern tube shot to her starboard beam. Set your radar range. Set target speed at one-half knot. To obtain target course, set angle on the bow at 90° starboard. Starboard back one-third! Port ahead one-third. I'm twisting our tail to point our stern directly at her."

"Set."

*Time 2349.* "Final bearing—mark."

"Set."

"FIRE 7! FIRE 8! FIRE 9! SONAR, STAY ON THOSE PICKLES!"

"Sonar reports all torpedoes hot, straight, and normal." Tuck and I smiled at each other, holding up crossed fingers in breathless suspense as the seconds ticked away.

WHAM! WHAM! WHAM! BRRROOOM! The three hits and the magazine explosion blew her sky high with a blinding flash. The *Barb* shuddered, and then we were alone again.

After setting course for a roving patrol of the convoy lanes toward Paramushiru, I took a turn through the boat offering kudos to everyone for their flawless performances. It was a long and rewarding day, but no one wanted to sleep. The boat was a beehive abuzz with cheers, laughter, and the recounting of the quantity of honey we had brought home without getting stung. Two trawlers sunk. Four ships sunk. Satisfying, but we hadn't forgotten our two wolfpack mates, sunk, the eternal price of war.

Still, our immediate response was one of jubilation. "Four down—one to go! All hands splice the main brace!" Cheers rang throughout the boat. I thought, "Today we live; tomorrow—who knows? Best we be prepared and rested to avoid being bested. Yet, maybe we can cheat a bit on this and celebrate."

Through diplomacy Kito maneuvered himself to the head of the line as our depth-charge rations were poured out. We were a happy family. Each one depended on the others for his survival.

About 0200 people drifted off to their bunks, expecting a peaceful night. I don't know whether it was the letdown of relaxation or the depth-charge ration that affected me. I stifled only half a yawn before I started snoring.

Something woke me at 0405. Quickly, I snatched the red goggles off my desk and slipped them over my eyes, forgetting that it was already daylight. I lay still a moment. The captain's call buzzer wasn't ringing. The boat was quiet. What had awakened me? I was groping for the cause when it came to me that it wasn't a noise, it was the lack of a familiar noise.

Just outside my cabin, above the forward battery that would fill a garage, was a deck hatch to the battery. Every hour an electrician opened it to take the specific gravity reading that indicated the battery's state of charge. It was something even my subconscious had become vitally concerned with. I grabbed the telephone and punched the maneuvering room button. "What happened to the 0400 battery reading?"

"Sorry, sir. Just finishing a cup of coffee—coming right up." Snoring again, I never heard him arrive.

"Sinkings cake time—five-minute call!" I opened my eyes. 0855. I hadn't slept this long since the patrol began. The officers were becoming more and more confident and sure of their responsibilities, so they didn't wake me as often. A lot of this was due to Bob's tutelage. How fortunate I was to have him as my executive officer; he was perfect in every respect.

The masterpiece celebration cakes became even larger and more masterful. This one was iced with icebergs, two trawlers sinking, and two ships sinking. Kito requested, in fair English, a larger slice because he was becoming such a good torpedoman. The message of survival had been driven home to him with the noise of the cannon firings close aboard. He said these sounded like drumbeats on the hull. As a matter of self-preservation, he now considered it essential that their source be eliminated.

Our celebration over, Tuck, Bob, and I broke out our ship identification manual to determine by consensus what two ships had been sunk. Due to poor visibility we could only approximate, yet we agreed that the last ship—which exploded and sank—was two-thirds the size of the other.

*13 June. Time 0230.* The destroyer *Hatsuharu* got under way from the wharf at the Kataoka (i.e., Paramushiru Island) Naval Base. Following her astern was the 5633-ton *Takashima Maru,* a large Army transport that was one of the very few icebreakers that had not been sunk. Built in 1942, she was heavily armed and carrying 600 troops who were being moved to the Philippines along with their equipment. The convoy commander had specific instructions regarding this important mission. Once clear of the harbor and the small islands to the westward, he was to increase speed to 14 knots and set his course northward, using Soviet waters along the west coast of Kamchatka. Thence the convoy was to cut westward, zigzagging, with the base course toward the northern part of Sakhalin Island. The lower half, Karafuto, had been given to Japan when President Theodore Roosevelt mediated the end of the Russo-Japanese War.

The basic plan to counter the wolfpack, which had been causing so many sinkings to the south, was to cross the Okhotsk in steps and stairs, using 90° base course changes every four hours. Superimposed on this base course was a zigzag plan wherein legs varied from 3 to 15 minutes. Some zig courses would be as much as 70° off the base course to befuddle the submarines.

*Time 1713.* "Captain, visibility is clearing. I have smoke to port and to starboard."

"Aye, don't speed up!"

On my way to the bridge I stopped in the conning tower to take a look through the high periscope. To starboard, a lone ship was heading for Paramushiru. To port, a larger ship with a destroyer escort was well ahead of us on a parallel course, heading for Soviet waters off Sakhalin Island. Lowering the scope to avoid our being detected, I had Dick put on four engines at flank speed and station the tracking party. I could see the superstructure of all the ships, so we had to keep the periscope down. My decision was to go after the larger ship and destroyer, broad on our port bow.

Tuck relieved Dick on the bridge so he could take his place with Al at the torpedo data computer. Bob arrived and Dave went to the radar with Brendle. John and Jim manned the plot in the control room.

"Bob, the escorted target must be more valuable. It's a transport. Speed will be higher and they'll have a more complex zigzag plan. Only take a peek with the periscope every 10 minutes until we have their masts in sight from the bridge. Then we'll give the bearings. End around as close as we dare. Our speed advantage is only a few knots. I'm going up."

"Aye, captain."

On the bridge I could see that we were still edging in toward the smoke. With our small relative-speed advantage, we needed to end around on the surface, unseen, as close as possible. I took a look at the color of the horizon's background against which the enemy lookouts might spot our periscope shears as we passed by. The clouds were pure white. The camouflage paint on our vertical parts was a light gray—not white enough. Our horizontal parts were painted black to avoid being seen when submerged.

I had an idea. In the torpedo rooms were some spools of strong, light line. "Control room. We need to improve the color of our camouflage on the shears. Send up four bed sheets from the wardroom to tie around them. Also, I need strong men—Laughter, Wierski, Williams, and Powell—with a spool of light line. Lehman can assist with the radar mast and Bochenko with the radio antenna."

"Aye, aye, sir."

John spoke up. "Captain, this plot is the craziest one I've ever seen. At times the target almost reverses course."

"John, he's the boss. Fortunately, we're on the surface and can work in and out with him. We'd never be able to keep up with his gyrations submerged."

The bed sheet end around was working. Instead of circling around the target at a distance of about 10 miles, we had shortened the radius to 6.

*Time 2036.* After three hours we were 12,000 yards ahead of the target, a position that every submarine would like to achieve before submerging and boring in for the attack. Yet I felt uncomfortable. Something was wrong. Tuck had his finger on the klaxon. "Are we going to dive, sir?"

"I don't know, Tuck. Take charge. I'm going below for a conference."

Bob and I dropped down to the control room to study the plot showing the target's positions since we first made contact. Jim and John had sketched in the approximate base course of all the zigs and zags, which were wild. "Captain, why don't we war-game the possibilities?"

"Great idea, Bob. I'll pick five submerged positions for the *Barb* 12,000 yards ahead, based on random points of his base course. You and I will play the *Barb* part. Jim and John will play the enemy movements, assuming that we remain undetected."

Ignoring the target's known past movements, I marked an X five times along the line. The targets followed the exact movements they had made on the plot at that time. The results: The *Barb* had three successful attacks and twice was way out in left field, and the targets escaped. "Not good enough, captain," Bob said. "With night approaching, he may change his base course and further reduce our chance of success."

"Concur. We must stay surfaced to preserve our speed advantage. Keep tracking for a night surface attack."

As twilight faded, the convoy changed the base course to the west, which left us with a short, high-speed end around. Darkness brought with it a moonless night, a light overcast, and the worst phosphorescence we had experienced. The loom of this on the steam from our engine exhausts gave one the impression that we had two searchlights trained aft.

I turned to the quartermaster. "Shankles, lend me your pack of cigarettes for a moment." Holding them over the side of the bridge rail, I read aloud the complete fine print on the back of the pack from the loom of the phosphorescence.

"Problem, captain?"

"Tuck, we are going to have to change our strategy and tactics. This is too much. We'll be sighted. I don't like the idea of submerging for a radar depth approach when we are ahead again. They might get away. I want to shoot for a sure sinking at about 1500 yards. The destroyer, at higher speed, is patrolling close ahead, working back and forth from beam to beam. He'll see us unless . . ."

"Unless what, sir?"

"Conning tower and plot, listen carefully. The phosphorescence is incredible and will ruin a normal attack. We have only six torpedoes left, four forward and two aft. When we're in position ahead again, we'll stop, lie to, secure the engines, and answer bells on the battery. Note the time it takes the escort to move from port beam to starboard beam. Now, so that we can escape from the stopped position if something goes wrong, I'll keep our stern pointed at her. This means we have no choice other than to use our stern tubes. Those two torpedoes must hit. Understood?" Both responded affirmatively. "Then make ready all tubes. BATTLE STATIONS TORPEDOES!"

*Time 2223.* "All stop. Secure the engines. Open the outer tube doors. Tuck's got the ship now on the TBT aft. Ship has a zero angle on the bow, heading right at us. Escort is just crossing her bow from port to starboard. Set torpedo depth six feet. If she doesn't zig, we may have to dive in a hurry!"

Her phosphorescent bow wave was tremendous as she bore down on us. "Come on, zig, please!" We stopped breathing.

*Time 2229.* "Captain, she's zigging to starboard! Thank God!"

"Tuck, steady on her. Al, she's still turning. Angle on the bow. Port— 10, 20, 30, steady. Spread your torpedoes from aft forward. Range?"

"1450 yards."

*Time 2231.* "Final bearing—mark."

"Set."

"FIRE 9! FIRE 10! SONAR! STAY ON THOSE TORPEDOES!"

"Sir, torpedo run 45 seconds, torpedo track 60° port. Sonar reports both torpedoes hot, straight, and normal."

WHAM! WHAM! The flash of each explosion lit up the ship for an instant, followed by the green phosphorescent cascade of the descending plumes of water.

"Bob, the first pickle hit in her stern and blew off part of the fantail. The second hit halfway between her stern and funnel. Target is stopped and settling aft with her whistle blowing. Sweet music." Cheers welled up from below.

"Bridge. Radar jamming."

"ALL AHEAD FULL! Here comes the escort!" The race commenced with her dropping depth charges. FLOOM! FLOOM! "Radar, keep on the destroyer."

"Range is decreasing rapidly; 3000 yards." Finally, at 2600 yards, she turned back when she was winning the race. We stopped to observe, two miles away.

The destroyer maneuvered alongside the sinking transport with her searchlight illuminating soldiers with emergency lanterns lowering the huge boarding ladder from the main deck. Immediately it was crammed with troops pouring into the destroyer.

In barely two minutes the destroyer suddenly pulled away as if she were escaping from something. Panic resulted. The pressure of the crush of men coming down the ladder was such that those on the bottom landing were forced overboard.

*Time 2241.* In the dim light that had been focused on the ladder, I could see men hanging outside the rail and dropping into the sea. Then I saw the escort heading for us again!

Range decreased as the destroyer, unencumbered by the *Takashima,* made her speed advantage felt. Again, she laid down a depth-charge pattern, turned, and went back to the ship. This scene repeated itself again and again. Five minutes alongside and 10 minutes chasing us. It was extremely nerve-racking for all.

Every time I moved the *Barb* in for a possible attack on the escort, the destroyer came after us. It was difficult to comprehend my target's motives. We had continuous radar jamming; his radar must have picked us up inside of 3500 yards. Why was he breaking off his chase and dropping a depth-charge salvo? He must think that we had submerged. It was very eerie.

In the black, yet clear, night the weird chases frothed the green phosphorescence, and the depth-charge explosions rumbled against our hull. It was a monstrous night. "Tuck, I'll drop down to radar to see how the sinking is progressing. Her pip should be diminishing."

As I talked with Bob, something caught my eye. Glancing down the hatch into the control room, I was shocked to see Reb Holt (a fictitious name), frozen with fear, gripping the sides of the ladder, his face stark white. "Bob, take charge of any evasion. I have to go below for a while."

Skidding down the ladder, I put my arm around Reb's shoulder. "Reb, let's go up to the wardroom and talk."

He was silent until we passed through the watertight door of the forward battery room. "Captain, don't dope me, please. Please don't dope me!"

"Reb, nobody's going to dope you. We're going to sit down calmly, have a cup of coffee, relax, and talk. Nothing else. Jackson, bring us a couple of cups of coffee."

"Captain, I'm scared. I've never had a night like this. It's horrible. Those depth charges pounding against the hull as if I was in a coffin. It's like thunder below! We'll be lost!"

"Reb, wake up! They're going to be lost, not us. This is why I brought you here. The enemy is sharp, but we're sharper. Until you come to your senses, I am not going to have you infect the crew with fear. I need every man on board to be at his best in combat. Fear breeds errors, and there's no room for that tonight. Our target is sinking. You stay here until it does."

"I'm sorry, captain. This submarining isn't for me."

"Jackson! This is an order. Reb is to stay here until we secure from battle stations. Reb, drink your coffee. This isn't the end of the world."

*14 June. Time 0003.* Bob yawned. "Captain, we've less than two hours of darkness remaining. I recommend we circle around to the other side of the ship and finish her off with one torpedo, since we can't get within torpedo range while the escort is alongside."

"Agreed. She needs a *coup de grâce* so they can't tow her away to come back again another day." We needed a battery charge before dawn after all the high-speed running we'd done without being able to use the engines. And search planes would be coming. I was sure they had reported us.

"Tuck, go ahead two-thirds speed. Open the outer door on tube one."

*Time 0016.* The *Barb* was halfway around when the lookout bellowed: "The escort's pulling clear again!" All binoculars were turned in that direction. The bow of the damaged ship slowly rose majestically in the air. Then it seemed as though an unseen hand grasped the tortured hull and dragged it to its frigid bosom. The death of a beautiful ship; the hideous side of war.

The destroyer headed down our old track, depth charging desperately. I called the control room. "Fill one of those empty five-gallon milk tins with oil, punch a couple of holes in the top, and send it up to the bridge."

Sam Turnage soon appeared, lifting the can gently through the hatch. "What do you want me to do with your oil, sir?"

"Sam, climb down on deck and toss it overboard. This will give our friend an oil slick to depth charge at daylight. Hurry before you freeze."

"Tuck, we haven't enough juice left in the battery to have a pillow fight with the destroyer. Let's get out of here. Secure the outer door on tube one. Crank up all four engines at flank speed. When we're well clear, slow to standard speed and put two engines on charge.

"Bob, secure from battle stations. Give us a course for lower Kamchatka. We'll make a sweep along the coast for any floating Japanese canneries and check our new-found convoy lanes. They seem to be avoiding subs by using Soviet waters. What's our depth-charge count so far?"

"Course 060, sir. Depth charges 29, still dropping. I hope his sonar isn't pinging on seals."

"Could be—or baby whales. Tuck, I'm going below. Have your relief watch our foaming-mad enemy."

Passing through the joyful control room, I cut on the speaker. "Hear ye! Five sunk—zero to go! Tonight the *Barb* put to sleep a large transport full of troops. Despite the worrisome and spooky circumstances that forced a change in normal tactics, all torpedoes struck home. Her fantail blew off so she wasn't going any place. To preserve our last four torpedoes, we had to wait almost two hours for her to sink. Meanwhile, the destroyer escort attempted to plaster us with ashcans. True, the rumbling echo on the hull sounds like 'thunder below,' as one shipmate described it. It can be unnerving.

"Listen! There, he dropped another one. But really, is there any sound more pleasant for a submariner than the sound of distant depth-charge explosions? For what the *Barb* did tonight, Well done!

"Now, until she stops dropping, it will be difficult to sleep. Let's splice the main brace!"

Cheers and more cheers filled the *Barb*.

No one was in the wardroom except Reb. Smiling, I looked at him. "Captain, I'm okay now. You're right. It depends on the way you look at it. It is a pleasant sound. I won't let you down again. Thanks."

"Forget it. Join the others."

Tomczyk, recounting the day's events, scribbled a few interesting quips in his illegal diary.

> Sighted smoke on the horizon. To port was a lone ship—a cinch. To starboard a larger ship with an escort. I guess you know the Barb picked the one with the escort.
>
> Both torpedoes hit the large ship in the belly. It started to sink. The escort came around and chased us. Kept closing the range, dropping

The *Takashima Maru,* a passenger-cargo transport and icebreaker, was sunk on 13 June 1944.

depth charges all the way. Well everybody was practically dropping it in their pants.

Instead of diving the Captain outsmarted him, thank God, and kept running on the surface. When the escort got to the spot where we should have been, but weren't, he dropped well over twenty charges. We hung around to make sure the ship goes all the way down. Escort kept pinging and searching for us—dropping charges here and there, but to no avail. Ship finally sank. It's a good thing the night was real dark or we might have also been tagged. Standing by for anything.

*Time 0315.* Final depth charge exploded. Total dropped, 38. Ah, sleep, that knits up the ravelled sleeve of care balm of each day's life, each night's oblivion.

*Postscript:* The Japanese reported the passenger-cargo icebreaker *Takashima\* Maru* sunk. The destroyer *Hatsuhara* reported that we had fired four single torpedoes at her while she was alongside removing troops. We did not.

---

\* "Falcon island"

# Chapter 4   Trapped

Late in the morning we were gathered around the cake. On it, the bow of the transport was above the surface. The destroyer was dropping depth charges as plumes of water surged up. The *Barb* was in a corner watching. I cut the first piece.

"DIVE! DIVE! PLANE HEADING IN!" As klaxons blared, the cooks grabbed the cake. I cleared the diving-planes bench because the bow and stern planesmen had to take refuge behind the conning tower ladder until the lookouts dropped down to take over the planes from the control room watch.

"Green board, sir. Pressure in the boat."

"Take her to 62 feet."

"Captain, we picked it up at eight miles. I don't believe we were seen."

The bow and stern diving planes manned, I climbed up into the conning tower. I swung the periscope around in a full circle; the plane was there. Gradually, it turned away. We stayed submerged so we could have our cake and eat it too—peacefully—as well as do routine maintenance on the four forward torpedoes.

After lunch we surfaced again, ever searching.

Toward evening, atmospheric lenses kept us fascinated. We sighted the island of Araido To clearly at 90 miles. The next day, off the west coast of Kamchatka, was an up-and-down day. The *Barb* dived to avoid numerous aircraft contacts, then identified them as Soviet. The same was true for our submerged approaches on various ships, all finally confirmed as Soviet. We found no floating canneries, but there were many ashore. Another day brought forth heavy fog as we searched off the west coast of Paramushiru. Finally, we contacted a ship steering various courses and went to battle stations, but we couldn't get close enough to safely identify her. We secured from battle stations and trailed her for seven exasperating hours until dark.

Then we were able to close to about 100 yards off the target's track in order to identify her and still be out of reach of her Y-gun depth-charge throwers and stern racks, should she turn out to be a destroyer.

"Tuck, secure the engines. Motors are quieter. We don't want her to hear us before we hear her."

"Captain, range is less than 1000 yards. Do you still want to close?"

"Bob, if she's Soviet, she should have her running lights on. This fog is very dense. I can barely see our stern. Of course, it's not lit. We'll close in more. Open the outer tube doors just in case she shoots first. I'm sure you have a setup."

"Sir, after nine hours of tracking, we have her speed accurate to one-thousandth of a knot. You can fire any time."

*16 June. Time 2200.* "Range 450 yards."

"Bob, nothing yet. Must be Japanese. Turn down your volume. I know it's hairy, yet I'll move in to 100 yards to identify. If we don't see any lights, we can move out to 400 yards on her beam to get enough run so the torpedoes will arm, explode, and sink her."

The *Barb* made approaches to firing position on 13 ships like this one before recognizing them by their side markings as Russian.

"Range 310 yards."

"Captain, I can hear an accordion." We listened.

"Tuck, that's a Russian balalaika, plus there are stern lights." We secured from battle stations.

At 2220 we picked up another contact that also turned out to be Soviet. It was most annoying to the whole crew. We wondered: Why don't the Soviets declare war on Japan and open a second front here—or at least help us? It would be such a tremendous assist if they reported Japanese ships from the sightings by their merchant fleet. From Kamchatka, land observation of Paramushiru and air observation of the harbors there would enable us to clear the Okhotsk of shipping.

So far on the patrol we had chased 13 Soviet ships and 6 Japanese. In the Pacific the Soviets were not only a zero, they were a negative force. It was another day of frustration.

We passed a week circling all of the islands, searching their harbors and conducting a photo reconnaissance of Matsuwa. The latter was requested of us after our small North Pacific fleet based in the Aleutians conducted a lightning night gun-bombardment strike.

Submerged, we moved to within 1500 yards off the air base and empty harbor for photos. I swept from one edge of the land to the other with the periscope at high power and clicked away with the camera.

"Bob, they won't be too happy with these. Most of their shells were overs; there's very little damage. They also requested us to capture a prisoner from a patrol boat if possible to assist in damage assessment, but there is no one here."

Our mission accomplished, we moved out, surfaced, and headed for more fruitful waters. As we entered a strait between two islands, Dick sighted a vast whirlpool like a giant saucer more than half a mile in diameter. I was on the bridge, hypnotized by the counterclockwise motion of the swirling waters.

"Captain, permission to steer around?"

Usually we avoided these, but I was hungry for action. "Now hear this. The *Barb* is going to show us what happens when we pass across a whirlpool. Don't worry, we've got plenty of power to pull clear if need be. As a precaution, however, stand by to close all watertight doors, or to dive if we sound the collision alarm. We'll start down one side of the saucer in about two minutes. Hold your hats!"

Dick piped, "This is gripping!"

The *Barb* tipped over and started down the shallow crater. What would happen?

The helmsman called up through the hatch. "Captain, I can't hold her on course!"

"Hofferber, take it easy. The current striking our bow should be faster than the current striking our stern. If you can't hold course with left full rudder, I'll speed up the port motor to assist, or back the starboard if necessary."

The *Barb* now listed to port some 5°. Evidently, the current on the keel was greater than that on the surface. "I'm okay now, sir. She's steady on course with 15° left rudder."

"Great, hang in there!"

Basking in the sunlight as we glided down toward the center of the crater with our 5° list to port, I felt very smug. So proud was I of having figured out the forces in the whirlpool that I cautioned the helmsman, "Ed, as soon as we start up the other side of the whirlpool, you'll have to shift to hard right rudder. The *Barb* will list to starboard."

"Aye, sir."

We leveled off at the bottom of the crater and pitched up. "Sir, I can't hold course with right rudder. She's going to circle out of control!"

"For God's sake, put on hard left rudder then."

Dick burst out laughing. "We've still a list to port."

"Sir, she's holding steady with 15° left rudder."

I *would* open my big mouth. What a perverse gal. She did it on purpose, just to make me look like a blasted "know it all." What happened? Up and over the hump, we secured from our adventure. The main benefit was that it gave the crew something to talk and write home about without it being censored.

From Tomczyk's illegal diary for this period:

*21 June.* Somewhere near Paramushiru on the surface as usual. Things are pretty quiet. Shallow water, 100 feet, the deepest up till now, and a helluva lot lower most of the time. The fun started. Had to dive four times for Jap planes. Guess we weren't spotted, no bombs. Thank goodness. Chased a ship. Manned battle stations. Identified as Russky, one of our Liberty ships. Secured and surfaced 1000 yards away from him. Boy—you should have seen the commotion and the faces of that crew when they realized how close they came to visiting Davy Jones. They went and we went about our business looking for RATS.

Captain Gene got some good pictures at Matsuwa. Came in so close that we can plainly see their 5 inch gun batteries. Boy they're mean looking.

*22 June.* Six miles off Paramushiru. Been on topside today, a real treat, which Gene permits. Witnessed one of the most beautiful sights I ever did see. High mountains reaching up 3000 feet, with their peaks all covered with snow. Volcanoes alive and breathing off their smoke and fire. Have seen over 25 planes practicing dog-fighting over land.

*23 June.* Paramushiru. No luck. No ships. Going in to the mouth of the harbor for our first full day submerged. Dove at 0200, daybreak. Destroyer Base on the stern of us. Submarine Base to starboard and airfields all around. Sonar contacts and periscope sightings—Patrol Boats. At least eight. Some fun, smoking lamp out, and with the heaters turned off it's as cold as a witch's lip. By 1800 air started to become very foul and breathing was difficult.

*Time 1953.* Everyone jumped up in alarm. I made a dash for the conning tower. The ominous sounds—continuous rubbing, clanking, and scraping noises against the hull—had everyone in a tizzy. Are we in a minefield?

"Dave, what have you seen through the periscope?"

"Nothing but a couple of sampans at a distance."

I took another 360° swing. "We're towing two red flags astern! Pass the word to the crew—before they use up all the remaining oxygen—that we're caught in a fishnet. Start the fathometer."

"Forty-six fathoms, sir."

"All stop. Starks, take her down easy to 150 feet. Maybe we can get out through the bottom of the net. At worst, we'll drag the red flags out of sight. With their movement some fishermen may think they've caught a whale." The clankings and scrapings continued. Bob and I conferred.

"Sounding 45 fathoms."

"Frank, ease her down to 200 feet. Maintain a slight down angle. We'll try backing her clear."

"Two hundred feet. Two degrees down angle."

"All back full. Try to keep her between 150 and 200 feet. We'll back for 10 minutes."

*Time 2020.* The clankings stopped, then the rubbings and the scrapings. We were clear. Sighs of relief.

*Time 2131.* What luck! We were caught in another net, the same noises grinding against the hull. "Stand by to surface. Arm the 4-inch gun crew with knives ready to come up to the bridge and cut us free."

*Time 2141.* Surfaced. The *Barb* was covered: bridge, shears, and deck were encased in a strong net of line as thick as your thumb. In the last of the twilight we lay to. The cutting party—augmented by the rest of the gun crews—climbed all over the topside unleashing the *Barb* once more. I noted that piles of clippings were passed down below as souvenirs. The material of the nets intrigued me because it was so difficult to cut. It was made of neither twine nor hemp and was three-quarters of an inch in diameter. Prying a bit apart, I found it was made of an exceedingly tough rolled wax paper. The whole net was buoyed with wood and weighted with lead pellets and rocks. Quite ingenious.

"Hey, Captain. We got a problem!" Swish called from the deck aft.

"Not so loud. We have four sampans at about 3000 yards. Come forward."

"Sir, this net is wrapped tight around the propellers. We can't pull it loose. Someone has to dive over the side, follow the net down, and clear the screws."

"Wow! In this 34° water. Who's our best swimmer?"

"Traville Houston."

"Has he done any diving?"

"Has he! You know that 100-foot diving tower at Pearl, where we train to escape from a submarine? Well, he's the only one I know of who's dived in at the top, swum down to the bottom, and back up again without any outside assistance."

"Sounds like a human fish. As he's the first loader for the 4-inch gun, he's somewhere down on deck with you. Both of you come up to the bridge."

They arrived. Houston volunteered. "I can do it, sir. I need to put on my flannel underwear and a diver's belt with a knife. Swish can hand-pump air to the face mask. The props are 15 to 18 feet below the surface. It'll be very dark down there."

We couldn't use lights because of the sampans, but we could bring the props closer to the surface by flooding the bow tanks. That should bring the props up to about seven feet below the surface. Fortunately, the *Barb* was only a 1000 miles below the Arctic Circle, so this dim twilight might last for another hour. "Houston, you should be able to see the shape of the props. Now let's get cracking and bring some towels and blankets up."

In jig time Houston was over the side and clawing his way down the net to the props. Five minutes later, his head bobbed up. He was hauled on board. Trouble!

"Swish, the air line is too cumbersome. Let's use the mask plug. I need a few more dives. Good God, it's cold." He dove in again.

In four minutes he reappeared, gulped a few deep breaths, and disappeared again. After several more performances I became worried and told Swish to bring him aboard.

Houston had now been in the water almost 30 minutes. Swish gave two jerks on the tether line and Houston's head bobbed up alongside. Swish wanted him out, but Houston objected. "Captain, he's almost finished cutting the net loose and wants to finish the job now."

"Okay. But he's lost a lot of body heat. It's dangerous. Tell him not to overdo it."

Down he went for almost five minutes, came up, gulped air, and went down again. I sent for Doc Donnelly. "Doc, get down on deck and get Houston aboard. He'll kill himself or get pneumonia."

As Doc headed aft, Houston emerged holding his fingers on high in a "V," but he was too far gone to climb aboard. The men on deck formed a human chain and lifted him up bodily. He was shaking so badly he could hardly move. Wrapped in swaddling blankets, he was passed down through the after torpedo room hatch and on to his bunk. Doc started working on him with medicinal brandy.

We cast the slashed net over the side and tested each shaft at slow speed. No noises. The *Barb* was ready to answer bells.

At dawn, before my customary trip to the bridge at the end of navigator's twilight, I dropped by Houston's bunk to see how he was doing. Doc was resting in a chair next to him and motioned for me to be silent. Then he gave a thumb's up as our diver slept. I tiptoed away, happy.

Two different sunsets fascinated us. The first caused me to ask our navigator, Bob, "What time is sunset?" The mirage split the sun into round suns. As we watched one sun set, the other suddenly dropped below the horizon, as if someone were drawing down a shade. The next day the sun elongated and set in the shape of a pagoda.

*29 June. Time 0030.* We entered Kunashiri Strait for a planned bombardment of the cable station. Heavy fog inside the cove at Kushibetsu, Etorofu Island, had defeated our attempts to bombard on 9 June. Tonight was different: we had a clear sky and a half moon. The landscape was startling, particularly the beautiful snow-covered, steaming volcano bathed in moonlight. The crew was awake, standing by for the call to battle stations.

As we neared the promontory of the cove, radar contacted a small boat. Recalling the patrol boat that we had found here before, we passed beyond the range of her automatic weapons. The cape rounded, we saw a white blob covering the cable station area. How disappointing! "All stop. Bob, maybe we can wait until the fog pocket dissolves. I'll wish it away."

"Captain, the patrol boat is blinking at us."

"Tuck, it's probably some sort of identification challenge."

"She's speeding up and heading this way."

"ALL AHEAD FULL! RIGHT FULL RUDDER!" We moved out, keeping our distance.

"She's turned around, returning to the cable station."

"Do likewise."

Back and lying to again, we saw the blinking once more. "Shankles, get the blinker gun and answer the skipper's challenge." Our reply resulted in a much longer message.

"Send him, 'Japanese submarine R-24' in English."

No reply. "Here he comes again."

"Tuck, evade." After three minutes he turned around and so did we. "Bob, that fog pocket is as thick as ever. We'll shove off in 15 minutes, but this patrol boat interests me. As soon as he chases us out into deep water, he leaves. I think he's scared."

"Here he comes again!"

"Okay. Tuck, I'll take over. ALL AHEAD FULL SPEED! Let's go right at him!" The cat-and-mouse game ended. The patrol boat spun around and headed for the beach. "All stop." I didn't want to join him. He drove the boat right up onto the beach. The crew jumped out and scurried up into the hills. "Bob, I give up. There's nothing left for us in this area, and we're almost out of torpedoes. Secure from bombardment. It's still a good target. Maybe the *Barb* will return another day. Let's go home."

The Okhotsk Sea, being under Japanese control, had other ideas. The last stretch of the strait brought tail currents of seven knots that rushed us out. To maintain our desired course, we steered 30° away from it. The sea blending into the Pacific Ocean had a peculiar chop: the barometer took a skid; waves bounced up to 35 feet; and the wind shot up to 60 knots. We slowed to six knots, the bridge looking like a bathtub. As cold water welled up inside my parka, I watched the bubbles oozing up by my chest. The Okhotsk was sending us off with a final warning never to return.

Yet submarines have an alternative to the mayhem of cruel seas: dive. Although wild and wooly topside, the sea was blissfully calm below. Men resumed their letter writing. At periscope depth we were still rolling 10°, so down we went to 200 feet where calm prevailed. Here we quietly said good-bye to the forever-fascinating Okhotsk Sea.

Leaving Imperial Japanese waters, we still maintained a taut alertness, though our possible enemy contacts consisted only of a chance encounter with a weather picket, a far-ranging patrol plane, or a returning Japanese submarine. With danger diminished, an aura of relative, restful peace settled on us.

Bob and I were taking a break from patrol report writing with a game of cribbage. Top Electrician John Hogan interrupted our respite. "Captain, we have a leak in our main induction line. It will have to be fixed."

"How bad? Can the pumps handle it?"

"Sir, it appears to be quite a lot, though that may be due to the pressure at this depth. The bilge pump is handling it so far." Bob was smiling.

"Hogan, in the middle of the *Darb's* last patrol, a similar problem occurred. Lieutenant McNitt volunteered to crawl through the length of the main induction air pipe from the engine room to find and fix the leak. Bob, what's it like inside?"

"As I recall, it's a 22-inch pipe with interior ribs, wet, smelly, and too small for someone to be on his hands and knees. When I got up to the vertical section, I had to crawl back for another wrench to tighten the packing on the valve stem while we were submerged."

"Enough." At this stage it was too dangerous for anyone to crawl through the main induction line while we were submerged in a combat zone. The only such case I knew of was that of a chief motor machinist named Earl Archer. They had to grease his body to get him through, the water surging around him. I had read his well-deserved citation for a Navy Cross. Bob certainly deserved a proper commendation for his risky undertaking. I regretted that I had only been the prospective commanding officer at that time and so had had no input.

"Captain, what do you want us to do?"

"Nothing other than plane up to 100 feet to halve the pressure. No one goes up in that line on our way home. The cause could be only a zerk fitting whose grease fitting was carried away by the pounding waves above. Keep me informed if it increases. We'll fix it on the surface or at Midway."

"Aye, sir."

"Bob, now that I have saved you from drowning in a pipe, stand by to be skunked in this game."


*30 June. Time 1030.* The wardroom table was cluttered with letters awaiting censorship. As usual, the officers were grousing about the length of the letters they had to screen. I reminded them how lucky they were that they didn't have to censor the incoming mail as well. Such was required while I was in Panama in 1942, when volumes came in from friends and relatives. I told them about one particularly entertaining moment.

"An outstanding first class machinist mate writing to his wife unknowingly had me as the censor. He hadn't left the Coco Solo Submarine Base on liberty since the war began because he didn't trust himself to be good. If only she were here, he wouldn't have this desperate desire for another woman. He just had to have one. Yet he wouldn't do such a dastardly deed without her permission. He hoped that she would think this through and let him know quickly. He was going crazy. Two weeks later I had the duty when her reply arrived."

> Dear Heart,
>
> I love you. Tears came to my eyes as I read your letter. You know how much I long to sleep snuggled up to you, but it cannot be. I know what a delightful "lech" you are and I never want you to change. You do have my heartfelt permission to go out and have a woman—provided you will accept these three conditions.
>
> First, please don't fall in love with her. Second, don't bring home something that you didn't go away with. Third, for God's sake don't pay too much for something you can't give away free up here!

"You know," I said, "he never left the base." Smiles and silence. "Now stop dawdling and get censoring—it serves a real purpose."

During the transit to Midway we busied ourselves writing reports and maintenance orders and sprucing up the *Barb* to enter port. Now we had a battle flag with the *Barb* emblem on a blue field, white flags with red spots for each of the five Japanese ships we had sunk, one for the ship sunk on her previous patrol, a Nazi flag for the tanker sunk in the Atlantic, and two smaller flags for the trawlers. This hand-made flag would fly from our periscope. By custom we would also fly pennants separately, indicating the ships sunk on this patrol.

*4 July.* Crossing the International Date Line gave us our second Fourth of July. We moored at Midway for our overnight stay to refuel. With all flags flying we received a heroes welcome. Kito, blindfolded, left with a Marine escort after I assured him that he would be safe, unharmed, and sent home at war's end.

*11 July.* Escorted by a destroyer and a dozen planes, we moored at Pearl Harbor to a tumultuous welcome with the band playing. Admiral Lockwood, Commander Submarine Force Pacific Fleet, came aboard and said, "Gene, you're the first skipper to tell me what he would sink and did it! Five ships sunk is enough. You can have whatever you want. My hat is off to you and your crew."

On 16 July I received the admiral's off-the-record, "eyes only" reactions and comments, which he had written after he had studied our patrol report.

> It is a pleasure to read this report. The initiative, determination, headwork displayed, and your expertness in torpedo and gunfire were all of the

highest degree. You have undoubtedly instilled a spirit of self-confidence in BARB which will produce many more successful patrols.

Charles A. Lockwood, Jr.

We were congratulated on our super patrol by others, also, and encouraged to have many more like it. Thus ended the *Barb*'s eighth war patrol, my first in command.

June 1944 shipboard calendar, drawn by a crew member.

# Chapter 5　The Royal Hawaiian

Through the good efforts of Fleet Admiral Nimitz and Admiral Lockwood, by 1944 the submarine force had taken over the best hotel on the beach of Waikiki as a recuperation center for the submarine crews returning from war patrols. Nothing less than the famous Royal Hawaiian (commonly known as the Pink Palace) would do. Our same endearing Uncle Charlie also arranged for relief crews to take over the two-week refit of the submarine the day she arrived.

Thus, within 24 hours the entire crew was on the beach at Waikiki soaking up sunshine, fresh milk, beer, and sports. Rest was important: each man had a bed with an innerspring mattress. Health was important: the best and freshest food, often unavailable to others, was channeled to the Royal.

We held no reveille, but quiet was required from 2200 to 0700. Barbed-wire coils—a fitting appellation to fence in the *Barb* crew—encircled the Royal grounds. Wartime Oahu was blacked out and had a 2200 curfew. The Royal gates were locked till dawn. Anyone out late was picked up by military or civil police.

Single females were booked for dates a month ahead. Company for submariners was rare. Those who managed to find a girl, however, were allowed to bring her into the ground floor during happy hour, 1700 to 1900. Military police watched the stairways and elevators to make sure that no one misbehaved.

Bob and I trekked to the submarine base often. We were losing one-third of our crew to new submarines being completed. Bob was selecting replacements. Men and officers with five patrols or more were to be rotated to less stressful duty ashore for at least a year's tour before returning to combat. While Bob was reconstituting the crew, I was tied up with conferences, operations, intelligence, training, gunnery, plans, strategy for the next patrol, public relations, and supply.

The "can-do" base supply officer was eager to assist until I told him I needed an innerspring mattress within two weeks for my *Barb* bunk. Astonished, he said, "That's impossible!" I explained my situation and offered to telephone Admiral Lockwood. "He told me I can have anything I want, so get cracking!" I said.

"Okay. I'll get it done somehow."

Returning to our recuperation paradise, we were met by Max Duncan, who told us that all *Barb* officers had been invited to a civilian party only a mile away at 1800. The party was fun, if a bit boisterous, and the 2200 curfew meant little for us guests living in the Royal. Nevertheless, roving patrols were outside. "How are we going to sneak back, captain?"

"We must beard the lion in his den. Bob, call the shore patrol and ask for a lift to the Royal from this address. Tell them we're just in from a great patrol." Ten minutes later the patrol wagon came to our rescue, bypassing the police, and the locked Royal gates opened.

While we were laughing about our good fortune at the hotel entrance, a night clerk gave John Post a letter from Admiral Lockwood. John had taken some pictures of him when he had ridden the *Barb* from Pearl to Midway, en route to her seventh patrol. John's wife had had them developed for Mrs. Lockwood in Coronado, California, who was ecstatic over the thoughtfulness and quality, passing this on to the admiral. His longhand letter thanked John and invited him to drop by his office. How thoughtful it was of Admiral Lockwood to take the time to write, when Admiral Ernest King had arrived the day before, General Douglas MacArthur was arriving that day, and President Franklin Roosevelt three days later. His gesture showed why he succeeded in making the submarine service what it was.

Since John was a bit awed, I suggested he join me for my appointment with the admiral. The next day, spit-shined and polished, we arrived at his office to an effusive welcome. After a few minutes of pleasantries, John excused himself so we could get down to business. As the admiral closed the door, he said, "Gene, you've turned *Barb* around and come up with one of the finest first patrols as skipper. You're only equalled by Klakring, Kirkpatrick, Cutter, Griffith, and O'Kane.* I've just recommended you for

---

* Commanders Burt Klakring, Charles Kirkpatrick, and Lieutenant Commanders Slade Cutter, Walter Griffith, and Richard O'Kane.

the Navy Cross, so work up your list of officers and crew for the other decorations resulting therefrom. Now, for your next patrol, would you like to have Australia after your last arctic work?"

"No, admiral. Without being presumptuous, I don't like the way our commanders operate down under. From study of the patrol reports of subs based there, it seems to me our U.S. skippers have little to say as to where and how they search for the enemy. There appears to be almost daily direction from their headquarters to transit to a certain spot or to stay submerged. That's not my modus operandi. Luck is where you find it— but to find it you have to look for it. During her seventh patrol *Barb* was submerged every day waiting for the enemy to pass her way. It's no good. The area of search is practically nil.

"There's a big ocean out there. I search it on the surface with our high periscope up and a wide, sweeping zig plan, using as high a speed as our fuel supply will allow. Now, I realize that we may be sighted, depth charged, and bombed more often, but we'll find a helluva lot more targets. On our last patrol, we spent only one full day submerged to check their biggest harbor. Admiral, I like your system. You give us an area and all the info you have on it. Then we do what we want with it. If we really try, but don't sink anything, you give the skipper another patrol to produce. The second time he comes back with an empty bag, you get a new skipper who may be luckier. I like this, and I want to stay under your command."

"Okay, Gene, that's fine. What's your choice then—back to the Okhotsk Sea?"

"No, sir, the *Barb* is ready for bigger things than coastal traffic. I couldn't ask for a better crew or officers. We're well trained and confident. Admiral, we'd like to tackle big-time convoys, including the Japanese Navy. You won't regret it."

"Wolfpacking?"

"No problem, sir. Since we lost both our wolfpack mates, the *Golet* and *Herring,* there was little chance to operate as such. The *Golet* we never contacted. I feel sure a mine caused her demise. The *Herring* worked well when the two of us eliminated the entire convoy. She must have caught it around Matsuwa. I prefer to have the wolfpack commander on another submarine, for he might not agree with the way we operate. With all due regards for the old-school methods of silent approaches, ours are very noisy. I require a voluminous amount of information, constantly, from all sources, fed to me on the bridge during surface attacks and when I'm at the periscope submerged. Weighing our chances, we look for every opportunity to engage. Some don't agree with this."

"Understood, skipper. I'll talk with Dick Voge in operations. You'll get your chance. I'm most pleased that you want to stay with me. How about dinner tonight at my quarters, six o'clock?"

"With pleasure, sir, and thanks."

The admiral's dinner was delightful and lightened by the presence of a lovely Army surgical nurse, Martha Hendrickson. Engaged to an Army lieutenant somewhere in the South Pacific, she never dated, but was often invited to parties.

Gregarious, she kept the three admirals, two squadron commanders, jovial Division Commander Roy Benson, and myself in stitches with operating room humor. By 2100 we said our good-byes, for working hours, seven days a week, started at 0700. On behalf of the *Barb* officers, I set up a date with Martha for happy hour at the Royal.

A civilian couple, Helen and Jan Hull, invited all of the *Barb* officers to a barbecue and swim at their home not far from the Royal. Knowing the difficulty of rationing, I telephoned Helen Hull two days ahead and offered to bring meat, lettuce, and drinks, including a gallon of fresh milk for their six-year-old daughter, sweet Leilani. Helen was elated, because they had been saving up their meat ration for a week to have the party.

Jan (an engineer with Castle & Cooke), Helen, and their daughter quickly became lifelong friends of ours. Their beautiful home, high on the side of Diamond Head, had a breathtaking view over Oahu. In 10 minutes we were all swimming in their large pool and frolicking about. Jan had once won a national cooking contest with his own barbecue sauce, so the steaks were unbeatable. The day ended too soon.

Driving back to the Royal, I thought of how much I truly admired the innumerable families on Oahu that frequently invited military people into their homes. No matter what Uncle Sam does for someone, there's no place like home. I wished the war would end. What a sweet dream.

During our second week of rest and recuperation, the crew was broken up. Torpedomen were sent off to school to train on the new type of electric, wakeless torpedo we would be carrying. Leading men and officers checked up on vital repairs. Bob was bobbing around on a myriad of details. I had been informed confidentially that the *Barb* would be in a wolfpack with the *Queenfish* and *Tunny*, with Captain Edwin Swinburne our commodore. As wolfpack commander, he would "ride" *Queenfish*. Good!

My days were filled with studying the charts of our area—the South China Sea between Formosa and the Philippines to the coast of China and Hong Kong. The area was to be divided among three wolfpacks. A hot spot to be sure, it was exactly what I had requested. I studied all of the previous war patrols until my eyes were bleary, and I compared notes with one of our greatest skippers, Sam Dealey, who had sunk several destroyers and had just flown in from Australia.

Martha arrived a bit late for happy hour, fresh out of emergency surgery. The officers were captivated with her ready wit. As we sat on the terrace enjoying our fancy drinks, a beautiful lady, and a painted Waikiki evening, our frigid patrol in the Okhotsk almost faded into dreamland. Yet, from the glances of the officers, I could sense something afoot. As we started our second round, it came out.

"Captain, you're real good in outsmarting the Japanese, but we'll bet you a dinner for Martha and yourself that you can't outsmart the military police."

"In what way, John?"

"We bet that you can't get Martha into your suite on the floor above without being caught."

"No bet, John!"

Martha interrupted. "Oh, come on, Gene, you can figure it out some way. After all, it's a free dinner at the Outrigger Club."

"Martha, we'd never be able to pull it off by ourselves."

"Well, you're the captain—order your officers to assist."

"Okay, you're on. If we get caught before Martha gets into my suite, I'll pay for two of your dinners. Agreed?"

"Agreed."

I laid out the battle plan. While Bob, Max, and I diverted the MPs by asking for directions to King Kamehameha's statue, the rest of the group would escort our convoy, Martha, up the stairs. I directed Miss Convoy not to click her heels or talk while we were sneaking her up, and told her to duck low so as not to be spotted.

"Understood, sir!"

"Thank you, Martha, for showing proper respect for your convoy commodore. Bottoms up! Battle stations in one minute. I'll take a swing around with my periscope now and eyeball the enemy. On my return, move out!"

The target MP was most obliging. At the entrance he answered Bob's questions about the way to King Kamehameha's statue as the convoy was swiftly tiptoeing up the ladder. Bob had him repeat the directions to gain extra time. Thanking the corporal profusely, we slowly walked back around the corner, took the elevator up, and ran down the corridor to my suite. I could hear the laughter from within. As I knocked quietly, the laughter hushed. Tuck opened the door a crack and peeked out, then let us in. Martha was so exuberant that she gave each one a kiss on the cheek. John complained, "Captain, this isn't fair. We took all the risk. You couldn't have been caught red-handed."

"Sour grapes over a dumb bet that you thought was a lead-pipe cinch, John. You'll just have to smarten up to beat your Old Man."

Then Martha dropped a bombshell. "Now that we're victorious, how do we get out of here?"

"Lads, it seems that it takes a nurse to give us a lesson in tactics. We all know well that when we attack, we must have a retirement course in mind. Start thinking!"

The dead silence was shattered by loud knocking on the door. "MILITARY POLICE! OPEN UP!"

I bolted for the door to open it. Martha sat frozen in her chair. Tuck grinned. "Well, that solves that problem!"

"Captain, sir, the receptionist reported seeing a lady going upstairs with your officers, probably to your suite, since you only took the elevator to the first floor. Is she here?" Turning around I found the officers lined up screening Martha from view.

"Why of course she is. What's the problem?"

"It's against hotel regulations, sir. She must leave at once."

"As you say, corporal. I suggest that you put up a sign to avoid this happening in the future. We'll take the elevator down, then you can return to your post."

Merrily we went along to the Outrigger Canoe Club to dine and take turns dancing in the dark by the light of the stars with Martha. Her hospital ambulance arrived on the dot of 2100 and whisked our Cinderella away.

Back at the hotel I had a message to call the submarine force medical officer who had examined me the day we arrived. This sounded a bit ominous and ruined my sleep. I did not want anything to take the *Barb* away from me. At 0700 I had him on the line.

"Gene," Doctor Walter Welham said, "I just wanted to give you the final results of your physical. Hold on while I get the report...."

"Walt, will you stop scaring the hell out of me and tell me what's wrong."

"Nothing, Gene, nothing. You're in great shape—physically, psychologically, mentally. You do have to watch your weight. You lost 14 pounds on your last patrol."

"Well, we were up on the polar circuit. You lose a lot of body heat, but we ate a lot of cake. Besides, my nights were messed up with frequent contacts and approaches on Soviet ships, not to mention the leather transom seat I inherited for a mattress. I'm having an innerspring one manufactured."

"That's fine. I've never seen one in a submarine. Good hunting!"

"Thanks, Doc." I relaxed again.

As the last few days of our recuperation waned, Bob set up a ship's picnic in one of the city parks on Saturday. "Gene, all the men are looking forward to it. We've arranged for food and drinks aplenty, music too, but only five or six men have dates."

I decided to see if I could round up some girls. One perennial trouble with stag affairs is that the stags tend to drink too much. Not infrequently, this leads to trouble. Ladies have a soothing influence.

The average age of the *Barb* crew was 23. At 30, I was the oldest person on board. En route to the submarine base I had a brilliant idea—the laundry queens (as they called themselves) who worked in the base laundry. Mrs. Chang, the manager, was most helpful. She was sure that many would volunteer, because on Saturday they only worked half a day. Could I have transportation there at noon to take them? No problem, said I. She then got on the loudspeaker and explained the *Barb's* invitation, asking all those who accepted to hold up their hands. She counted the washer-dryer side, while I counted the pressing and bundling side. Our total was 39. Hooray! With the other dates, that made 45. Enough to hold our 76 men at bay—or baying.

Transportation was not quite as simple. Other subs were arriving, and the area was chock-a-block. Not a chance for anything.

"Now look here," I told the tranportation officer, "I can't let those girls or my crew down. You've got to find something. Don't you have any trucks with benches?"

"No, but I do have a stake-bed truck that I can shake loose. It's not too large. There'd be no room for benches with that many people."

"That'll do. Many thanks."

Saturday was hilarious. The truck arrived, bursting with waving girls and some of their friends. Like a magnet they drew the men to help them alight.

The party was an immediate success, everyone talking at once. Beer in hand, I sidled over to the bar. From the adjacent group, I heard a strident female voice ring out. "Rig in your jib lad, before I deck you with a belaying pin!"

I just had to meet this old salt. Tapping Chief Frank Starks on the shoulder, I asked, "Who is she?"

"Mother Reilly, captain. Would you like to be introduced?"

A few minutes later, I had taken her away from the boys. "Mother Reilly, I have to leave shortly, but I must know where you picked up such salty language."

"Sure, skipper, but first I want to tell you what a wonderful crew you have—they think the world of you. I've had groups out to my quarters and shown them around the island. As for me, I am married to the sea. My husband was skipper of the *Empress of Asia,* lost when the Japanese sank her. I then shipped out as a stewardess on the Murmansk runs. Caught German torpedoes and sank twice. The last time I was picked up by a tin-can—my leg and ankle were broken. I have to use a cane because my leg is not quite right."

"What are you doing now?"

"Sitting on my backside waiting for a berth. They stopped letting women serve aboard in the Atlantic, so I came out here. Some women are on runs to Australia. Cane and all, I'm ready. As long as there's an American flag flying over the waves, you can bet your bloody seaboots Old Reilly will be in there pitching!"

"Mother Reilly, you're a perfect gem. I wish everyone in America could hear what you've just told me. I must run, but may I kiss you for good luck?"

"Come here, skipper!"

Regrettably, I had to depart for a skippers' conference with the wolfpack commander.

Tomczyk's diary for this period is brief:

> *12 July.* In the barn. Getting ship ready for refit. Working like a madman so that I can leave with the rest. Received my bundle of mail, but didn't have a chance to read it. At 1300 left ship for the Royal Hawaiian Hotel. OH BOY! Upon arrival took a good long hot shower and hit the sack. Slept till noon the following day in a BIG bed.

> *24 July.* Moved back aboard ship from the Royal. Got plenty of rest, exercise, and fresh food. Also got good and stinko.

The *Barb's* chief of the boat, "Chippy" Brendle, received orders to a submarine tender. Bob came to me on his behalf. Brendle, crying, had pleaded with him to keep him on board. He had been with the boat during her construction and all eight war patrols. Having done a lot to mold the crew into a fighting team, and believing the *Barb* had an aggressive skipper and was sizzling for action, he hated the idea of leaving his mates. "Captain, what can we do?"

If the problem had simply been that Brendle had had over five patrols, we could have kept him. But he was basically a chief fire control man, as scarce as hen's teeth and desperately needed by major warships or in shop repair. We were not even using him properly as the boss of all our enlisted men. I was indeed sorry, but there was nothing I could do, much as I wanted to. I asked Bob to make Brendle understand my position and tell him that I had recommended him for the Silver Star Medal, the third highest award for combat. "I'll see him before he leaves. Now, for his replacement, we've agreed on Swish Saunders. Have the watch send him to my cabin."

Shortly thereafter, Swish entered. "Swish, how would you like to relieve Brendle as chief of the boat?"

Swish looked startled. "Not me, captain, no way."

"Brendle, all the officers, and the crew consider you the best choice. Be frank. Why not?"

"Captain, all the men are my friends. As chief of the boat I'd have to tell them off and discipline them? How could I do that?"

"Swish, I don't want a bastard, I want a leader. We don't drive men on board the *Barb*. We lead them. From my experience with bastards, they achieve about equal results. But there's one big difference. When you lead men, they ship over and want to stay with you. Anything else?"

"Sir, there's all that responsibility. What if I goof?"

"On responsibility, you'll grow with it and enjoy it as you shape things and people. On goofing—so you goof. Don't hide it or cover up. Do your best to correct your mistakes and don't be afraid to ask for help from anyone from top to bottom. You'll find people are complimented when you ask for help. In a few days you'll be advanced to chief gunner's mate and be the junior chief on board. The other chiefs understand that as chief of the boat you become the senior chief on board. In submarines we hang our rates on the gangway when we come aboard. It's what you can do that counts with me."

Swish thought for a moment, then nodded. "Sir, I accept!"

"Good, tell the exec and Brendle."

Repairs checked, individual sea trials and an overnight exercise with the wolfpack completed, we had two days of loading before our morning of departure. The reconstituted crew was shaping up. Max, sharp as a tack, handled the essential torpedo data computer with perfection. Because Bob was scheduled for orders after this, the *Barb*'s ninth patrol, his replacement arrived.

Lieutenant Commander Daniel "Chic" Baughman, fresh from being a skipper and a division commander of torpedo boats (PTs), had no previous submarine experience. Yet he had two Silver Stars and one Bronze Star from action with the PTs. It would be a tough problem to qualify him as a submarine officer and make him an executive officer in one patrol. Such had never been done. Submarining required much more experience and knowledge than handling small boats on the surface from close-by land bases.

Upon completion of the second day of loading, Swish checked with all departments, then reported to me. "Captain, the *Barb* is completely loaded in all respects and ready for sea."

"Not quite, chief. I've a few things for you to load."

"What, sir? There's no space left."

"Don't worry, we'll make some. Listen carefully, because this is not legal. Our crew of 23-year-olds definitely do not enjoy our Schenley 'Black Death' depth-charge ration, which I use to celebrate our sinkings. So, I have convinced the submarine base welfare officer who controls the beer ration for ships' parties to loan me 24 cases of beer until the *Barb* returns to Pearl in about six months and has a party. When we sink a ship, we'll have beer."

"Sir, where can we stow such a quantity?"

"Easy. First, we'll load the officers' shower to the top. This will give a stimulus to the officers to find and sink ships versus using their Pullman type wash basins as a bathtub. Any cases left over go into the torpedo room bilges. After dark, take a few trusted men to the service door of the officers' club. The beer is just inside. No one will stop you or be around. The officer has arranged this, and a jeep will be at your disposal at the head of the dock. Be quiet, act normal, and warn our shipmates that anyone who reveals our beer cache will be keelhauled."

"Understood, sir."

This last evening ashore, the Hulls had invited us, including Martha, for a barbecue. Only a few of the officers were free from their final chores. Al was present, telling sea stories of how the fishnet caught the *Barb*.

Everyone marveled at Houston's diving achievement in 34° water. Leilani, my young sweetheart, then asked me if I had done any diving with a mask. I replied, yes, I had, but only twice in my life, and not voluntarily.

"On the *Bonita* in November 1941 I was the engineer and diving officer. We were the first sub to set up the Bermuda Lend-Lease Base. Entering the very tight harbor at Saint George, our skipper, Shorty Nichols, struck a rock with our port prop. A resulting loud squeal meant we had to check the extent of the damage.

"On board we had only an antique, brass diving helmet, with three glass faceplates, which rested on the diver's shoulders. A rocking hand pump on deck provided air through a rubber hose to the helmet. We had no instructions. None of the crew on board had ever made a dive.

"Shorty called the crew to quarters and asked for all volunteers to take one step forward. Dead silence—no one moved. Annoyed, he said, 'Gene, you're the diving officer. Dive!' I lost a short debate as to what diving officer meant. Yet, I had lots of help to put me under with a couple of deep-sea leads tied loosely around my ankles to hold my legs down. It really wasn't too bad. I dropped down about 20 feet and sat on a prop. Light was fair. I found the tips of two blades curled over about two inches.

"Suddenly, by peripheral vision through the sideplate, I saw a big shape coming toward me. I didn't wait to identify it. Kicking off the deep-sea leads, I ducked out from under the helmet, grabbed the tether line, and scrambled up and on board. Everyone was astonished at my flight. 'What's the matter?'

" 'I don't know, but there's something big down there besides me.'

"Again Shorty called the crew to quarters and asked for volunteers. None. Again he ordered me to dive. This time down I looked around and found a big jewfish, acting like a sick calf. It was probably about 150 pounds and just plain curious. It would come near my faceplate, opening and closing its mouth—wonk, wonk. Not vicious at all. It came so close I could push it away. After that dive one of my engineers went down and cut the curled parts of the blades off with a hacksaw.

"Leilani, my second dive was a beaut, but it's time for us to shove off. I'll save it for the next visit."

With departure set for 1330, the morning was a madhouse. I finished 20 love letters to my beloved wife, Marjorie, and five for my daughter, Barbara. Impishly, she had been telling cronies that my sub was named after her. I explained to Marjorie that we hadn't been able to go on patrol yet due to machinery trouble. I took the letters to a friend in headquarters operations who would mail one every three days to my wife and one a week to my daughter.

While I was there, Captain Voge came in. "Gene, there's been a slight change in plan. Elliott Loughlin has been in to see Lockwood. Since this is *Queenfish's* first war patrol, they've got to shakedown. The skipper doesn't want anyone looking over his shoulder. He requested that the wolfpack commander ride another sub. As for George Pierce in the *Tunny,* this is his first patrol in command. Reluctantly, the admiral has ordered Captain Swinburne to ride the *Barb* with the most experienced skipper, in spite of his promise to you. I agree, for he hasn't had a combat patrol and he'll need your help.

My heart plummeted. "He'll be welcome aboard, sir."

Hurrying back to the *Barb,* I could recall Swinburne only as the regulation exec of the Submarine Base New London who insisted on a hair length less than two inches. What about the beer?

On board I informed the crew that the commodore would ride the *Barb* instead of *Queenfish*. He was very strict. No one should mention the beer. I would disclose its presence to him at an opportune time.

At 1100 Commodore Swinburne came aboard. The officers, chief petty officers, and leading men—Tomczyk included—were paraded. I presented them one by one, with a brief description of their duties. Chief of the Boat Saunders stood proudly in his brand-new chief's uniform. They were then dismissed to ready the *Barb* for the wolfpack's departure.

Below, I offered Ed my cabin, without mentioning my brand-new innerspring mattress, hoping he would decline. He did, choosing a berth in the four-officer bunkroom.

Loughlin and Pierce, the other skippers of our wolfpack, came to eat lunch and receive any final orders.

# Part II The Ninth War Patrol of the USS *Barb* in the South China Sea, between Formosa and the Philippines, 4 August–3 October 1944

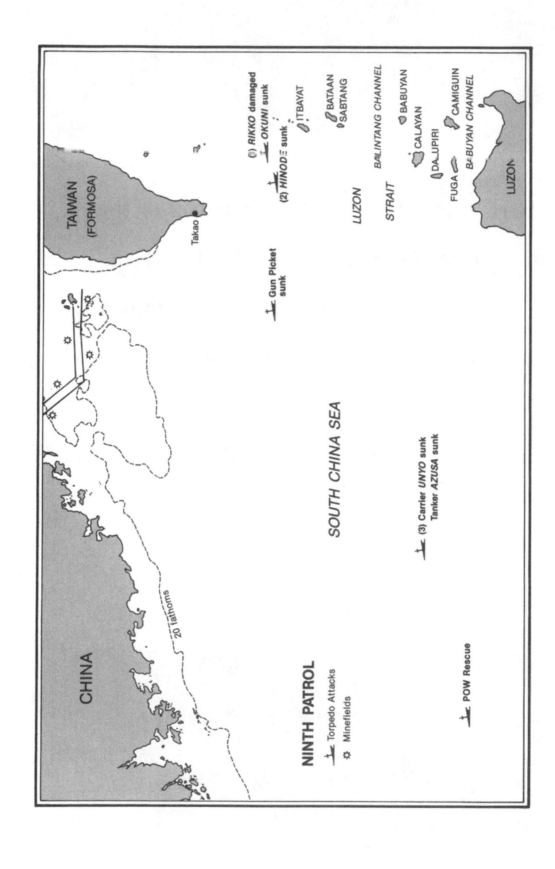

# Chapter **6**   From the Frying Pan

*4 August 1944.* Lunch was a surprise. Paul Ragland, our ever-gambling steward, had been in a crap game with a loyal Nisei fisherman. Rolling his bones successfully, he had cleaned him out of cash. When the Nisei offered to bet five langoustas he had caught on the northern reefs of Oahu, Ragland's lucky streak won those too. Proudly he presented his "lobster newburgh." Swinburne knew he was riding a winner.

I looked forward to working with Elliott Loughlin and his *Queenfish*. He graduated from the Naval Academy in 1933, and I (class of 1935) had idolized him as the greatest athlete in his class. He was awarded the top Thompson Trophy in football, basketball, and tennis. A natural athlete, he could have starred in any sport. I was sure Elliott would be a superb team player and that I could depend on him in a crunch. I barely knew Pierce, the *Tunny* skipper, but he seemed to fit.

Following our lobster lunch and a last-minute conference, the wolfpack got under way with a destroyer escort, headed for Midway. As we followed in single file past the battleships and carriers moored at Ford Island, many sailors waved farewell and wished us luck. Swinburne and I stood on the bridge, waving. He was proud and happy with his assignment, even chummy, for he insisted that we be on a first-name basis. Still, I wasn't sure enough of Ed's attitude toward regulations to inform him about the beer. Would he order it off-loaded in Midway? Fortunately, as we cleared Oahu the ocean roughed up, which meant that no one could possibly shower. The beer was still safe.

Days of wolfpack exercises went by and still the weather deteriorated, until we were rolling our guts out. That helped, however, for it taught everyone to secure all loose articles that might otherwise become dangerous flying missiles when we were being depth charged. In off hours we tested and retested everything on board. Casualty drills were frequent as we honed our skills to fight the ship. An electrical fire brought realism to these exercises. We tuned up our diving times to escape being sighted or bombed: 70 seconds from surface to 175 feet just wasn't good enough.

For the dawn training-dive we increased our diving angle from 15° to 25-30°. Dick had the bridge watch. I instructed Torpedoman Third Class Clarence Byers, a lookout, to cry out, "ENEMY PLANE COMING IN FAST, PORT QUARTER!"

Dick responded with alacrity, diving the boat. Taking over the dive when he bounced onto the control room deck, he had the *Barb* responding nicely as she increased her downward incline. Passing 20°, he ordered, "Take your planes off full dive. We can't exceed 30°."

"Bow planes are jammed on full dive, sir."

"BLOW MAIN BALLAST!" I could hear the command up in the conning tower, but no air blast occurred.

"ALL STOP! ALL BACK EMERGENCY!"

The scene below was wild. Edward McKee, the petty officer on the main ballast blow manifold, slipped on the wet deck with the down angle and fell all the way to the forward bulkhead of the control room. Our baker, Russell Elliman, who watched the diving evolution from the watertight door to the galley, saw McKee sliding and swung himself through the door, hung on to the valve handles, and blew main ballast. The *Barb* reversed her descent as we neared 300 feet. A dizzying rise to the surface followed. We were able to rig the planes in, but they would require repair at Midway. In the meantime, we dived using stern planes alone.

On arrival at Midway, Shorty Edmonds greeted us and invited Ed and all the skippers to his quarters for lunch. He commanded the island as well as the submarine squadron. During lunch I mentioned to him that the average age of the *Barb* crew was 23. The majority didn't like the depth-charge ration of booze, which I offered to everyone when we sank a ship. I hoped that the regulations would be changed to permit us to carry beer.

"Skipper, you can carry beer! Regulations state that all ships are allowed to carry a half case per man for picnics."

"Commodore, you mean that I could order beer right now and legally put it on board?"

"Absolutely!"

"Excuse me, sir, for a moment. May I use your phone to start loading some now?"

"Sure. The phone's in the hall."

Leaving the table, I hurried to the hall, picked up the phone while holding the button down to void any transmission, and said loudly, "Connect me with the *Barb,* please. *Barb,* is Saunders around? Swish, Captain. Drop whatever you're doing, take our jeep, and draw 24 cases of beer from supply. Sign my name on the order. Stow them on board. What! You have no space? Forget it! Stow them in the officers' shower. Don't worry. It'll encourage us to sink some ships. Swish, get these cases on board and stowed before I return from lunch. Understood? Good, move out!"

The party was laughing when I returned. "Gene, you must sure want that beer bad."

"More than you know, commodore."

After lunch, Shorty brought us up to date with information on the war. Then word came that our bow plane repairs would not be completed until after midnight, which meant that our departure would be delayed until dawn. Shorty invited Ed to stay at his quarters overnight.

I used the bow planes job as an excuse to shove off. On board the *Barb* I clued in everyone regarding the beer—and the time warp—just in case Ed asked, which he did when he returned to pick up a few items before returning to Shorty's quarters. "Saunders, did you get the beer stowed in the shower?"

"Yes, sir. To the top."

"Do you think *Barb* will sink as many ships as you did on the last patrol, so the shower will clear?"

"She'll sink more, sir."

The next morning, "Ed's Eradicators" cleared the tight, man-made channel of Midway as the sun was peeking over the broad-daylight horizon. Now at sea, with no further land contacts to possibly compromise our mission, we could reveal information to the crew. "Now hear this! Our area for this patrol lies 3600 miles from Midway in the South China Sea. We are fortunate to have such a hot area with the best wolfpack. No Soviets are here, so sharpen up your cutlasses! The *Barb* is going to cut a wide swath."

On Bob's suggestion, I returned to the intercom. "With good weather we should be transiting Luzon Straits in two weeks. This doesn't mean we can slack off strict vigilance. We are in enemy waters now and can expect an attack if we are sighted."

En route to the area, the three subs formed a scouting line on the surface

20 miles apart. We passed by Wake and Marcus islands, both Japanese strongholds. Watches, trim dives, drills, studies, and maintenance gave everyone a 12-hour-plus workday. For 10 days there was little action. Small-arms fire holed a floating mine, yet it failed to explode. Twice we avoided periscopes, for we were in a restricted area used by our own returning subs. The ocean was calm and the weather hot. Water was in short supply. Tomczyk recorded in his diary:

> *16 August.* Water shut off, except for drinking, cooking and face washing. Boy do I stink—sweat keeps pouring like out of a faucet. Here's one for the books—It's started raining like hell. So almost all hands are going topside, one at a time in their birthday suits, soap and washcloth, and taking a nature made shower. Believe you me, it's swell. Short day only ten hours work. All hands can't wait until they get into action, especially the new fellers. Wonder how they will feel after the action is over!

Two days away from Luzon Straits things started heating up, with patrol bombers forcing us down. One bombed the *Tunny,* but missed. The night of 23 August the pack transited. Tomczyk wrote:

> Tonight will be very interesting. As usual the Straits are very narrow and well protected by radar stations. Got a funny feeling we're going to run into something.—Listening to some good records, boy what I wouldn't give to be with a certain pretty girl right now—that's right—the Jersey coast. Sighted 4 heavy bombers on patrol. Captain Gene has his eye on them. Bataan, Diogo, and Itbayat Islands are in sight, but we aren't diving as yet. Just started thru when we picked up 4 Patrol Boats at the mouth. Lots of interference from radar on land and the boats. They've picked us up! We're going thru regardless at full speed with the patrols chasing us. Thank God it's a dark night. Lost the boats. WE MADE IT! We're now in the South China Sea bagging ass. Gene wouldn't put up with that word. I mean bending on all engines and really making knots. He wants to be well inside by daylight so we can search on the surface—not held down by flyboys.

For the next week our pack, surface patrolling, found nothing but occasional pesky aircraft. The total area for the three different wolfpacks here, operationally known as Convoy College, is divided into four quadrants: Northeast—south of Formosa; Southeast—off Luzon; Southwest—east of Hainan Island; and Northwest—off Hong Kong. One wolfpack, Captain G. R. "Donk" Donaho's "Donk's Devils" (with *Picuda, Spadefish,* and *Redfish*), was off Luzon. En route to our area, an ULTRA message told of a convoy passing through his area, but Captain Donaho never sent us

A patrol plane is sighted on the South China Sea.

a contact report. They attacked, sank three ships, then left to reload torpedoes from our submarine tender at Saipan.

*25 August.* Ed and I were sunning on the bridge when Paul screamed, "MIDGET SUBMARINE — STARBOARD BOW — 1000 YARDS!"

"RIGHT FULL RUDDER! ALL AHEAD FLANK!"

"Gene, what are you doing?"

"We've got to ram her, Ed, or she'll sink us! She's at the firing point!" The midget sub downed her periscope and disappeared. We lacked the speed to catch her. Close call. Ed warned the *Queenfish* and *Tunny.*

*30 August. Time 1215.* Received an ULTRA message on a convoy departing Takao, southern Formosa, heading for Manila. Ed swung into action, ordering the pack to an intercept spot. At full speed we headed for the Bashi Channel, 50 miles south of Formosa. All of us were forced down a few times by planes. At 0120, 31 August, we sighted smoke and received a contact report from the *Queenfish* up ahead. We were receiving radar interference from three other subs.

Tuck took the *Barb* to flank speed and headed for the smoke until plot could give us a better intercepting course. The commodore soon joined me on the bridge. "Ed, *Queenfish* is some six miles ahead. With this three-quarter moon, she will attack submerged. *Tunny* is about five miles on our starboard beam. From radar interference, we have one of 'Ben's Busters' subs up ahead. She has probably sent a contact report to bring her sisters into the ballpark. Stand by for a circus with two wolfpacks jumping on the same convoy. This is an open area with Donk's Devils gone, so it's free for all."

"Concur, Gene. I can't coordinate the other pack. This one isn't in the books. But based on our positions, have *Barb* be the starboard flanker."

*Time 0143.* "Radar has a night flier coming in fast."

Ed dropped down with the speed of a gazelle before Tuck's voice boomed out, "CLEAR THE BRIDGE! DIVE! DIVE!" Other gazelles—quartermaster, lookouts, myself—followed, with Tuck last, snapping the hatch shut.

No bombs fell, but our plane was circling. Radio reported, "*Queenfish* attacking."

Forty minutes later, four torpedo explosions brightened the sky, setting the 4700-ton tanker *Chiyoda Maru* on fire.

*Time 0240.* At moonset the tanker sank. The *Barb* surfaced, for we had no contact with the nine-ship, five-escort convoy. Then we sighted a plane three miles astern after another sub, but we did not dive because we had to move fast on the surface to regain contact. From listening to some 25 to 30 assorted depth charges and bombs being dropped, we knew the *Queenfish* was being held down and the convoy evading northeastward. Several planes were working within three to four miles.

*Time 0526.* We sighted ships and went to battle stations torpedoes. The gongs silenced the distant depth charges. Ed joined me and I explained the setup. We had unknown subs ahead—left, right, and astern—so were in prime position. There were eight ships left, five escorts, and eight planes patrolling. We were a starboard flanker, but the enemy had changed his base course from 90° to 160°. Zigging, they were an hour away. We'd go deep and fast to work into a center position.

*Time 0624.* In the starboard column were a freighter and a tanker; in the center column, freighter, tanker, freighter; in the port column, tanker, tanker, freighter. A minelayer was the lead escort ahead of and between the starboard and center columns. Our position ahead of and between the center and port column was a dream. We made ready all tubes.

"Bob, a few more minutes and we can wipe out the center column and get the leading tanker in the port column."

*Time 0625.* Signal flags going up on the escort: one of the subs had been sighted.

"Sound reports torpedoes running—headed toward us!"

"Bob, keep JP sonar on those torpedoes. I'm sure they're from a sub, not the escort, so they won't be set deep."

*Time 0626.* WHAM! WHAM! "Two hits blew up the escort just as she was running up signal flags for a turn movement."

"Sound reports torpedoes coming right at us."

I kept the periscope down so we wouldn't be hit and told sound to let me know when they passed over us or if any sounded erratic. The whole formation changed course to the east. Damn! If only we could have shot first.

WHAM! WHAM! Two more torpedo explosions. Then through the hush came the eerie, high-speed whine of torpedoes passing directly overhead.

Ed said, "Gene, thank God we're not on the surface."

"Ed, those must have come from my classmate, Eli Reich, in the *Sea Lion*. He always shoots a nest full—with success."

*Time 0630.* "Bearing is on the largest ship leading the center column. Angle on the bow is 10° starboard. The lead ship of the starboard column is missing. Max, mind your torpedo data computer. We'll shoot the stern tubes at this ship in a minute as soon as we pull out. Use the same ship speed as you had before. Range will be about 750 yards.

"Phone talker; tell Jackson and Ragland to take four cases of beer out of the officers' shower and put them in the cooler." Muffled cheers were heard throughout the boat; smiles were seen all around the conning tower.

*Time 0633.* "Up scope. Bearing, mark, down. Angle on the bow, seventy starboard. Range 800 yards. Final bearing and shoot!"

*Time 0634.* "Final bearing—mark."

"Set."

"FIRE 7! FIRE 8! FIRE 9!" The *Barb* jolted as the torpedoes whooshed out on their lethal path. "JP sonar, get on those torpedoes!"

"Sonar reports all torpedoes hot, straight, and normal."

"We had a nice overlap astern of the freighter, with the tanker in the port column about 500 yards beyond, so if they had increased speed, she'd catch the first torpedo . . ."

"Ten seconds to hitting!"

*Time 0635.* WHAM! WHAM! "Beautiful. The first hit the stern of the freighter, the second amidships, at the after end of the superstructure." WHAM! "The third hit caught the tanker amidships."

Suddenly, we heard a click—bang! as a depth bomb went off astern. A steady string of explosions followed. "Depth-charge indicator shows depth-charge pattern astern and explosions all around." Then came a muttered, "It's a wonder it doesn't show tilt."

"Rig ship for depth charge. Stand by watertight doors."

*Time 0640.* "Higgins, rudder amidships. Steady as you go. Up scope. The freighter is upending, sinking stern first. Quick, Ed, take a look!"

"Gene, she's going down, going down, gone! Down periscope. What's left of the convoy has swung around to the northwest, well beyond the range of the torpedoes. Looks like they're returning to Takao."

We couldn't surface with the air cover peppering the area with bombs, but we secured from depth charge and battle stations. We would splice the main brace at 1100 to give the beer a chance to cool. We set course westward in case the convoy turned south again. Before we turned in, Ed and I went to find out from the *Ship Recognition Manual* what ship we had sunk and what we had damaged. It had been a long night, and we needed to rest. Anything might happen.

Unbeknown to us, the *Tunny* was making an approach on the same freighter and was almost ready to shoot when our torpedoes slammed into her. What a disappointment for them.

I took a swing through the boat to have a few words with those members of the crew who were not lined up at the heads holding their water. It had been too long a night. In the after torpedo room cheers rose as I entered; the place was humming with activity and noise as they reloaded torpedoes in the tubes. Buel Murphy and Sydney Shoard were directing their huskies, Houston, Richard Maxwell, and Roark, while Gunner's Mate Joseph Petrasunas was already changing the empty torpedo skids into bunks. As I went forward into the maneuvering room, Ezra Davis lifted one finger. "Captain, one down, four to go?"

"Well, if that damaged tanker sinks, it would help." Turning to Dallas Bowden and Norman Larsen, I asked, "How's our battery?"

"Almost a full can, sir. We topped off until we dove."

The after engine room was blessedly quiet submerged. Chief Franklin Williams, Walter Price, and Clarence Spencer were busy replacing an injection valve, so I didn't tarry other than to tell them planes would probably keep us down for at least four hours. Knowing that Ed was waiting, I buzzed through the forward engine room where Elliman, Norman Wearsch, Rudolf Schmitt, and William Whitt were standing easy. "Whitt, we want to make your first patrol in the *Barb* exciting for you."

They all laughed at his response. "Thanks a heap, captain. Will it be like this every day?"

The *Okuni Maru* sinking stern first after the *Barb*'s two hits, 31 August 1944.

Back in the wardroom, Ed was leafing through the manual. Using bow and stern shapes, superstructure, and positions of funnels, kingposts, and masts, one could find the class of ship. The freighter we sunk turned out to be the *Okuni Maru* (5633 gross tons), built in 1936; the tanker we hit was the *Rikko Maru* (9181 gross tons). The *Okuni Maru*'s lifeboats had yellow-and-black striped sails.

Exhausted, we all decided to skip breakfast in favor of sack time and sweet dreams; if only there weren't that frequent drumming against the hull of bombs and depth charges.

*Time 1037.* A closer bomb explosion banging against the hull brought us all out of our bunks. Max, the watch officer, passed the word that air cover had intensified. None were on us, yet all were dropping bombs at intervals and returning for reloads.

*Time 1115.* "Now hear this! In honor of your splendid performance last night and this dawn, which resulted in the sinking of the freighter *Okuni Maru* and the damaging of the tanker *Rikko Maru*, which I hope has since sunk, I beseech all hands to follow our ancient naval custom and splice the main brace! Your most capable chief of the boat, Swish Saunders, with the doughty assistance of Turnage and Jack Kerrigan, has released four cases of beer from the cooler to be slaughtered. Doc Donnelly is offering his medicinal whiskey to anyone who prefers such to beer. For the beer I propose three cheers. Hip-hip-hooray! Hip-hip-hooray! Hip-hip-hooray!" Tongues hanging to their knees, the crew made a mad dash for the control room.

According to Tomczyk:

> This beer is sooo good I'm taking a sip every couple minutes to make it last. All hands in high spirits and the sea stories are really flying. I'm wearing boots, but will soon get into my bathing suit if they keep it up. Boy oh Boy what a show. They got hit from all sides and didn't know what way to turn, so they went home. And they're still dropping bombs & charges. One came steaming by and dropped 18 in nothing flat. Then a squadron of planes swooped down and let go their load. Counted 96 so far near us. Gene says distant "thunder below" is a very pleasant sound. I'm telling you that even a near one that just shakes us up a bit sounds just like putting your head under water in a bathtub and somebody hitting the side with a sledgehammer. It's hot as hell, with sweat bubbling out. The air is foul and getting hard to breathe. I feel like I just ran 5 miles. All winded, and so sleepy that I could fall asleep standing up.

The Japanese post-action report from the captain of the *Rikko Maru* shows that he capably handled the damage control on his crippled tanker. The *Barb's* single hit amidships blew a gaping hole 20 feet in diameter, but fortunately his ship was in ballast. There was no explosion or fire. He quickly righted her by counter flooding. With her engines aft, he had lost none of his power, even with the drag due to the yawning hole. The big *Rikko* had more speed than the smaller or older freighters.

Messages reveal the scramble returning to Takao. Their convoy commander was lost when the *Sea Lion* sank the lead escort, the minelayer *Shirataka*. The *Rikko Maru* radioed her need for immediate dry-docking on arrival. The Takao Naval District requested her location: they had received a message at 0700 stating that the MI-15 convoy was under submarine attack at latitude 21-11 north, longitude 121-11 east. A hunter-killer group of six ships patrolling off Takao had been dispatched to sink the sub. This group was expected to be on the scene by afternoon.

Just before sunset the *Rikko* arrived at Takao. The dry dock was open and pilots moved the *Rikko* in. On the blocks, with the dock pumped dry, floodlights revealed that the damage was more extensive than they had thought. Many spaces were leaking through cracked plates. Repairs would take over three months.

*Postscript:* The air raids of 10, 11, and 12 October 1944 caused no damage to the *Rikko Maru.* Due to more urgent work, a temporary patch was installed and the captain was ordered to take his ship up Formosa Strait to the shipyard at Keelung, Formosa, to complete repairs. He left Takao at full speed without an escort, his ship leaking. The 150-mile trek was a continuous fight to save the ship. Just before dawn, as she was gradually sinking, he drove her hard aground on the rocks outside of Keelung. She was never bombed after she was beached. On 6 March 1945 the Japanese gave up all attempts to salvage her.

After World War II, the Joint Army-Navy Assessment Committee assessed the *Rikko Maru* as a *marine casualty,* instead of giving proper credit to the *Barb* for damaging her. A reassessment in 1956 credited aircraft with sinking her, though she had never been attacked or damaged by aircraft after the *Barb's* torpedo hit.

# Chapter 7 Into the Fire

*31 August. Time 1349.* "Captain, smoke sighted to the southwest. I'm changing course to head that way, but haven't speeded up. Lots of planes still scouring the area. Sea is calm, visibility perfect."

Hooking my left arm over the periscope handles, I made a fast swing around the sky searching for nearby planes. Bob, Ed, and I estimated the black smoke to be about seven miles away. We could tell from the smoke trail on a flat sea with no wind that his course was a bit south of west. At the torpedo data computer, Max inserted an average target speed of approximately nine knots. After whirling a few dials he calculated that the *Barb* should head southwest for five hours at full speed to close to a firing position. Minor adjustments would be made as we refined target course and speed.

That would leave us with a flat battery if we were depth charged. A risky situation, but nothing ventured . . . "Oh, hell—let's go for it!"

We went to a depth of 150 feet to avoid any planes spotting our wake in the glassy sea. I told the crew we had a far-off target, that we were all tired, and that we would slow once an hour to come up for a look. Except for the watch, they were to stand easy on all stations and try to sleep at their posts. I told the helmsman to wake us in an hour. Ed went below to the wardroom to flake out on the transom. The tracking party curled up on the steel conning tower deck and slept like babes.

*Time 1500.* Reveille for the tracking party. We went up to 60 feet to check, then went back down. We were plugging along. Next check—1600.

"Gene, this is the screwiest approach I've ever seen in all my years as a training officer."

"Ed, chalk it off to the unusual; it may get screwier."

As the remnants of the MI-15 convoy returned to Takao for protection, counteraction against our two wolfpacks was under way. Captain Kouzou Suzuki, commander of the 45th Minesweeper Squadron, happened to be aboard the *Hinode Maru* (minesweeper #20) patrolling off Takao. The naval district commander relayed the MI-15 convoy's 0700 message—"Under attack by submarine at 21-11N, 121-11E"—and ordered Suzuki's hunter-killer decoy attack group to sink the submarine.

He headed his force toward the location using a formation that had been extremely successful in luring submarines into a hunter-killer trap. His tactics were to station two special, low-profile antisub escorts 300 meters broad on each bow, one of his sleeper antisub escorts five kilometers on the port beam of his decoy flagship (the *Hinode Maru*), and the other two sleeper escorts five kilometers on each quarter.

The *Hinode* was a huge, wooden, oceangoing trawler that had been built in 1930 with a forecastle, mast, welldeck, high bridge and funnel, and poopdeck. She gave the appearance of being a much larger ship. Converted to a naval minesweeper in 1941 after Pearl Harbor, she had her crew augmented to 28 men when losses due to subs became serious. Assigned as a hunter-killer decoy, her extra crew became lookouts. Thus she was a perfect convert to mine sweeping, and her shallow draft and coal-fired boilers made her perfect for use as a smoking decoy.

*Time 1230. Hinode Maru* lookouts sighted three vessels with black-and-yellow vertically striped sails. These were lifeboats from the *Okuni Maru.* The *Hinode* took 57 men on board.

*Time 1330.* The *Himode Maru* sighted minesweeper #21—one of five escorts for the 10 merchant ships of the MI-15 convoy—picking up survivors. The 57 men were then transferred.

*Time 1615.* The *Sea Lion,* still submerged, noted smoke growing larger. Then a mast appeared coming directly at her. Eli Reich started the approach, thinking she was a merchantman.

*Time 1707.* "It's an antisub vessel," reports Captain Reich. "High forecastle, poop, bridge structure, and stack. Deck gun mounted forward. A range finder located atop the pilothouse. Steady course."

*Time 1719.* "Fired three torpedoes at 1500 yards. Port track 135. At about hitting time, a lookout pointed. Target turned left and headed at us. Went deep and rigged for depth charge. He passed down our port side, crossed our stern, and dropped two depth charges on our starboard quarter, then went away. Maybe he's not sure he had a sub."

*Time 1800.* "Heard 10 underwater explosions."

*Time 1700.* Reveille for the *Barb* tracking party. The target had changed course to northwest, which would help us close. We would still have to hurry to reach a firing point. Next look was in half an hour.

*Time 1730.* Our target was a small freighter with two escorts on his quarters: mast—funnel—mast, three-island superstructure. "MAN BATTLE STATIONS TORPEDOES! All ahead full! Make ready tubes four, five, six. Angle on the bow is about 90°. Not sure how many escorts, for we can't waste time looking. After four hours of chasing, this is going to be a photo finish—I hope!"

*Time 1740.* "Have the electricians get me a battery gravity reading."

"Maneuvering room reports the battery gravity has dropped from 1250 to 1105. At 1080 the battery is flat and we stop, sir."

"Okay, hang in there. Max, how's your setup checking?"

"TDC and plot are right on. Target course of 252 is correct within 1°. Speed jibes to within one-tenth of a knot. In 10 minutes we'll be at our closest possible shooting position, 1200 yards. After that the range opens and the firing bearing worsens."

"Sound reports that pinging well astern is diminishing."

*Time 1750.* "Excellent. We've passed under the starboard quarter escort. The coast is clear. We have met the enemy and he is ours. Shift four cases of beer from the shower to the cooler. Open the outer tube doors. Set torpedo depth at four feet. All ahead one-third."

"Time for a look. Final bearing, captain."

From my crouched position, I flopped the scope handles down as they cleared the deck, glued my eye to the eyepiece, and rode it up till it was clear of the water. Something was wrong! I swung the scope around from side to side, withdrew my eye, blinked, and looked again. I banged the handles against the scope in vain. "Damn! Down scope."

"Gene, what's wrong?"

"Ed, I can't see; there's something covering the scope. I don't know what it is or what's wrong!"

*Time 1753.* "Captain, range is steady. We've got to shoot or it'll open."

"Listen, Bob, same thing again, but this time raise the scope and dip it on my signal. Maybe we can shake whatever it is loose. Up scope. Nope, still there. Try again."

"Sir! Range is opening slowly. We're losing bearing."

I burst out laughing. "Hey, we've encountered the latest fiendish, antisub

*Top:* This is the bird that kept interfering with the final periscope bearing at the time of firing torpedoes at the hunter-killer decoy *Hinode Maru,* causing both scopes to be raised to fire. *Bottom:* The *Hinode Maru* (# 20) converted to a minesweeper XAMS, was sunk by the *Barb,* 31 August 1944.

weapon of the Japanese—a patrolling bird! As the scope clears the water, he or she is there and lights on it, draping his or her tail over the slanting exit eyepiece. Let's foil it. Ed, control one scope while Bob controls the other. Bob, have your scope clear the water, as a feint, a fraction before Ed's scope, which I'll use for shooting. Understood?"

"Range is opening, 1260 yards."

*Time 1759.* "All ready. Up scopes. Camera, quick!" Slapping it on the scope, I took a photo of the bird, took off the camera, and swung the scope to the target. "Final bearing—mark. Down scopes."

"Set."

"FIRE 4!" Whoosh—Jolt. "FIRE 5! FIRE 6!"

"Dick, what's the hitting time?"

"Seventy-five seconds, sir."

"Sonar has all torpedoes hot, straight, and normal!"

"Ed, Bob, same deal on the scopes; I can't miss this."

"Fifteen seconds!"

"Up scopes. His guns are manned fore and aft! There are about 15 lookouts, dressed in white, on a catwalk above the bridge. An officer is looking . . ." WHAM! "My God! Right under the bridge; bodies are flying through the air." WHAM! "Waterspout at the welldeck!" WHAM! "Under the forecastle. The gun crew's been blown overboard. The ship's breaking into a V! Quick, Ed, take a look. Everybody in the conning tower take a quick peek through one scope or the other."

It was soon time to stop being spectators and check on the escorts—they were coming fast! We had no time for a down-the-throat shot. I had Paul take her deep—to 340 feet—using a 25° down angle. We moved at flank speed with full right rudder, rigged for depth charges, silent running, with all watertight doors closed tight. I had the sonars keep on different escorts and give us the bearing quickly whenever it was constant, meaning it was heading in to drop on us.

It looked as though the crew of one of the subchasers was greasing up their depth-charge racks. There was nothing to do now except have the yeoman break out our depth-charge forms, which bureaucracy required.

*Time 1804.* Thunder below! None too close—100, 150 yards.

*Time 1824.* "Sound reports help is arriving from the east and northwest, all pinging."

Ed realized, I knew, that I had goofed and misjudged this one. That had been no small freighter: that was bait! I had never picked up any escorts but the two subchasers. The rest of the escorts were sleepers and probably first team. We lacked enough juice in the battery to fight them. I told Ed, "I plan to tiptoe and get the hell out of here with a whole skin."

"Skipper, you do just that!"

We crept out at minimum speed, heading northwest on the assumption the first team would not search where they had just come from. I had the word passed for all hands to be quiet—no unnecessary talking—and to take off their shoes. Paul eased her down to 375 feet, concerned about the pressure on our thin-skinned sub, but we had been that deep before. We watched for leaks and any signs of the decks starting to rumple.

Gradually, we pulled clear. The antisub group found something to their liking astern of us and pounded it with 58 depth charges. Darkness fell with a full moon and no clouds. As we surfaced with the group five miles astern on the horizon, two engines started charging the batteries. The other two on propulsion sped us westward—evading various small patrol craft—to a new scouting line that Ed set up.

*1 September.* After some 200 bombs and depth charges the day before, we looked forward to a peaceful interlude well clear of any enemy bombsights. Studying the charts spread over the wardroom table, Ed wondered whether we had a good spot or were jumping from the frying pan into the fire.

"Commodore, only test and time will tell."

Test and time told us to act like a yo-yo, evading air cover by diving and surfacing some nine times during the day before dinner. It was exhausting being the fox for the hounds. Taking our leisure, we unfolded our napkins as a platter of steaks was served. Suddenly, "LEFT FULL RUDDER—ALL AHEAD FLANK! CAPTAIN, PERISCOPE STARBOARD BEAM."

Dashing to the bridge with my napkin tucked in my belt, I met Tuck, who had the situation well in hand. Chic, the officer of the deck, looked a bit nonplussed, almost shellshocked. As he was the only officer not yet qualified as submarine watch officer, we put him on watch with Tuck as a backup. Thank God. I don't understand the psychology of it, but from my experience in peacetime battleships and destroyers, I became convinced that three-dimensional operations are more demanding. Perhaps it's a David against Goliath syndrome, or "God helps those who help themselves," or the loneliness of knowing there's no outside help that quickens one's reaction time, but it's there. No one, regardless of rank or other experience, becomes qualified in subs until he achieves that, for one hole sinks a submarine.

"Good evasion, Tuck, return to course."

After dinner Max relieved Tuck and Chic. Dusk was approaching. I wolfed down half my dinner before joining Max on the bridge. Dawn and dusk, the most dangerous times for a surfaced sub due to the fading visibility, normally found me sitting topside searching the sky and nearby waters for floating mines or periscopes. That evening the sky was sprinkled with cumulus clouds, much like a field of cotton.

*Time 1848.* Max's strident, "PLANE ASTERN! CLEAR THE BRIDGE! DIVE! DIVE!" had me landing in the conning tower before the klaxons sounded. Lookouts came tumbling down, and Max deftly slapped the hatch shut and gave a powerful spin to the wheel, tightly engaging its locking lugs.

"Up scope. Left full rudder." As the scope came out of its well, we heard a tremendous roar as the plane passed over us, but no bombs fell. Peering through the eyepiece, I saw that the plane was heading straight for the *Tunny* four miles ahead. I could see the *Tunny* still on the surface! She hadn't seen him!

"Gene, can't we warn her?" Ed asked.

It was too late: he was dropping bombs already. As the *Tunny* submerged we heard the rumble of explosions, and I saw the *Tunny*'s stern rising. I told Ed to take a quick look. A bomb must have exploded right under her tail.

Ed backed off and shook his head. "It looks bad, and that plane is still around."

*Time 1952.* I watched through the periscope as another salvo of bombs landed on the *Tunny.* Enemy aircraft had marked her position with float lights, and another string appeared to indicate her course on diving. Then they dropped a green magnesium parachute flare on her position.

"Gene, try to contact the *Tunny* by sonar," Ed said. "Tell her to stay deep, and that she's marked with float lights and flares. We'll let her know by sonar when the coast is clear and it's safe to surface. Meanwhile, *Barb* will stand by, nearby, to render assistance."

Kerrigan failed to make sonar contact.

*Time 2200.* Air cover cleared out; no lights around. We surfaced. Couldn't make radar contact on the *Tunny.* We also tried radio, but no contact.

*2 September. Time 0003.* Dark as pitch. Radar receiver showed enemy radar-equipped planes approaching. Two minutes later, as the bridge was cleared, the plane passed 100 feet overhead. The roar brought everyone on board to his feet. Barefoot and in my undershorts, I made it to the conning tower in eight seconds, yelling commands for maximum speed, left full rudder, 300 feet, and rig ship for depth charge with watertight doors locked.

One minute later two bombs hit very close to starboard while we were passing 100 feet and swinging hard left. Light bulbs shattered.

"EMERGENCY LIGHTS ON! Bob, he must have dropped on his second pass. He'll be back. Level off at 375 feet."

"Captain, suggest only major damage be reported to keep the phone lines clear."

"Good idea. Check to see if the commodore is okay."

Two more bombs close to starboard as word was passed that Commodore Swinburne was stuck in the control room. The lower hatch closed before he could reach it.

*Time 0008.* Another bomb. "Depth-charge indicator shows above and on starboard quarter about 400 yards."

"Good. Stand easy; watertight doors may be opened. All ahead one-third. Shankles, make your course 250. Have the electricians send some light bulbs to the conning tower."

Tuck was making himself useful, as usual. Borrowing a dustpan and foxtail from Shankles, he looked like a frog, sweeping up glass from the shattered light bulbs. Grinning at me, he asked, "Sir, with all the interruptions, have you forgotten the beer in the cooler?"

"Blimey, Tuck, I had. Bob, how are your damage reports?"

Bob took off his phones. "No major damage. Bow plane shafting is now squealing. We can fix it submerged. Gauges, light bulbs broken. Minor high pressure air leaks. There'll be a lot more of this type that can be fixed while the bow planes shafts are being taken down. Topside is an unknown and may be worse."

It could have been worse. I announced to the crew, "Now hear this. I must apologize for having forgotten the beer left cooling after our sinking of the hunter-killer group decoy at 1800 yesterday. No excuses. Lieutenant Weaver has properly upbraided me for this. Wherefore, he will gently lift the four cases of beer out of the lower cooler into the cradling arms of Chief Saunders, to be deposited without shaking in the control room. This beverage is far too precious for a drop to perish in foaming spills. Secure from depth charge. Splice the main brace!"

*Time 0030.* With a sigh of relief, I turned the ship back over to Dave, telling him, "Great work and quick thinking, Dave. You saved us." As Bob and I descended into the control room in our undershorts, Ed, in trousers and shirt, was grinning up at us. "Skipper, I've learned a lesson. In the *Barb* you don't have time to get dressed or you'll be locked out. Does this happen frequently?"

"Not on your life, sir, but yes—for your life!"

Men were busy sweeping up the glass fragments, paint flakes, cork dust, and pieces of insulation that had blown off the hull as Tuck and Swish arrived, each hugging two cases of beer. All hands drifted in, talking about our hair-breadth escape. We almost got it like the *Tunny.* I could see that the men were visibly shaken. Our good Scot Bob McNitt took the lead in bringing them back to battery. "I know a beautiful old Scottish prayer that is apropos to this little squeak a half hour ago. Please be quiet and bow your heads." All hands dutifully complied.

"Good Lord, do deliver us—from all the ghosties and ghoulies, and long-leggity beasties, and things that go boomp in the night!"

Tension turned into gales of laughter, and the beer was passed out. Tomczyk recorded in his illegal diary:

> *2 September.* Boy did this day start with a bang and more. The lookouts and O.D. saw a plane 50 feet above them. We dove and got to about 100 feet when he circled, came back, and dropped 4 bombs. God it seemed like the whole overhead was coming down. Lights blew out, fuses jumped out, some valves were fouled up, cork and paint was coming down like a snowstorm. The *Barb* jumped up and down like a piece of paper in a strong wind. So help me, I honestly thought it was the end of us. I *still* don't believe that we *still* exist. A fraction higher to the surface and I won't get to read this in my old age. When I said this was a hot area, I wasn't kidding. Old "On-the-Ball" McNitt's prayer sure fits us.

Relaxing in the wardroom with the officers, sipping beer, I mentioned to Ed what Dave had told me. He had dived when radar intercept was steady on. We all agreed that these were new Japanese tactics.

"Tell me something," Ed said. "Dave Teeters' real name is Robert, and he says he was always called Bob. Why do you call him Dave?"

"Commodore, we have a Bob—Bob McNitt. All the other officers, you and I included, have different nicknames. In emergencies we can't afford any possible confusion as to whom we are giving orders. When Teeters came aboard, we christened him 'Dave.' He answers to it."

"Sounds like you're naming a dog!"

*Time 0100.* While work progressed on the bow planes, we planed up to periscope depth to take a look. Float lights marked our position several miles astern. Half an hour later, with the bow planes fixed, we surfaced. In the zodiacal light we assessed our topside damage. Our port antenna was blown off. SD air radar search supports were bent back. The deck was littered with bomb fragments, and the tail vanes of one bomb lay just forward of the conning tower. "Send Tomczyk and a bag to the bridge."

"Tomczyk, I won't risk more than one man on deck. You're quick. There are no planes around. How about dropping down on deck to pick up the tail vanes and bomb fragments for souvenirs. We won't dive without you."

"Yes, sir."

Returning a few minutes later, he took his booty below for distribution and added a couple of lines to his diary: "The bomb fins were *stuck* into the deck. Boy that's *close* when you find these reminders. Incidentally, I have a few pieces."

*Time 0150.* At long last a message came in from the *Tunny* asking permission to return to Pearl Harbor due to critical damage. Her after torpedo room hull was dished in. All four after torpedo tubes were out of commission. All three radio antennas were blown off. She had minor hull leaks and rudder damage, et cetera. Ed sent her on her way and offered assistance and escort, if required, but none was needed. They had made it through Bashi channel into the Pacific before they could rig a temporary antenna.

Fortunately, after dawn, aircraft contacts ceased, enabling us to charge our batteries, patrol on the surface, and get some badly needed sleep. With dusk, however, the night radar planes came after us again, forcing us to yo-yo, diving and surfacing.

*Time 2314.* Later this same day, three bombs struck close—above and to starboard—while we were passing 200 feet and swinging left at flank speed. Cork and paint flakes flew as a few bulbs shattered. Two electric cables were pushed in through their hull glands, showering men nearby with a stream of salt water, similar to a fire hose. The leaks were quickly brought under control by tightening the gland nuts. The only damage was to our nerves!

Safe, but held down by circling planes, we set the regular watch and the officers drifted into the wardroom. I noticed Bob shaking his head and daintily caressing his forehead. "What gives?"

He removed his hand, revealing an emerging red lump. "I was on the bridge when that Betty bomber came at us. As we dove I zipped up my windbreaker to keep my binoculars from striking the ladder. Sad to say, my beard got caught in the zipper, and I dropped down through the hatch with my head bent forward. Naturally, my head rattled the ladder rungs like a xylophone until I landed on the conning tower deck. The beard comes off in the morning. You know, it did pound some sense in my head. We should set up an early-warning system in the radio shack to save precious seconds when the search plane's radar signal peaks at some level."

"Bob, magnificent idea. Let's tell Hinson."

Leading Radioman Hinson joined us and immediately agreed. "Captain, we can place a strip of adhesive tape on the scope that shows enemy radar signals. When his signal is strong enough for the spike to reach the bottom edge of the tape, we dive."

"Perfect. Put the tape on before you turn in. The radioman on watch now has the authority to order 'Dive!' Bob, spread the word throughout the boat."

The planes held us down till dawn. Again, we remained on the surface all day.

*4 September.* We dived and commenced an approach on a ship contact. It turned out to be a large four-masted picket, not worth torpedoes. Since our basic orders were to sweep the seas clean of all fishermen, pickets, weather vessels, et cetera, we decided on a gun attack.

I decided to use only the 40-millimeter gun on the afterdeck of the bridge and the twin 20-millimeter guns on the bridge deck forward. The 4-inch gun crew was to stand by below. Ed wondered why we didn't use all of the guns, and I explained that with heavy air cover it was too dangerous to have all the gun crews on deck at once in case we needed to dive in a hurry.

*Time 0906.* "BATTLE SURFACE!" Tuck, Max, and I were followed up by the two gun crews and lookouts. Speedily they unlimbered their guns and opened their magazines. In less than 20 seconds I ordered, "COMMENCE FIRING!" The 40-mm gun jammed after the third shot, with one hit. The 20-mm kept the Japanese crew from manning their guns forward.

*Time 0910.* "PLANE COMING IN, STARBOARD QUARTER!"

All of us tumbled down into the conning tower as we dove. Bob was at the scope. "Bearing 150, range five miles. I don't think he's seen us, but he's heading this way."

No bombs. We went back up to periscope depth. The plane was circling about 50 feet above the water. The picket crew was jumping up and down, waving their arms wildly. Some were pointing down into the water. I could actually see the pilot's face. He was grinning and waving back. He must have thought they were just lonesome sailors happy to see a plane. He shoved off.

I told the 4-inch gun crew to prepare for action and waited for the plane to disappear over the horizon. John Lehman, our crack radar technician and battle station radar watch, interrupted my thoughts. "Sir, don't forget that the cooks have our celebration cake in the oven for the 31 August sinkings. Do you think the 4-inch gun will make the cake fall?"

"John, having a cake in the oven during a gun shoot is a new experience for me. This may not solve the problem, but I have a solution that may ease the pain of the bakers, if the cake falls flat. Tell Ragland, Jackson, and Bentley to take four cases of beer from the officers' shower and put them in the cooler."

"Problem solved, captain."

The sky was now clear, but I had Bob ease us out to 1000 yards. The picket crew had manned its forward gun, so I moved out of range. I would not have any Purple Heart medals awarded to the *Barb's* personnel.

The gun crew scrambled up the gun-access trunk just forward of the bridge while the ammunition train formed to pass the shells up from the magazine below the radio room. Tuck, Max, and I—with the lookouts Don Miller and Byers plus Quartermaster Francis Sever—went up from the conning tower. Ed came later.

The fire from the picket was short in range. In 20 seconds Houston rammed the first shell home and jumped clear of the recoil as Swish fired. Emil Novak caught the ejected hot shell in his gloves and heaved it overboard. The first shot was short due to the cold barrel, the second one over. Max spotted the range down and six hits finished the picket. She sank stern first.

*Time 0944.* Ed was watching from the afterdeck of the bridge as the gun was secured. "Good show, skipper. Now I know why the *Barb* survives. That gun crew performs like a ballet. When we return I'll sure be a better training officer."

I had Bob send Tomczyk up to test-fire the jammed 40-mm gun and went below to the galley. Russell Elliman, our baker, and Dougherty, our cook, were hoisting the huge cake out of the oven. "Did it fall?"

"A bit, captain. We'll just make the icing thicker. Should be ready at 1100. We made the decorations yesterday when things relaxed a little. This is the freighter sinking stern first, and the lifeboats with the vertically striped black-and-yellow sails. Here's the decoy sinking in a 'V' shape, broken in two."

"Well done. Can you add the picket?"

"No problem. We'll have her bow upended with three of her masts showing as she sinks into the cake."

*Time 1100.* Ed asked to announce the celebration. "Now hear this from your wolfpack commander. This is my first combat patrol. From the experience I've had in the *Barb*, I could not have found a better flagship nor a more courageous crew. We've had our ups and downs fighting the enemy. Realize that you are making history in the best traditions of the United States Navy. I am sure that John Paul Jones would be as proud of you as I am. I also want to thank you for drinking the beer. Your red-headed skipper has informed me that the remaining cases of beer in the officers' shower will be moved to the empty space in the magazine left by this morning's outstanding gun action. For that I thank you. Now I can take a shower. Splice the main brace!"

After the photographs were taken, Ed gleefully cut the first slice of the majestic cake. The cooks then took over to slice another 85 pieces as the beer was doled out.

We gathered in the wardroom and the sea stories erupted. Ed joked about Paul's rapid recompensation of the *Barb* to the neutral buoyancy required for diving following the weight loss of the shells fired. Surely he must have been thinking of Tuck's story of being sunk off the Kurile Islands due to the failure to recompensate following a gun shoot and the resulting inability to dive. "Sir, none of us will ever forget it!"

Ed turned to me. "Gene, Shorty Nichols told some of us at a dinner party before we left Pearl about having you as his diving officer on board the *Bonita*. It seems that the division commander was on board after you had just completed the south transit of the Panama Canal with the *Bass* and *Barracuda*. In Panama Bay, as soon as the water was deep enough, the subs would dive on the DivCom's signal to periscope depth with their periscopes constantly raised. Having to answer a call of nature, he went to the officers' head, only to find you squatting beside the toilet bowl holding up a battery hydrometer.

" 'Fluckey, what are you doing?' he said.

"You answered, 'Compensating the boat, sir.'

"The DivCom thought you were crazy and told Shorty so, but Shorty said, 'Just watch him.'

"A few minutes later you appeared, hydrometer in hand, picked up the phone, and told the control room to pump out 8000 pounds of water from the auxiliary tank. Then you announced, 'Captain, we're ready to dive.'

"At the raised periscope the DivCom executed the signal. The *Bonita* dived nicely, never dipping the periscope. The other subs disappeared. It was over two minutes before the periscope of the *Barracuda* appeared and the *Bass* broached. The DivCom made no comment until the commanders were ashore later having Cuba Libres. Then he gave the other skippers hell for not knowing how to compensate their boats. The hydrometer was not mentioned. Now, Gene, please elucidate."

I explained that keeping a sub in neutral buoyancy is not difficult. "It's something like balancing your checkbook monthly. You take the last balance, add deposits, subtract expenditures, and you have your new balance. In compensating a submarine in the Atlantic Ocean, follow what you did with the checkbook. Dive and get a good neutral trim, or balance. Before your next dive, regardless of time, note the weights you've removed and the weights you've added. Figure your balance and that's compensation. There's no problem in the same ocean.

"Now, move out of the Atlantic into Panama Bay. Your exchange is different. A lot of fresh water is dumped in from Gatun Lake and the surrounding jungles and mixes with lighter, warm salt water from the Humboldt Current. Overall, this mixture is lighter in weight than normal ocean salt water. The specific gravity has decreased.

"All this, of course, was before the days of the bathythermograph that we have today in *Barb*. The only available instrument we had was the hydrometer, used for taking the specific gravity of the acid in the batteries to determine the amount of charge. Yet the float was too heavy for water; it sank to the bottom. By drilling a hole in the float, removing some of the tiny lead ball bearings, and resealing with wax, it was light enough to use for brackish and ocean water. Reading the comparison from the Atlantic to Panama Bay, I could compensate accurately. The officers' head, where salt water was used for flushing, was the handiest place for me to take my quick and essential gravity.

"So," I concluded, "what Shorty said actually happened."

*Time 1229.* "DIVE! PLANE ATTACKING!"

This time Tuck had sighted him diving at us out of the sun at half a mile. "Captain, it's silver colored. We are lucky our zig plan was just giving us a big swing to the right. The plane missed his bombing position on the first run, passing 40 yards off the port side. I could see the faces of the fliers. I sure blew this one!"

"Forget it, Tuck. This could happen to anyone. The rudder indicator below showed you had some right rudder on, so I increased it to full. Normally, I always put the rudder left full because the Japanese may have been trained that most Americans are right handed."

*Time 1232.* Two bombs exploded extremely close by—above and on the port quarter—as we were passing 300 feet. Light bulbs shattered, cork and paint flew. Bob was on the phones below getting damage reports. Four more bombs exploded, not close. A couple of cables were pushed in, showering some of the crew. Nothing more.

At lunch, Ed let go a calming one-liner. "If this keeps up, we're going to have to return to Saipan for a reload, not of torpedoes, but of light bulbs."

Surfacing, we yo-yoed with patrolling planes. Once we had a double-barreled threat, having sighted a plane and a periscope at the same time. We avoided both.

The *Queenfish* reported she had battle surfaced to shoot at a Japanese flying boat that had landed on the water, but it revved up and ran away from her while they were exchanging fire. She had also dodged two torpedoes running on the surface. The *Queenfish* paralleled one at flank speed; the other she avoided as it crossed her stern. Thereupon, Ed moved the pack to patrolling submerged during daylight for the next three days, reconnoitering the islands.

*5 September. Time 0203.* En route to his new station, Elliott scurried to the *Queenfish* bridge, reacting to a radar contact at 3400 yards. With his alert peripheral vision, he spotted the impulse bubble from a firing submarine and took split-second action to avoid the oncoming torpedo. Speeding up to 19 knots and deftly paralleling the churning torpedo, he watched it scrape past at about 50 yards, with all hands on the bridge holding their breath—or something.

Simultaneously, radar picked up a new contact 1000 yards ahead. Elliott instinctively reacted in the third dimension and dived, passing under the midget submarine at 200 feet.

Ed decided that there were too many midget submarines in this area and plucked us out of the fire. We moved west, away from the proximity of close-to-shore bases.

# Chapter 8 Singapore Slaves

Far far away in Singapore, history was on the move. Still-proud remnants of the Australian and British divisions that had been surrendered with the fall of Singapore crammed the docks—slaves-to-be for the mines and factories of Japan.

Two Australian Army buddies, Neville Thams and Jack Flynn from Queensland, and Alfred Allbury of the Royal Army all worked as slaves and lived the tortures of the damned, building a 265-mile railroad through Thailand and Burma. Captured in Singapore on 15 February 1942, they were sent to one of the many construction camps of 1000 men each. Over 61,000 Allied prisoners from Singapore, Java, HMAS *Perth,* and the USS *Houston* labored for the Japanese; 22,000 of them died. About 250,000 Asians also worked at bayonet point from separate, but nearby, camps; 85,000 of them died.

The three men gathered one night to discuss their situation. Neville Thams spoke first. "Al, after finishing that gawdam 'Railway of Death' last March, we were brought back to civilization. Six times since then, after being picked as the most healthy, we've been herded to the docks to embark, then returned to dig that dry dock. Are we going to shove off from this barbeque pit for something worse or not?"

"Neville, yesterday a Frenchie slipped me a newspaper. It says the American Navy is pushing the Japanese back toward Japan with a ring of steel around the whole perimeter. They quote from Chinese geishas: 'Japanese sailors say you can walk from Singapore to Japan on American periscopes.' I think they're frightened. How about it, Jack?"

"It's a cinch, we'd be safer here. If we get to Japan, these fanatical bastards will fight an invasion to the last man. By then every last one of us will be killed—if we're not already dead from overwork, starvation, or allied bombing. I'm not looking forward to a sea voyage. We'll be treated even worse than when we were locked in those solid steel railroad cars going up the line. I'll never forgive them for that ride—standing room only—baked by the sun like you're in an oven—no air other than the crack under the door. Men clawing at each other to breathe. Over a third dead and the rest of us with heat stroke were beaten by the Korean guards because we didn't have the strength to pull the dead out."

"Look, blokes, there are about 30 ships in the harbor, and lots of escorts. Should be a dinkum trip—if you can swim. Remember when they paraded us and counted our ribs to select the healthiest for this trip and that Jap colonel made that speech—'You POWs are the remnants of an inferior and dying race. You are most fortunate to be in our hands, your yellow benefactors. Fortune has smiled on you that you go to Japan. Our homeland is the center of the Co-Prosperity Sphere. There everybody is happy and contented in the land of plenty.' Then he stuck out his swollen, red, raw tongue—'Here, we too, have diseases.' "

Thams added, "Remember the unspoken part? On leaving, his retirement was made conspicuous by two prominent patches in the seat of his trousers. Mates, this time we go!"

"Guess you're right Neville. I've been loading this bloody *Rakuyo Maru* with copra, tin, rubber, and scrap iron for the last two weeks. Every hold is filled except the forward one. That must be for us, and it's just not big enough."

A few staccato barks from an officer and the Korean guards—snarling, swearing, and poking with their bayonets—herded 1350 POWs up the gangway. The guards lashed out savagely at any stragglers, bringing occasional yelps.

Topside, guards removed the hatch covers to reveal that platforms one meter high filled the hold below from top to bottom. All prisoners were to travel inside the hold. The men rebelled and wouldn't enter, certain that once inside they'd be locked in. Bashings commenced. A British officer intervened, and a heated conference ensued with the ship's captain and a Japanese officer, followed by an inspection of the hold.

Returning topside, the British officer announced that a compromise had been reached. Nine hundred men would have to be in the hold; the other 450 would be allowed on the deck forward. It was the best he could do. Looking at his swollen eye, the men knew he had done his best, and started below. Each had only three square feet of space. It was impossible either to stand or to lie down. They sat, leaning on each other. The litter patients, walking sick, and those with dysentery would be on deck. The rest would rotate. This victory took its toll—broken ribs, head gashes, cuts and bruises. A cotton blanket was given to each man, but they weren't fed until the next day.

Another group of 750 POWs was embarked in the *Kachidoko Maru.* The two ships then moved out and anchored in the roadstead, where the rest of the convoy had assembled.

*6 September. Time 0500.* The convoy of six ships and three escorts got under way. The POWS on deck had their first view of a Japanese aircraft carrier. The mighty *Unyo,** escorting six huge, empty tankers, entered the harbor as their convoy headed for the Inland Sea in Japan.

*9 September. Time 0119.* The *Queenfish* reported radar contact on a convoy at 20,000 yards; she would attack submerged. The *Barb* was on the port beam with a long way to end around on a brilliant moonlit night. Air cover flew 300 feet above the water, forcing us to dive. We kept plugging along submerged with that plane circling overhead. At 0213 we heard four torpedo explosions as the *Queenfish* sank two ships. They were followed by 11 depth-charge explosions. An hour and a half later the convoy passed ahead of us moving southeast at a range of 4350 yards, outside our electric torpedo range. Sadly, we turned south to open out so we could clear the aircraft and surface.

*Time 0432.* Still submerged, we sighted a sleeper escort. Maybe we could catch this frigate, but as I called the crew to battle stations torpedoes, something bothered me. "Bob, I don't like this. If we hit him, they'll keep us down and the convoy will escape. If we miss, we get the same results from end-of-run explosions."

"Right, sir."

Nevertheless, I raised the scope, checked range and bearing, and ordered tubes 1, 2, and 3 made ready. I guess once in every skipper's life there comes the urge to let fish fly at anything that moves, especially after a convoy slips by his nose unmolested while his hands are tied.

Bob said, "Captain, it's a real long shot."

We took a final check and fired three fish.

"Sonar reports all torpedoes hot, straight, and . . . Wait a minute! Number 2 is moving off to the left. She's circling!"

---

* "Falcon of the Clouds"

"Down scope. Switch sound to loudspeaker." All was absolutely quiet as we followed the erratic torpedo, which circled around and soon roared overhead. Our sighs of relief were cut short by three different pingers approaching. All torpedoes had missed. We went deeper and found a temperature gradient that gave us some protection. No depth charges were dropped after the end-of-run explosions, but the pingers kept searching. I told Bob to kick me in the shins if I ever pushed another lousy shot like that one.

The next morning was filled with planes and subchasers as we moved away. At about 1100 we commenced chasing the convoy, now 70 miles distant. Half an hour later a plane drove us under, bombs shattering more light bulbs. Through the afternoon we yo-yoed up and down as we gained on the convoy at flank speed. Another bomber caught us at 1621, coming very close. Lots of cork and paint flew, more light bulbs shattered. I mused at how strange it was that no one's eyes got cut by flying glass. Perhaps it was because we involuntarily blinked our eyes at the big bangs.

At 1900 the planes departed and we surfaced. The *Barb* had muffed, and the convoy escaped. The *Queenfish* must have made them mad. Ed moved us to a new area. We exchanged signals with the *Pampanito*.

Tomczyk wrote in his diary:

> It will be a miracle if we come back from this patrol. Boy they're sure out to get us. We dive. We surface. We chase. We get bombed. Paint and cork and light bulbs fly. We tried our best, but bad luck today after going at it for over twenty hours. I believe that convoy is out of our reach, darn it! Boy am I tired! These planes are right on our tail. Up and down. Some fun. We call it KEEP-SLIP. One slip and it's for keeps.
>
> It's hot as hell being submerged this way and twice as stinky, with the batteries throwing off gases mixed with that world known 'B.O.' Temperature is a blooming 94°. Surfaced at 1900 and the fresh air was something awful. What a peculiar smell. Tonight we laid dead in the water while a couple of heroes greased the ballast tank vents and the exhaust valves on deck. Two other heroes repaired our antenna that the Japs blew off. You should have seen their faces. DEAD WHITE! [scared that a plane might force the sub to dive while they were working in the superstructure].

*11 September. Time 0300.* An ULTRA message was received from SubForce Pacific for the *Growler, Sea Lion, Pampanito, Barb,* and *Queenfish.* A convoy from Singapore was headed northeast through the Paracel Islands to the Formosa Straits. All wolfpacks were to stay on radio frequency 2006 for contact reports while patrolling this lane. Ed's Eradicators immediately headed westward at full speed to the designated track. Ben's Busters, the other pack, would be 60 miles south of our waiting position. With five subs attacking this convoy, Ed, Elliott, and I looked forward to a clean sweep.

Al Allbury's odyssey on board the Japanese merchant ship continued.

> *Time 1200.* The Singapore convoy: 5 ships and 4 escorts carrying the 2100 Australian and British prisoners of war passed the Paracel Islands opposite Luzon. Another convoy of 5 ships and 2 escorts from the southern Philippines joined them on that sun-kissed sea. To the POWs on deck it was an impressive sight. Al Allbury said to his closest friend, Ted Jewel, "This convoy now has far more escorts for far fewer ships than we had crossing from Britain to Singapore. Guess we're well protected." Then Al leaned over the air opening to the hold where one hatch cover had been removed and described the scene to the 900 men still stuffed below.
>
> Neville Thams' eyes were transfixed on a plaque screwed to the bulkhead just above the lower section of the hold, where he squatted in the squalor. Waving to Al he shouted, "Tell the captain, the more the merrier. Our probability of getting sunk decreases with each added ship. He's a good bloke and speaks English. Also tell him there's a plaque down here that reads in English, 'This space is designed for 187 third class passengers.' We'll be happy with 300 here so we can lie down. The stench of sweat, puke, and crap is awful."
>
> Trying to reassure the prisoners, the captain had addressed them in English on departure. He told them he had made the voyage successfully 18 times. At that an Aussie voice piped up, "Yeah, matey, but Ajax went down on the 19th!" Ajax, a famous Australian race horse, had had 18 straight wins, but lost the 19th race when he fell.
>
> Al passed the word to an Aussie officer, who talked to the captain. Another hundred men were allowed topside.
>
> Drinking water was in short supply. The ration for each POW was one-half liter per day. Those with dysentery could obtain half a cup more each time the medical officer wrote out a special chit. There was plenty of rice, but it was cooked in half ocean water and half fresh water. Jack Flynn found that the naval gun crew manning the cannon on wheels at the bow had their own water line. For a guilder they would fill a canteen. Soon an active trade started, bartering photos of loved ones—anything— for water. After dark, men drained oily water from the winch engines and drank it.

It was the sixth day the convoy had been under way.

*12 September. Time 0007. Growler* made contact with the convoy, but she didn't send a contact report on the prescribed radio frequency for both wolfpacks. Twenty minutes later the *Sea Lion*, then the *Pampanito*, made contact. The *Barb* and *Queenfish* were on the surface eagerly awaiting a contact report during the entire time the Busters were engaged with this convoy.

*Time 0057.* The *Growler* sank the frigate *Hirado*.

*Time 0425.* The *Sea Lion* put two torpedoes into the *Nankai Maru* and two into the *Rakuyo Maru*, which carried 1350 POWs.

*Time 0553.* The *Growler* sank the destroyer *Shikinami*.

*Time 0700.* The *Nankai Maru*, previously hit by the *Sea Lion*, sank.

*Time 2140.* The *Pampanito* sank the *Zuiho Maru* and the *Kachidoki Maru*, which carried 750 POWs.

Chief Pharmacist Mate William Donnelly (*left*) and Motor Machinist Mate 2d Class Traville Houston, 12 September 1944, South China Sea.

From his view on board the *Rakuyo Maru*, Al Allbury saw the events:

*Time 0057.* We were asleep, oblivious of the submarines that had for hours been stalking us. We had, as the days and nights passed with a placid, uneventful serenity, almost forgotten our earlier trepidations. The bow of the *Rakuyo Maru* was simply a crowded transit camp. We sat and sweated and waited for the rice tins by day. We curled up in whatever space we had grabbed by night.

There was a vast roar, and the crash and thud of two explosions. The sudden shock, and the fear, set my heart hammering madly. One tanker was ablaze from bow to stern, a fierce, glowing torch of flame. Alarm bells rang. Hysterical screams and shouts came from the passenger deck above us. It was a night gone mad. Our ears were filled with the booming thunder of depth charges, the roaring crackle of flames, and the soft hiss and bursting of distress rockets as they patterned the sky.

On the bridge gun platform, 12 September 1944, South China Sea: (*row 1*) Chic Baughman, Bob McNitt; (*row 2*) Dave Teeters; (*row 3*) Dick Gibson, Paul Monroe; (*row 4*) Tuck Weaver, Max Duncan, Ed Swinburne; (*top row*) Gene Fluckey, Jim Lanier.

The hubbub coming from the hold increased. Those still hemmed in between the shelves were in an agony of fear. The dull reddish glow that suffused the well of the hold lit the struggling mass. With the shuddering concussion of the explosions that rocked the sea around, what I had always feared was happening. The result, the wild-eyed scrambling and trampling down of men who had lived too long as animals. There was a terrible cry of pain from the hold. Panic, in which a thousand, struggling, fear-crazed men fought like beasts in the dark.

"TAKE IT EASY, YOU CRAZY BASTARDS. TAKE IT EASY. WE HAVEN'T BEEN HIT!"

The young, hard-faced Australian officer sprang up on the hatchway. Leadership!

"Not yet, you mean," came a Cockney voice from the hold. That brought laughter that broke the tension. A quiet, shame-faced calm spread across the decks and down into the hold.

"WE ARE DRAWING AWAY. I THINK WE'LL BE ALL RIGHT IN THE SAFETY OF THE DARKNESS. I ADVISE ALL TO SIT TIGHT AND SETTLE DOWN AGAIN."

A muttering peace settled over the deck. Curled up in my blanket, I soon fell asleep.

*Time 0425.* A blinding shuddering crash and a sheet of acrid flame brought me up on my elbows with my stomach churning. The *Rakuyo Maru* faltered and leaped up out of the water. A solid wall of green pounded me flat as the giant sea boiled about me. For one terrible moment I thought we were plunging straight down. My friend, Ted, was gone.

Stumbling towards the thick cluster of men at the port rail, the young officer said, "Get out as quick as you can." The men paused and then vanished from sight. I could hear their continuous splash hitting the water. Two mounds of rafts were heaved over. The Japs had gone with the boats. Uncertain, I saw a knot of men scuffling, dived into the scrum and grabbed two kapok life jackets. In bright moonlight the ship was a perfect target. Crossing to the rail, I gave one jacket to an Aussie and jumped. The sea was warm.

I swam away from the sinking ship to a raft. Clinging to the rope, feeling sick, breathless, and glad of the security, a near voice muttered, "Can't I go anywhere without you taggin' on!" It was Ted. What a glorious watery reunion.

Ted had been squatting in the slatted box that the Japs hung on the port rail to serve as a latrine for the POWs. A Jap scream in terror from the bridge caused him to see the white streaks of two torpedoes approaching the *Rakuyo Maru*. One struck her amidships, the other just missed her stern. He had been washed on deck by the column of water. Hardly able to swim, he jumped overboard, and paddled to a boat. A Jap threw something heavy at him and threatened, so he struggled to a raft. Humbly he muttered a prayer.

"Reckon maybe I ought to try it other times, when I haven't got the wind up."

"Ted, d'ya want my life-jacket? I swim like a fish." He took it gladly.

## Jack Flynn's story provides another perspective of the initial attack:

Jack Flynn and Dick Heywood had a sense of being inside a kettle drum when the two torpedoes smacked the frigate not far away. Standing among the winches where they had been sleeping, they watched the ensuing panic knowing the *Rakuyo Maru* had not been hit. They watched the other escorts milling around dropping depth charges. The night, that had been torched off by the blazing frigate, resumed its moonlit tranquility—the South China Sea smothered the flames as the frigate sank.

They lay down again among the winches, wishing they were wenches and their bed was on terra firma.

Four hours later they were wrenched from their hopeless dreams by the crash of the torpedo striking aft. Gaunt eyed, Flynn struggled to rise as thunder below smashed an ear drum.

"Up I went in the air in a great squirt of water. If this is death, I wonder what it's like. Not so—down I came head first onto the steel deck. I was bleeding from my left ear and a chunk out of the back of my left hand. I had a concussion. Major Chalmers, a doctor, came along. He said, 'You're fit enough. Throw over anything that will float and get away from her before you go down with her.'"

Hatch covers were tossed over. Flynn slid down a rope and climbed on one with two others. "I got violently sick and lost consciousness. When I came to, I was on my own. I paddled over to a big group. Among them were three Japs. We were all for drowning them, but a warrant officer said to hold them so when they were rescued we'd do all right."

Another aspect of the attack comes from Neville Thams:

He was sleeping in a sitting position on the port side of the lower hold when the two torpedoes hit fore and aft of the hold. All were banged around, but no one was killed.

"The order to abandon ship was bellowed down from above. Some sailors from the HMAS Perth had briefed groups for escape from the hold, by setting up long planks with ropes thrown down from above. We moved to get out, but we had to wait until the hundreds on the top tiers jumped overboard. They blocked our exit.

"After an eternity, I pulled myself up a steep plank to the deck. An unforgettable sight with a tanker burning. The broad flames lit up an eerie sky. Guns were firing. High drama. Explosions. The decks seemed almost deserted. I heard a plop in the sea below. With the next plop, I climbed the rails and jumped the thirty feet from the deck to the water.

"Earlier, I had found a life jacket though a third of our mates had none. There was no way my life jacket and I would part. Down, down deep into the murky depths before I surfaced with my lungs bursting. I lost my Dutch straw hat. My water bottle was empty. A raft was close by.

"There were 19 of us under and around the raft. It had no buoyancy. Two sick lads we put on top. Unanimously, we kicked away from the ship and nearby burning tanker. Fear stayed with us from abandon-ship, death by fire, drowning, sharks, suction as the ship sank. If we only had had a little Navy know-how, we should have scrounged for food and water before we plunged into the sea. I was thirsty.

"Some of the Japanese did not escape the ship. They were waylaid by POWs who seized the opportunity to settle old scores. Others in the water suffered also.

"When the *Rakuyo Maru* didn't sink immediately, several groups tried to paddle back. The wind moved her away faster.

"The second escort was torpedoed. Another frigate raced over to attack the sub, dropping depth charges. These caused brutal and terrifying shock

waves in the area. On the port side of the ship they felt them like punches
in your stomach. On the starboard side they suffered casualties, doubling
men up with pain as their bowels blew."

The sun rose at 0700 on a dismal scene. Jack Flynn saw the frigate that
had stayed behind to pick up survivors.

> Out came the lifeboat alright, but one of the Japs in front of the boat
> stood there with a German Luger revolver and took the other Japs away
> from us, leaving all the POWs. Then as they went by us, they gave us
> the thumbs down sign. We gave them the old upya sign.
>
> Things got really tough from then on. The escort was still racing around
> dropping depth charges. Water hammers; it was like being hit in the
> stomach with a pick handle.

Al Allbury recalled:

> Eighteen men clung to our raft. Over a thousand were out there in the
> darkness with dawn near. Ted looked old, with no guts left in him. We
> had no reserves to fall back on. Fish plucked at my legs. A nearby voice
> said, "Bleedin' fish. Won't even wait till you're dead."—Silence.
>
> With the dawn came hope. The *Rakuyo Maru* was slowly settling a
> mile away. Her boats filled with Japs, close by. Closing fast was a large
> Jap ship. It wouldn't be long now!
>
> A sudden reverberating explosion, a geyser-like column, and before our
> eyes the oncoming ship, struck by a salvo of torpedoes, disintegrated.
> Then came the concussion. Great waves of almost unbearable pressure,
> hammering our bodies. The walls of my stomach collapsed. My bowels
> emptied.
>
> Then we heard the singing. Softly, almost shyly, it grew in volume,
> swelling into a great triumphant surge, flooding across the sunlit water:
> "RULE, BRITANNIA!"—This was retribution. At last, after three years, the
> Japanese were on the receiving end of the boot.
>
> The singing died away. A great silence crept slowly over us. Our moment
> of victory was past. We became a rabble of tired, hungry and thirsty men,
> faced with the prospect of approaching death.
>
> With our raft down to fourteen now, we held an apathetic council of
> war. We must reach the *Rakuyo Maru* for food, water, more rafts. Kicking
> and pulling together we set out. After an hour, exhaustion defeated us.
> Due to the currents or light breeze, the ship was farther away. Drifting
> aimlessly, we got covered with thick, scummy masses of oil.
>
> Afternoon now, with a harsh and blinding sun, I reached down for my
> shorts to cover my head. They were gone. I felt horribly exposed. Soon
> the raft was surrounded by a shoal of tiny silver fish. A soft mouthing

against my thighs served to bring home my loss. Ted grinned at my predicament, "Start worrying when something bigger comes along."

Before sunset a Jap destroyer and a small ship bore straight towards us. A ragged cheer swelled into a full roar. The ship picked up the Jap boats, now distant. The destroyer circled in wide sweeps, then headed our way at high speed. She passed by, thirty yards away from our gaggle of rafts without slowing. The wake washed over us. Stopping alongside the *Rakuyo Maru,* a boarding party clambered aboard, opened her seacocks, and returned. She shoved off. Suddenly, quietly, the great ship slid beneath the waves.

The POWs, close to the life boats pulling toward the small ship, had been held at bay by rifles and pistols. Once the Japs were aboard, we thought the boats would be used to collect us, but they drifted away empty and beyond our reach. Then the ship with nets still over her side turned our way. With a thousand men watching, waiting, and praying, they changed their mind and sped off. We knew—never to return.

Poor deluded fools! Even the "Red Cross Camps" in Burma had failed to teach us that there is no limit to man's inhumanity. Nobody spoke. No hysterics. No defiant oaths. Men, almost at once, began to die. After all we had gone through to be left to a death such as this. . . .

And Neville Thams remembers:

The *Rakuyo Maru* remained afloat for almost 12 hours, finally going down quietly at dusk. One frigate had remained through the day keeping a watchful eye on their men in the water with the POWs. Under its very guns at close range our raft was commandeered by the Japanese. We had to clear out. I swam to five other blokes on a hatch cover. We were up to our waist in water. A second frigate and freighter appeared at dusk rescuing their own men. The Japs in the lifeboats that had kept close to the *Rakuyo* were now picked up. Surprisingly, they did not sink the abandoned lifeboats which were soon filled with about 350 POWs, including the three senior officers. With this rescue freighter now close, our hopes of rescue soared. Sadly they made no attempt. When the English POWs realized this, they started singing, "RULE BRITTANIA." The Australians joined in. It rose to a great crescendo. Words cannot describe this act of defiance. Why they did not fire on us, I will never understand. The three ships turned away. We watched them go over the horizon. We were alone. All hope of rescue disappeared. Men began to die at once.

Al Allbury describes the nightmare: "That first night passed slowly as I hung on to the raft in a stupor of exhaustion. Vaguely it seemed men had gone berserk crying out in pain and fear, clawing, fighting, then muttering.— What was left of our world had gone mad."

# Chapter 9 Save the *Queenfish*

*13 September.* As Al Allbury recounts, the nightmare continued:

> The sun flushed a cloudless sky. Many had disappeared. As the rafts gathered together once more, there was room for everyone to get out of the water and sit. The redistribution worked out to about six on each raft and two to a hatch cover, with legs hanging over the sides.
>
> Hallucinations and dramas unfolded. Some saw ships coming, only to end in bitter disillusion, sobbing, and despair. On the next raft a vicious, lunatic struggle of four crazed men, tearing, hitting, and biting each other. One of them had caught a fish with his hands. Two men tumbled, still fighting, into the water. Too weak to swim, we saw the agony on their faces as they threw up their hands and disappeared, ten yards away. We sat and watched, feeling nothing.

*Time 0600.* Commodore Ed Swinburne gave orders to the *Barb* and *Queenfish* to head east to Bashi channel, since the Singapore convoy had failed to materialize. Three days and 400 miles had been wasted in this location, for Ben's Busters had failed to radio their contact reports to us.

Speeding back to the hot-spot plane area, the day was uneventful. Planes caused us to dive five times. No bombs were dropped.

Jack Flynn recalls:

> This second day, we ran into oil on top of the water from the tanker that sank with us. Some three inches thick, it was terrible. The boys were dwindling down by now, some by drinking salt water, and some by drinking their own urine. Peter Stokes disappeared. We don't know when.

114

Alby Meridith was rinsing his mouth with salt water. I was blind in one eye from the oil. I saw Alby drifting away. I swam after him. Just as I got to him, I lunged, but could only touch the top of his head. He went down feet first with his arms outstretched, perhaps just as well. Everyone had hallucinations. Only three of us left now.

Al Allbury continues: "I awoke to the roar of thunder. With the first spots of rain, our lips cracked apart in prayers of thankfulness to God. Hands cupped. Heads back. The drumming water stung my eyes, streaming down my cheeks and throat. All too quickly it was gone."

*14 September. Time 0003.* "Captain, radar contact 9000 yards. It must be a small ship. I'm heading for it at full speed."

"Good, Max. Station the tracking party. Send a contact report to the *Queenfish*. I'll be up in a jiffy." Slipping on my red goggles and khaki shorts, I grabbed a shirt and headed for the radar scope in the conning tower.

Lehman pointed out the target on the scope. "Sir, there may be another ship that's just coming in, 6000 yards on the port quarter of the first."

Bob was just climbing the ladder. "Bob, tell radio to make that contact report two ships. I'll be on the bridge. Max, let's slow to one-third until plot can find out what gives. I don't like the 6000-yard distance between the two ships. As soon as Tuck's eyes are night adapted, get below on the TDC." Tuck took the watch.

*Time 0023.* "MAN BATTLE STATIONS TORPEDOES! Tuck, let me know when your young eyes can detect what type of ships these are. Their movements are very odd."

"Aye, only shapeless blobs now."

After half an hour of tracking, Ed joined us on the bridge. "Gene, I've come from plot. There is no zig plan and their speed changes from 12 to 16 knots at intervals. I think you've got a bear by the tail."

Tuck called, "Got them, sir—two frigates."

"Right on the nose, Ed. Now let's twist it. Bob, we'll work in on the lead Chidori for a stern shot. Poor visibility may help. A hunter-killer group again. Make ready tubes 8, 9, 10."

*Time 0138.* Fired stern tubes 8, 9, 10. Depth set three feet; torpedo run 1580 yards; port track 70°. We had a calm sea. Sound reported all torpedoes hot, straight, and normal. Target did not change course until end-of-run explosions. All missed as we withdrew to check our complete system. This should have been a snap.

The *Queenfish* commenced her attack on the second ship and inquired as to types. Somebody replied, "One destroyer and one cargo vessel. The *Barb* missed and will re-attack from same flank."

*Time 0322.* Now, with a faint sliver of a new moon, the two subs neared firing positions on the separate targets each had chosen. Suddenly, our target blinked a recognition signal. We did not reply, as it was probable he didn't know our exact location. Still, I turned our tail his way, in case he started a chase. Again he challenged; again no reply. We played a waiting game.

*Time 0326.* The *Barb* became the target as our enemy flooded us with a brilliant searchlight! BOOM! Simultaneously, his forward guns opened fire. Before the five-inch shell roared past us at shoulder level like an express train, Tuck had pressed the diving alarm. Lego, Miller, Ed, Tuck, and I set a new all-time record in clearing the bridge. We commenced swinging right at flank speed for a down-the-throat shot while watching his shooting through the periscope. His stern gun fired at something on his port flank. In a couple of minutes he turned off his searchlight, whereupon the *Barb* came to radar depth and went after him again.

The *Queenfish,* seeing the *Barb* illuminated and fired upon by the frigate, took this opportunity to catch what she thought was the slightly larger cargo vessel. Seconds before skipper Elliott fired, however, his target turned away. Chasing up his tail at full speed, Elliott planned an up-the-kilts shot. At 1000 yards he was again ready to fire when his target turned right 90°. Only then did he realize he was not chasing a cargo vessel, but a destroyer. He got a quick new setup at a range of 900 yards and let fly four torpedoes. The destroyer kept her course and speed, but challenged the *Queenfish* by blinker. Not daring to turn, thus exposing her broadside, she dived. At that the destroyer reversed course and commenced depth charging.

*Time 0350.* The *Barb* was nearly in a firing position again. "Sonar reports torpedoes running . . . broad on the port bow. They're crossing the bow now . . . fading out broad on the starboard bow." No explosions.

"Ed, *Queenfish* must have fired and missed. The other frigate is depth charging her. Our target is turning that way, or doing something."

"Sonar reports another set of torpedoes running, headed our way."

"Sonar, watch the bearing of those. Is it steady?"

"Sir, bearing is moving slowly right." Through the hull we could hear them churning down the starboard side.

"Captain, from plot, somehow we goofed on our last contact report to *Queenfish.* We told her that one was a frigate and the other a small cargo vessel."

"Okay. We got her into this, it's up to us to get her out of their clutches.

If these two get together, one will stay off *Queenfish's* beam pinging her position while the other runs back and forth up and down her spine dropping depth-charge patterns. We must save the *Queenfish*. Let's go!"

*Time 0359.* It took a few precious minutes to track our target exactly by radar. We made ready our bow tubes and opened the outer doors, our course steady, speed 13.5 knots, range 1500 yards. "Up scope. Final bearing—mark."

"Set."

"FIRE 1!" Donnelly pushed the plunger for tube 1. At 10-second intervals tubes 2 and 3 were fired. As sonar tracked the torpedoes, we waited, praying.

"Hitting time 15 seconds."

"Up scope! Bob, the Chidori is circling right like she's spinning on a dime and coming after us. Paul, take her deep! LEFT FULL RUDDER! ALL AHEAD FLANK! Rig ship for depth charge. Rig ship for silent running. We missed, darn it!"

"Sonar reports depth-charge explosions ceased."

The *Queenfish* had a rough working over. Thirty close depth charges caused damage throughout. Now she crept away. It was the *Barb's* turn to be the target. As soon as I heard the two pinging on us, I knew I had overdone my good Samaritan act. My intention was to eliminate one frigate, or if that failed, to keep one of the two off the *Queenfish's* back. Surely I hadn't intended to end up with both. Yet, from a submariner's viewpoint, both heaven and hell are paved with good intentions.

"Paul, level off at 375 feet. All motors dead slow. Lock all watertight doors and the lower conning tower hatch. Right 15° rudder. I want to stop that pinger from staying on my beam. Perhaps we can maneuver them into a collision."

*Time 0404.* A series of explosions rocked us! Depth-charge indicator showed the shots to be very close, a perfect straddle and above. First team stuff, bouncing the *Barb* around with the now-customary mess. Bob, cool as usual, was on the phones. "Report only major damage."

"Right full rudder. All ahead two-thirds. Ed, here he comes again. I'd like to get him on our port side. If he drops another string of 10, we'll make a high-speed run to break loose on his third string as soon as the first charge explodes. The turbulence may screen us. Cross your fingers."

*Time 0409.* "Sonar reports shift to short-scale pinging. Screws coming up port side. He's dropping!" The second string of 10 were close, but high to port. Again he circled.

*Time 0415.* "Sonar reports short scale. Screws to port. First charge hit the water."

"All ahead flank speed. Maneuvering room, give her every ampere she'll take. Rudder amidships." After a series of click—bangs, we felt like we were a pin in a bowling alley. Men were knocked flat. The charges were set deeper, at about 300 feet. The *Barb* was pushed sideways and deeper. All the lights went out. The thunder below was enough to jar our fillings loose. The charges were so close we could hear the click of the detonator before the explosion.

With our emergency lights on, we sped away from the trap. This deep string had helped us. The deep turbulence spread and expanded as it climbed slowly up, forming great welling gushers on the surface. We profited not only from this three-dimensional screen but also from something more important: the fourth dimension—time! We had burst clear of their tightening explosive bonds bent on our extermination.

"Ed, we're free at last, and I, damn it, have wasted nine torpedoes on these friggin' frigates. Incidentally, did you know Frigga was the Norse goddess of marital love and the hearth, married to Odin?" Everyone relaxed in laughter, the great escape from stress.

Still smiling, Bob, added, "Either these frigates have flat bottoms or these new torpedoes are running deep." Ed nodded.

"Captain, no major damage. May we secure from depth charge and silent running?"

"Sure, Bob. Breakfast time."

*Time 0652.* Periscope depth. I sighted the tops of the frigates astern, searching and playing their sweet music, distant thunder.

Tomczyk recorded the latest encounter in his diary:

> At last a contact on two ships. No, 1 ship, 1 destroyer. Nope 2 screwy Chidoris. Very dark. Fired at one, missed. He dropped a couple. We're going full speed. This tub is really shaking. We're slowing and heading back in. Told to stand easy although we're not what you would call calm. Made ready the bow tubes, starting the attack. Astern is Queenfish. The Chidori is challenging us with blinker. We ignore him. He's turned on his searchlight, caught us full in his beam, and started shooting. We dove. Secured tubes. He's coming this way. Made ready bow tubes again in nothing flat. What a scramble! Fired 1, 2, 3 and missed. Going down deep with the charges raining down. Twenty on his first run. He's darn good though I hate to admit it. Captain Gene says only ten are close. Blew the lights out. All loose gear went flying around—paint and cork steadily in the air. I honestly thought this was it. Sprung more leaks. It looks pretty bad.—He's circling and coming in again. Rigged for silent running. Here they come—24 of them. They sounded so close we thought they tied two or three together. All lights are out except the emergency

system. I could fall asleep standing but they bounce us around like a fly. No sleep for the wicked. If we get out of this mess, you can bet your last dollar, I'm going to lead a lot cleaner and better life. I promise!—Hey, whatcha know we've scooted out from under them going like a bat out of hell and reloading torpedoes fast. Only nine left with seven days left on station.

# Chapter 10 Mission of Mercy

Al Allbury's third dawn adrift was bitter with the realization that the vast armada of human flotsam of three days earlier had been decimated. Hundreds had died, and the rafts were now widely separated.

Of the five of us left, the others were asleep, Ted across my legs. I spotted an oasis of rafts tied together, woke the others, and we commenced paddling in shifts to join them. It took half the day. It was the Aussie contingent led by the same young officer. He figured the stronger survivors could paddle west and reach the Paracel Islands in two days. Four or five of the wooden hatch covers would be hitched to each raft that opted for the journey. No sick could be taken. Each man had to pull his weight or be left behind. Those to be left gathered around a center raft on which they had erected a spar with a distress signal. Ted, I, and another elected to share a smaller raft that would go, the last in a string of five.

Shoving off from the incapacitated group was heart rending. Some wanted to come with us as a last hope. Soon we were moving along in the dark, paddling and resting on orders from ahead.

I never knew when I ceased to function. The next dawn I sat up and looked around. Ted was still there. The other man was gone. No longer were we attached to the group. They were not even visible. Others like us had been dropped.

*15 September.* Only Ted and I now, silent, huddled together, in this fourth dawn. Two empty rafts trailed along beside us, like riderless horses that clung for company. Other rafts with men sprawled across them dead. Twice, dirty-yellow triangular fins were circling our raft. That stopped further dangling of legs over the side. Later I spotted the doctor drifting by on a hatch bale. We pulled him, delirious with dysentery, onto our raft.—We tucked him between us. Thus, we could hold him during his ravings, so he wouldn't roll off.

*Time 1500.* The *Pampanito,* passing through the general area west of where the *Sea Lion* had sunk the *Rakuyo Maru,* was surprised to come upon a crude group of rafts and hatch bales loaded with lots of men. Coated with oil, they appeared more Japanese than Caucasian. She closed in cautiously, with small arms ready if there were any opposition. One of the survivors called out, "First you bloody Yanks sink us and now you're bloody well going to shoot us." This convinced Pete Summers to take them on board.

Quickly they revealed their sinking on the twelfth. So many were in the vicinity that Pete didn't have space for everyone. At 1610 he contacted Eli Reich in the *Sea Lion* to the north. The two swept the area until dark and no more could safely be accommodated on board the overcrowded subs. Sadly, they departed for Saipan: the *Pampanito* had 73 survivors, the *Sea Lion* had 54.

*Time 1930.* Al Allbury scanned the ever-empty horizon in despair, not knowing the rescuing submarines were just beyond it. Ironically, they might have been among those rescued if they had stayed behind with the large raft group of the sick. "The doctor's ravings were becoming infrequent. As darkness came, he lay quiet. In the torpor that came with the night, we forgot he was there. I kept falling across him. His body was just something to rest my head on."

*16 September. Time 0300.* Noting the *Sea Lion*'s report of the sinking of the *Rakuyo Maru* and the POW rescue, Admiral Lockwood in Pearl ordered the *Barb* and the *Queenfish* to the pick-up area. Both subs moved westward at flank speed.

Ed had just been exchanging information with Donk's Devils on the prescribed joint wolfpack radio frequency. They had now returned from reloading torpedoes at Saipan, and both our packs were searching for a convoy coming our way momentarily. Our loss of a crack at this convoy was not regretted, but it still irked me that we had to retrace some 450 miles to the area where we had waited vainly on 12 September, never having received one message from Ben's Busters.

*Time 0600.* On this fifth dawn the sun jumped over the horizon, rousing

Al Allbury from his dreams of home. He raised his head slowly off the doctor. "Ted was still there. Then I looked at my pillow; his mouth agape, his face a thousand years old. He was dead, his eyes still fixed with pain and delirium. I remembered all he had done. I became a human being for a moment filled with sorrow and compassion. The finest man I had known. We rolled his body off the raft."

*Time 0700.* That same day on the aircraft carrier *Unyo,* the dawn air patrol had been launched earlier, providing antisubmarine protection for this valuable Japanese convoy of their largest tankers. The largest, the *Azusa\* Maru* (11,177 tons) had been built the year before. The pride of the tanker fleet, she was loaded with 100,600 barrels of oil. Convoy Commander Rear Admiral Yoshitomi Eizo of the 5th Guard Fleet had placed her and the *Unyo* at the rear for added protection. The formation—starting counterclockwise from astern of the *Unyo*—was frigate *21,* then to starboard frigate *19,* inboard the *Otowayama Maru,* and dead ahead the *Harima Maru,* the destroyer *Kashii,* Eizo's flagship, and frigate *13.* To port Admiral Eizo placed the destroyer *Chiburi,* the *Omuroyama Maru,* the *Hakko Maru,* and frigate *27.\*\**

This HI-74 convoy had departed the Seletar-Singapore area at dawn 11 September for Japan. The 1st Marine Guard Headquarters sent a submarine alert on the twelfth informing them of the attacks on the POW convoy. To bypass that area Admiral Eizo had moved the convoy east 60 miles, even though there would be less air support from shore-based planes in Indochina and Hainan Island. This long-range coverage well ahead of important convoys was vital. The *Unyo* carried 48 planes, 12 of which were on deck and flown for local convoy coverage. The remainder were army aircraft cargo being returned to Japan for overhaul and major repairs.

The *Unyo* had been the *Yawata Maru* (22,500 tons), a new passenger ship capable of 25 knots. After the Pearl Harbor attack she was converted to an aircraft carrier. With the long reinforced flight deck, the added repair shops, cannons, and military equipment, her tonnage as the *Unyo* was now much greater. Also greater was her complement of 781 men: 45 officers, plus 24 pilots and 110 aviation crew men, and Inouye's staff of 20. Passengers pushed the total over 1000.

*Time 0700.* Breakfast over, the *Barb* was her usual beehive. Men off watch busied themselves checking, cleaning, and repairing the myriad mechanisms that stand between man and eternity in a submarine. I picked

---

\* Meaning the beautiful bell-shaped flower of the catalpa tree.

\*\* This convoy entered Singapore to load at Seletar as the POW convoy was leaving on 6 September.

*Top:* The *Yawata Maru,* 22,500 tons, before conversion. *Bottom:* The aircraft carrier *Unyo,* formerly the *Yawata Maru.*

up my plate of half-eaten, somewhat questionable cold-storage eggs, set it in the pantry, and broke out the chart of our present area.

Ed leaned over my shoulder, the fragrance from his mug of coffee erasing the sulphurous odor of the eggs. Ruffling my hair—a bit envious, for he was growing a beard to compensate for his thinning scalp—he said, "Don't you redheads ever sleep?"

"Lots to plan, Ed, since your wolfpack message. You've added a new dimension to submarine warfare, which we hadn't anticipated."

"What do you mean?"

"Simply this: Our normal mission is detect-attack-destroy, whether it's among icebergs or under a broiling sun. Now we add rescue of what was inadvertently destroyed. Not a simple Red Cross exercise, either. This mission includes danger, crew health, medical support, nursing, feeding, clothing, strict water conservation, and the maximum quantity of men we can accommodate in our *Barb* submarine hotel. I have been conferring with Bob, Max, Swish, Doc Donnelly, Tomczyk, and Shoard for the torpedo rooms, along with chief motor machinist mates Thomas Hudgens and Franklin Williams. We realize there may be additional unexpected problems. Medically, our supplies are limited and inadequate. You know how one cold sweeps through a sub. Yet we can't isolate new diseases that come aboard. If the POWs can put up with them, so can we. Ed, we're preparing to pick up 100!"

"Congrats, skipper! Make it 101. I'll give up my bunk and sleep on the transom in the wardroom."

Al Allbury recalls the fifth day on the hatch cover:

> Ted's plight was pathetic. Both eyes were sealed tight from the thick oil we had drifted through during the night. Trying to part his eyelids with his fingers made him shake his head with pain as some oil got on his eyeballs. My hands were coated too. My left eye kept sticking. After a half-asleep spell, I couldn't open it. Between the two of us we now had only one eye. Nothing to see, only drifting corpses.
>
> As darkness fell, Ted was delirious. I knew he was gulping sea water whenever I dozed. I swore at him as he drank in sly, greedy mouthfuls. In delirium, he kept plunging off the raft. Drifting in his lifejacket a few yards away, he would cry out in fear. I'd paddle the raft and pick him up.
>
> The trance of exhaustion into which I had sunk made me hardly care. Then sickened with the fear of finding myself completely alone, I would shout and shout until I saw the dark blob that was his head. I knew I couldn't swim now.
>
> As if in a dream, I realized that he had left the raft. He was floating somewhere near, but I could not see him. I shouted and shouted. His cries came back fainter and fainter. More piteous, more urgent, more knowing what was happening,—his cries died away, and fear crept over me. Would I follow?

*Time 1100.* The Japanese convoy HI-74 was now abeam of the Paracel Islands. The convoy commander's new flagship, the destroyer *Kashii,* had all of the latest German radar and sonar equipment. Submarine warning messages had been received concerning the attacks in Luzon Straits on 31 August and just north of the Paracels on 12 September. A base course between these locations had been chosen for the next 36 hours, then the convoy would be safe in the well-mined bastions of Formosa Straits. Air coverage from the *Unyo* was increased to four planes during daylight hours.

**Time 1234.** In the *Barb's* after torpedo room everyone was preparing for the rescue. Murphy, Shoard, Irving Greenhalgh, Roark, Joe Salantai, and Wells each led groups. Their last two torpedoes were in the tubes. The crew had decided to turn over all their bunks to the POWs. The *Barb's* men would hot-bunk in eight-hour shifts. Four torpedo skids were being padded with rags and burlap bags to make 12 bunks. Three shifts at each bunk would take care of 36 men, almost half the crew.

"Murph, this is a tight squeeze."

"Tight squeeze, sir—why that's more room than my old lady gives me at home."

In the wardroom Bob and the commodore were huddled over the charts, estimating the most probable position of the rafts. Ocean currents, tides, and probable winds for six days entered into the calculations.

I climbed up to the bridge. After an hour I felt jumpy. No reason. No planes. Just couldn't sit still. Dave, dropping his binoculars, chided, "What's the matter, captain? Nervous in the service?"

"I don't know. Something's wrong. I can't explain what."

Instead of going below I stayed there for another hour scanning the horizon, the sea, the sky. It was unusually clear, the horizon too distinct, the air hot and oppressive. "Quartermaster, what's the barometer doing?"

"A little higher than normal, sir."

The *Barb* carved the oily waves. The light breeze that had been slowly backing counterclockwise died. A long, low swell came from the southwest, barely rolling us. As if someone were pricking me, I suddenly realized we had a far more potent threat than enemy aircraft. The air and the sea had that uneasy feeling that gets into the bones of any sailor. TYPHOON!

"Dick, a typhoon is building up. No one on a raft will survive if it breaks. I'm going below."

I told Ed we were going to be shipmates with a typhoon. He dropped his pair of dividers. "My God! Haven't we enough trouble? Are you sure?"

"No. Let's take a look." Ed, Dave, Bob, and I went topside and returned in a moment.

"Your honest opinion, Bob, please."

"Well, the barometer is not following its regular diurnal oscillation. I've no idea how far off it is. Since the war started, we've had no weather reports from Japanese-held regions. Maybe tomorrow, maybe the next day, maybe it will turn off south of us. If it hits us, we'll be in the dangerous semicircle. But for the POWs, we should reverse course."

"Gene, can you crank her up a bit?"

"No use. Bob says we'll arrive two hours before dawn tomorrow. We need light to find them."

"Well, I'll warn Elliott that we'll plow on through with a prayer."

Tomczyk wrote that evening in his illegal diary:

> Heading to rescue survivors the Sea Lion unknowingly sank. Gene figures on taking a hundred. All skids fixed for bunks. Spare clothing, cigarettes and stuff donated by the crew are set and waiting. Probably drop them off in Saipan. Sighted two periscopes, but swerved and kept on going. Expecting anything and everything, we're prepared for all kinds of trouble. Crew has been assigned jobs to do. Imagine 17 lookouts topside plus me on the high periscope. We all hope we get there in time.— Everybody is GUNG HO!

*Time 1600.* The aerologist in the *Unyo* informed the admiral that weather reports indicated a typhoon had formed 160 miles southwest, heading up the convoy's track with winds building up from 60 to an expected 100 knots. The speed of advance was 20 knots. The full force should arrive in 24 hours and might stop planes from flying the next day as gusts and seas picked up.

*Time 1846.* Toward sunset I was on the bridge with Max. The breeze was becoming a wind; the sea was waking from its sodden slumber. The men on watch swung their binoculars incessantly. They combed the blue vault of heaven, the fine hairline that divided the blues of sea and sky, and the deeper blue of water close aboard.

Klinglesmith, the starboard lookout, called down, "Plane bearing 030, altitude 5°."

I said, "Don't dive, Max."

All our binoculars centered on the plane far away.

"Okay to stop, captain? He may not see us if he doesn't spot our wake." We stopped and watched. With a sigh—thank God—we watched him depart.

*Time 2110.* The *Queenfish,* on the ball as usual, made contact on a convoy at the astonishing range of 34,000 yards. Ever a team player, Elliott quickly sent us a contact report. Since the *Barb* was miles astern, he crossed the northeasterly base course of the zigzagging ships to attack from the port flank.

Forty minutes later the *Barb* made contact and moved in to attack from the starboard flank when the *Queenfish* finished.

*Time 2231.* The *Queenfish* fired four torpedoes, range 3200 yards and 95° port track. The *Kashii* sighted the torpedo's wake passing to starboard and ordered a 45° course change to port toward the *Queenfish*. One torpedo hit in the *Omuroyama Maru*, which the *Queenfish* saw and the *Barb* heard. The *Kashii* fired a red rocket flare signaling submarine attack. A second explosion followed.

*Time 2250.* The *Queenfish* reported her attack completed, fresh out of torpedoes. Ed directed her to trail. Now it was the *Barb*'s turn.

*Time 2254.* "BATTLE STATIONS TORPEDOES!" I waited for the gongs to stop, then announced to the crew: "Men, we have a nice fat convoy of five big tankers and six or seven escorts. *Queenfish* hit one that hasn't slowed, so they're alerted. We may have a bit of tangle, but oil for the lamps of Japan makes the *Barb* thumb her nose at anyone. On your toes! We're heading in!"

I told Bob to make ready all six tubes forward and the two after torpedoes, leaving only one torpedo forward for a reload, which we might need. "If we do, make it quick."

"Understood."

"And tell Jackson and Ragland to put four cases of beer in the cooler." Muffled cheers were heard from below.

"Bob, if you can spare Broocks, send him up. With Tuck on the target bearing transmitter, I need a quartermaster up here to help me keep track of this flock of escorts. I'll cross between the starboard ones to shoot the lead tankers."

*Time 2323.* The *Barb* bored in at full speed, with the convoy heading north and zigging. We were almost sandwiched between two escorts. As we were working out the setup to shoot the after tanker in the starboard column and the big one just ahead of her in the center, Tuck yelped, "That big ship between the columns is an aircraft carrier!"

"Ye gods! A flat top!"

A submariner's dream, the chance of a lifetime appeared before us. One of His Imperial Majesty's carriers was protecting this important convoy.

Tuck shouted, "We've got to get closer. It's hard to line her up in this poor light."

"Tuck, the wind's backed around to the east. We're riding in like a whitecap. With a course change, we'll be spotted."

Broocks tugged at my trouser leg. "Sir, the frigate has speeded up. She's got a bone in her teeth heading this way."

There was only one thing to do: "ALL AHEAD FLANK SPEED!"

The *Barb* surged forward, shook her tail, and flauntingly crossed ahead of the oncoming escort. Seconds jelled.

Miller, the port lookout, eased the stress. "Golly, if mother could only see her little boy now. I promised her I wouldn't do anything rash."

*Time 2331.* "Bob, I've got a perfect overlap of the carrier ahead of the tanker's bow. Tuck will stay on the tanker's bow constantly. Spread the six fish across it. We won't have time for a second setup. This damned frigate is coming in to ram. He's on our port quarter. Give me a quick range to him, then the tanker and carrier, and we'll shoot. ALL AHEAD ONE-THIRD! OPEN THE OUTER TUBE DOORS!"

"Ranges: escort, 700 yards, tanker 1800, carrier 2100."

*Time 2332.* "FIRE!" The first torpedo whooshed out with a jolt and broached. "QB sonar, get on that torpedo! First looks erratic; watch her for circling!"

Broocks grabbed me. "Look! The frigate's going to ram!"

A single glance was enough. "Take her down, Tuck!" Then I jumped down the hatch.

"CLEAR THE BRIDGE! DIVE! DIVE!" The lookouts, plus Broocks and Tuck, came dropping down. Tuck pulled the hatch shut while Broocks spun the locking wheel. The torpedoes were still being spurted out of the tubes as Doc pushed the firing plunger at 10-second intervals, so I couldn't change course.

"Sonar reports first torpedo erratic to port, did not circle."

"Rig ship for depth charge and silent running!"

As we went under, the frigate filled the whole glass on the periscope. Ed was handling its raising and lowering. Bob was orchestrating the torpedo firing, a perfectionist maestro. "Paul, level off at 365 feet. Down scope. Ed, house it all the way. He's sure to drop. No use looking at the targets. Torpedo run time's not up. Should be about two minutes."

As Bob shot the last torpedo, I shouted, "There goes Annie B! Go clobber her Annie!"

"Sonar reports erratic torpedo steady on southerly course. The rest are all hot, straight, and normal." Then, "High-speed screws coming in pinging! Short-scale on port quarter. Bearing is steady!"

"Lock all watertight doors and lower conning tower hatch! Hang on!"

*Time 2334.* BOOM! Dick called out, "Hit in tanker!"

*Time 2334.09.* WHAM! "Hit in tanker! My God, she must have exploded!" The *Barb* bounced around with the concussion. Our grins were cut short by the nerve-racking sound of propellers passing directly overhead. Stand by! Dead silence.

On board the surfaced *Queenfish*—trailing the convoy—Elliott and his executive officer had their eyes glued on the convoy. Harry Higgs, after careful study, announced, "Captain, there's a carrier smack in the middle of the convoy." BOOM! "*Barb*'s hit the trailing tanker!" BOOM! WHAM! "Good Lord, that second hit—she's exploded! Look at that giant fire-ball! Gad, it's over 500 feet in diameter! Must be a million gallons of gasoline going up."

On the *Kashii,* Admiral Eizo, having settled the convoy back on its prescribed track without a loss after the *Queenfish* attack, was startled by the explosion of the *Azusa* and signaled "Emergency turn starboard."

On the *Unyo,* the officer of the deck logged: "Recognized a column of water starboard side *Azusa;* judged to be torpedo damage. *Azusa* blew up on second torpedo attack."

On a raft 125 miles to the southwest, Al Allbury felt the great loneliness of the night and the sea: "Soon I will drink the seawater and know the terror of insanity. The spectre of the choppy waters, the last living gasp, the drowning, haunt me."

On the *Barb,* breathless seconds passed as we looked at each other. "Sonar reports high-speed screws reversed course."

"What happened? He didn't drop!"

*Time 2335.* Cheers from stem to stern—three hits in the carrier!

"Shankles, take your rudder off. Steady as you go."

"Gene, I think we were saved by the tanker exploding just as the skipper was ready to let go. We don't know what's happening on the surface, but from the jolt we got down here, it must have been enough to distract anyone."

When things had calmed down, Ed added, "By the way, Gene, what did you mean when you said, 'There goes Annie B. Go clobber her Annie'?"

"Ed, it's the skipper's traditional privilege to name one torpedo. The men do it, too. Annie B. is Ann Brindupke. She and her husband, Brindy, are old friends from *Bonita* days. She lives near us in Annapolis. After Brindy shoved off on his last patrol, I noted they changed his orders to refit at a forward base on his return. Knowing she's like a hen on hot bricks when he's on patrol, I dropped her a line. You know—big-help me. I told her not to worry. With the slowness of mail, she might not hear from him for a long time. What a boo-boo! The *Tullibee* was sunk on that patrol. So I printed Annie B. on that torpedo. When she slammed into the carrier, I said a silent prayer of sweet revenge. God bless."

"Sonar reports high-speed screws have swung around and are headed for us again, pinging short scale!"

"Dave, what's the bathythermograph doing? Have we a temperature gradient?"

"Isotherm 84° from surface to 40 feet. Down here 65°. Good thermocline—19°."

"Great! That should deflect the accuracy of his attack."

The sonar operator took his earphones off and bellowed, "HE'S DROPPING! HANG ON! HERE THEY COME!"

And come they did, his unwelcome calling cards too close for comfort. The indicator showed them deep and to starboard. The cloying cork dust was a nuisance, filling the air and sticking to the cold sweat on the men's bodies. Amongst all the hell there was a cheering undertone: breaking-up noises, hissings, underwater explosions, whistling, crackling, crunching— sure signs that a ship was sinking. Caught in the cavernous maw of the depths, bulkheads collapse and boilers are crushed even before the ship accelerates and crashes against the ocean bottom.

How great can personal and impersonal emotions become? Depth charges bursting all around, yet my mind was keenly alert as I directed evasive courses, speeds, feints, coaching the well-honed *Barb* team so the ship and her men and I would survive to rescue others. Yet a small part of my subconscious pondered the fate of the carrier that, head high, proudly steamed along just four minutes ago. Now she was writhing in her death agonies, her planes slithering over the side. The ugliness of destruction was tempered by the sobering knowledge that such planes could deal death no more.

Random depth charges were felt far off, not close. "Sonar reports destroyer turning and heading back at us."

"Dave, give me a course from plot to pass within 50 yards astern of the tanker's position when hit. Bob, I'm cutting as close as I dare to the stern of the tanker. There should be something burning up above. I hope she can't follow without getting singed."

On the *Unyo* after the *Azusa* exploded the hydrophone room detected a torpedo sound just abaft the starboard beam. The emergency alarm was pressed, indicating the sub's direction on the bridge battle board. Captain Ikuzo Kimura ordered "LEFT FULL RUDDER! BATTLE STATIONS!"

The *Unyo* had turned less than 10° when the torpedo hit below the steering room on the starboard side. Two seconds later a second torpedo hit the main engine room. Captain Kimura was furiously doling out myriad orders over the damage control circuit that was feeding information to the bridge, fighting to keep his *Unyo* alive. Reports kept jamming in: "steering, main engines, auxiliary machinery stop functioning and are unfunctional. All lights are out. Fires are extinguished in all boiler rooms and all main valves closed. Eighteen dead in engine room, one in steering."

Just then the third torpedo hit near the stern. The *Unyo* heeled over 5° to starboard, righted herself, and began to sink stern first, slowly. To prevent further attacks, Captain Kimura ordered, "Commence volley firing starboard cannons at the prearranged range. Point abaft the beam."

"Bone in Teeth," by combat artist Fred Freeman. A Japanese destroyer tries to ram the *Barb,* whose periscope is seen going under on the getaway. The *Barb* had just torpedoed a tanker and the escort carrier *Unyo.*

On the *Kashii,* Admiral Eizo canceled any attempt to assist the *Azusa,* burning ferociously from bow to stern with a tremendous lake of fire around her. There could be no survivors from that holocaust. He radioed the stopped *Unyo* on the voice circuit. No response. He then spoke in uncoded plain language to all ships on the voice circuit to continue with Z Plan. "Tailing escorts—frigates *21* and *27*—to assist *Unyo,* the vice admiral, pick up survivors and pass this message: 'Land-based air cover is requested.' "

Upon receiving this message, frigate *21* passed it by blinker to the *Unyo* and notified the *Kashii* of her depth-charge attack, believed to have sunk the submarine as it disappeared off sonar near the *Azusa.*

The *Barb* was deep now and running fast, cutting across the formation close astern of the tanker. A second set of noises—mixed with the rumble of the depth charges astern—tolled the death knell of the second ship as she began to break up.

"Bob, secure from depth charge and silent running."

Paul poked his head up through the lower conning tower hatch. He was soaked to the skin. "Had another cable pushed in down here, causing a fire hose at this depth. Tight now, but control room bilges are flooded. We're a bit heavy. I'll start pumping now. One air compressor foundation is cracked. We can weld it later."

"Good work."

Deep inside the *Unyo* a courageous damage control battle was being waged with a limited number of emergency battery lights. Her main emergency lighting system had been blown out. Bulkheads were being shored up. Working in waist-high water at times, sailors repaired leaks. Electricians, knee-deep in salt water, risked their lives connecting live substitute cables from a gasoline generator topside. Lighting was desperately needed.

Depth bombs for the planes—stored in ready lockers on the hangar deck—had broken loose, rolling around, banging the elevators. Three dropped into the sea and one exploded.

Smoke enveloped the entire deck leading to the main engine room, and carbon monoxide gas was being generated. All gas detectors were useless without electricity. Watertight doors were reinforced.

*Time 2350.* The *Azusa* sank with all hands.

On the *Unyo,* the auxiliary generator began working. One radio message was sent from the after transmitter room. Then gas forced the radiomen to abandon the room and set up a portable telegraph set.

*Time 2355.* The *Unyo'*s sinking stopped. This was confirmed and spurred all hands to extra efforts. Sea level aft was at the hangar deck. Frigates *21* and *27* circled, depth charging.

*17 September. Time 0005.* The *Unyo'*s reinforced bulkheads collapsed. Sinking increased; carbide in the damage control warehouse was underwater and began emitting gas. Brave volunteers carried it out and threw it into the sea. Crew members began pumping water out of the passageways in the sealed compartments. All secret documents were locked in the vault.

In the *Barb,* Ed, Bob, and I conferred over the area chart.

"Bob, we've a torpedo forward and two aft. I'd like another crack at those last four tankers. How much time can you give me?"

Ed looked at me, "Not satisfied yet?"

"With three fish left?"

"Captain, at 19 knots we should have shoved off 10 minutes ago. With the vagaries of wind, sea, tides, and currents—plus this threatening typhoon—our search area is enormous."

I couldn't deny the itch for action, but neither would I deny priority to our mission of mercy. "Ed, we don't need orders. All our hearts bleed for the poor wretches, wherever they may be, imprisoned on their flotsam for this, their sixth day." Ed nodded approval.

*Time 0040.* At periscope depth, a quick swing around revealed only one escort in the light of the new moon, which was very faint. Random depth charging continued.

"Bring her up to 45 feet, for radar."

Two pips were spotted at 11,000 yards, probably picking up survivors. With no burning oil, we surfaced and took off at flank speed. We were submerged only one unforgettable hour.

"Now hear this: *Barb* teamwork tonight was perfection. I feel sure we sank a 22,500-ton carrier of the *Otaka* class and a large tanker of the *Itukusima* class—10,000 or more tons. I am proud of each and every one of you—you know how I feel. I know you're exhausted, but we must be ready at dawn, so now I ask you to turn to for half an hour and make the *Barb* shipshape for our guests. At 0130 we'll splice the main brace!— then hit the hay until dawn, when the rescue lookout stations will be manned."

Secreted away, Tomczyk wrote:

> Made ready bow and stern tubes. Fired all six forward as we were diving with destroyers coming right down our throat. Five hits, three on a flat top and two on a tanker that blew up. We heard them breaking up. WAHOO! Always wanted to hit a carrier. Went way deep and the ashcans commenced. They're still dropping and coming closer. Counted 58 so far. We've lost them. They've stopped dropping. OOPS! I'm wrong—there's another and another. Surfaced with all 4 generators on the line breezing out of here. Two destroyers astern. Total ashcans for that lovely picnic— 63. Squared things away in the Room. At 0130 interrupted again, but this time it was fun—our ration of beer and a shot of whiskey for good measure if wanted.

*Time 0150.* On the *Unyo* the brave salvation battle continued. Damage control parties discovered much leakage coming through air ducts between reinforced bulkheads. They cut and sealed these and sinking stopped completely. Large gangs of men, in desperation, began draining water from passageways and compartments using pots and pans from the galleys. Ship trim was 20° stern down. The ship was bow high, pushed by wind, so the stern was up wind and sea.

*Time 0355.* Wind and waves became severe, hitting into the rear of the *Unyo*'s middle deck, flooding into the torpedo hangar. The stern was gradually sinking.

# Chapter **11**   Search for Survivors

Morning twilight did not bring joy to our world in this area of the South China Sea. Ed's Eradicators had formed a scouting line four miles apart. Speed had been reduced from 19 to 15 knots as the seas picked up. Our search commenced in earnest with the stinging salt spray flogging our faces. The *Barb's* 17 lookouts scrubbed the surface. No flotsam.

The *Unyo's* crew—exhausted and without sleep—continued their gallant struggle. Waves turned over the depth bombs on her rear deck. Torpedomen heaved them overboard. By 0530, the pounding seas caused water to penetrate into the elevator room, making it impossible to dispose of the remaining depth bombs.

Elsewhere that morning, Al Allbury recalled:

A brutal wave slapped my head hard against the raft enticing my clawed fingers to let go and end this misery. End the drugged exhaustion that would find relief only in the pleasure of death. Opening my one Cycloptic eye, still stinging from the saline water, I surveyed my heaven and my hell. The first streaks of dawn, the start of my sixth day, broke through my haze. I reached for Ted, but he was gone. I could still hear his last muffled cries for help. My raft slid up and down the steep-sided waves. I could see nothing. Tiny speck that I was upon the endless water, I knew that nothing could see me. But, GOD, CAN YOU HEAR ME?

Victorious from the last affray, the *Barb* heaved as if sighing as she surfboarded along the following sea, eager to get on with a more important mission. Her men brought her alive as they scrambled up the periscope shears swinging their binoculars. The bridge-level fore and aft gun platforms were filled. Every single pair of binoculars on board was at the ready as the first fingers of dawn splashed the east.

"The first person to spot a POW gets the honor of cutting our celebration cake for last midnight's work. Plus a beer!" I added. Louder cheers rocked the *Barb*.

*Time 0630.* On the *Unyo,* Captain Kimura finished a conference with his senior officers. They were in unanimous agreement: the vice admiral must leave the ship while lifeboats could still be lowered. The stern was sinking; the *Unyo's* demise was inevitable. The admiral agreed, sent his flag lieutenant to take down the Emperor's portrait and bring it to him in the lifeboat, then ordered Captain Kimura not to go down with his ship because aviation needed him.

*Time 0645.* Captain Kimura passed the word to stop emergency operations, all hands on deck, man all lifeboats. With his final order to abandon ship, the Rising Sun naval flag was hauled down as another sun rose over the horizon heralding the new day.

*Time 0655.* The *Unyo* sank!

*Time 1130.* Frigate *27* picked up 761 men; over 200 were lost.

Back at the Naval District Headquarters in Pearl Harbor, in a carefully guarded, super-secret enclave known as the "Black Chamber" (where Japanese messages were intercepted, decoded, and translated), Captain Jasper Holmes, a former submarine skipper, was at work there in cryptography. As a liaison with the submarine force, he furnished selected information for the ULTRA messages sent to submarines.

Looking through a batch of Japanese naval messages, his eye caught a garbled one. The best that he could make of the decipher was this: An air vice admiral had shifted his flag from (blank) ship to (blank) ship in the South China Sea and the imperial portrait was safe.

Getting on his private secure line to Captain Voge, Admiral Lockwood's operations officer, he asked, "Dick, I have a garble here that's elusive, but significant. Can you shed some light on what is happening?" He then read the message.

"Jasper, why is it important?"

"Well, they don't put air vice admirals on anything but carriers. Shifting flags is quite common, but they don't take the imperial portrait as baggage unless a ship is sinking. We have no other U.S. naval forces engaged in that area. I'll bet you a buck that one of your subs has bagged a carrier. Who's in that area?"

"Let me check."

"Donk's Devils are in Luzon Straits. I believe they would tell us if they sank a carrier. *Barb* and *Queenfish* arrive about now to rescue POW survivors sunk by Ben's Busters. They're in a dangerous situation now; they'll have a lot of people topside. These POWs have been drifting for six days. I'm sure the subs won't take a chance on being picked up on a direction finder by sending any high-powered transmission. Probably be four or five days before they clear the straits for Saipan. We'll hear then. Yes, I'll take your bet and hope that I lose, only because I won the last bet. I'm off to dine at Uncle Charlie Lockwood's quarters. I'll pass your estimate on to him, but I think it's too optimistic and a bit iffy."

In the *Barb*, passing the barometer, I tapped it twice just to make sure. No doubt now that it was in a steep slide. The typhoon could break within a matter of hours. The sun rose red and angry, in spite of the ever-increasing wind. Our eyes reddened from the unrelenting binocular search. A veil covered the sky, but not the sun. I expected to see the black mass of clouds come scudding over the horizon ahead at any moment. From time to time I caught myself mumbling little prayers. Button, button, who's got the button? Where are they? Are six days of starvation and dehydration beyond their physical limits? Probably.

The morning trickled away vainly as tempers shortened.

*Time 0956.* Tomczyk, the high periscope lookout, yelled up the hatch. "Wreckage, broad on the port bow! Do I win the prize?"

"Providing there's somebody alive. Coach the helmsman left until it's dead ahead. Good work."

In a moment we had it in sight from the bridge: a large area of flotsam. Swish called up from below, "Captain, do you want the rescue party?"

"Not yet, Swish. No sign of life, just floaters."

"Aye, sir. Some of our first patrolmen ask permission to see what Japanese wreckage looks like?"

"Granted—come up!"

Trailing down the wake of flotsam, there were numerous floaters identified as Japanese. The bodies were ghastly—bloated and blistered by the sun. Some were slightly gnawed by the more carnivorous species of fish. Their skins, grossly inflated, were stretched as tight as drums. At last I knew we were in the vicinity. The men gazed, amazed and awe-stricken by these absolute horrors of war. Thoughtful Bob had scheduled an early lunch to clear the mess tables for medical use.

"Lunch is served. Chow down," the intercom squawked.

One man vomited. Others had had their fill and went below, having decided they weren't hungry. Me too.

Slowly the sea increased, the wind whistled through the antennae.

"More wreckage ahead, sir."

Everyone strained, prayerfully.

"Ship bearing 300, on the horizon, sir."

Hearts pounded. We couldn't lose time to dive. Our rope was almost at it's bitter end. Max waited for my signal to dive.

Tomczyk called, "Looks like the *Queenfish*."

"Give her a recognition signal." Prompt reply, affirmed.

Ed formed a scouting line abeam, separation four miles. The seas were too rough to permit a wider search. Now we were running into Allied floaters as well, some with life jackets.

*Time 1255.* Tomczyk called again, "Captain, I've got two rafts, one with three men, one with two men."

"Great! Put Shankles on their heading. ALL AHEAD FULL! AWAY ENTIRE RESCUE PARTY! Notify *Queenfish*. Bob, for God's sake, in these seas they're out of sight so much of the time we may have passed some by. Have your deck rescue group be careful. Watch they don't get caught between the raft and the turn of the hull. In these seas a leg can easily be broken."

As the deck group scrambled up, they waited for the ship to slow because waves were breaking over the awash deck.

To our dismay the three survivors, though sitting up, clinging to the hatch grating, evinced no interest in our approach. Jim Lanier, one of Bob's group, said, "Their complexions are almost black. They're not British nor Aussies. They must be live Japs."

*Time 1303.* "ALL STOP! ALL BACK FULL! Rescue party on deck. Warning! One keeps peering this way cautiously. He may be armed." We drew alongside as Tuck, conning the *Barb*, stopped.

The same survivor stared at us. Then with his last ounce of strength, he got to his knees and screeched, "Hey, Yank!"

"Did you hear that, Max? No mistaking that cry."

Our deck rescue party went into action. Heaving lines had been made up with a loop in the end. Swish heaved one across the raft. "Slip the loop over your head and under your arms!" he shouted. No response. Two just stared; the other was past movement. Bob dropped his heaving line, slipped Houston's loop over his head, and dove into the four-foot waves. At the raft he pulled a man off with a cross-chest carry and hauled him back to the *Barb,* where ready hands lifted Harold New aboard. Bob pushed off from the surging waves around the curve of the hull to pick up another. With another line, Houston dove in for the third man.

*Above:* A raft comes alongside as McNitt and Houston rush for heaving lines, then dive in to rescue those too weak to grasp the lines. *Below:* The rescue party brings a POW aboard. Lanier, Swish Saunders, McNitt, and Houston were scraped and cut by barnacles as the typhoon broke.

*Above:* Covered with oil, the POWs are given a quick cleaning and passed up to the bridge, to be lowered down the hatch. The decks were awash much of the time. *Below:* The rescue continues.

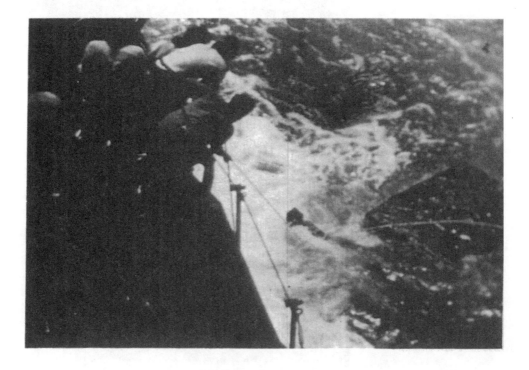

Bob was hauled back with the Aussie, Jack Flynn, who had yelled, and both slopped aboard exhausted as the *Barb* lurched.

Houston, our Floridian fish, was in trouble. Aussie Murray Thompson was unconscious and couldn't hold his head up. With the waves, it was a real fight to keep his head out of the water. Yet, as I watched, Houston did it by holding his man aloft while his own frame was completely submerged. Once he was on board, Tuck headed the *Barb* toward the second raft with two men.

Meanwhile, others on deck were stripping and cutting off the Aussies' oil-soaked garments. The featherweights were then carried to the gun platform at bridge level. Here the transportation gang took over. On to the bridge and gently down the ladders, by way of the conning tower to the control room, the survivors were then carried into the crew's mess.

"Tuck, this is horrible; I'll take the conn and watch for planes. You help this next lad waiting to go below."

Tuck looked at him. His oil-covered skin was so soft from immersion that his rescuers feared to handle him in case it would fall from his body. He could barely move. Yet he looked up at Tuck and said, "If I can't do anything else, I can still eat. I'd give a lot for a slice of a roasted prison guard whom I will never forget, but no rice, please."

Tuck laughed to see such wit, such humor, such spirit in this guy. It impressed and left a stamp on him that would last forever.

Pharmacist's Mate Donnelly and his nurses bathed, doctored, and brought life anew to these battered skeletons laid out on the mess tables. At the end of this production line were the sleepers who would carry the patients off when more room was needed, tuck them in a regular bunk, and guard them in their delirium.

*Time 1317.* Jim Lanier dove in to assist the two on the second raft, Cecil Hutchinson and Ross Smith. Swish was in the water alongside, holding their flimsy flesh off the rasping barnacles. No more rafts were in sight.

"Bob, as soon as these are below, clear the decks so we can speed up, or someone may get washed overboard."

"Aye!"

*Time 1429.* We picked up Lloyd Monro—a sailor who had been on HMAS *Perth*, before she was sunk in 1942 by the Japanese—and our first Brit, Jimmy Johnson. Both were so far gone that Bob and Jim dove in for them. These two looked as if they were sitting back to back on top of the water with their Jesus shoes on. Their hatch bale was so small we couldn't see it. Even with their light weight, it was always four to six inches under water.

Monro, the first one aboard, collapsed like a beanbag. "Sorry, matey, I can't seem to stand up."

Four men lifted him. "Take it easy, Mac. We've carried five of your pals down already. Gad, you're like a greased pig."

"Bunker oil from a sunken tanker. It saved our lives, for we'd have been fried by the sun."

As he passed by on his way below, he stared at me with haunting, sunken, pus-filled eyes. I smiled at him. Reaching out, he plucked my sleeve with a scrawny hand and whispered, "God bless you, captain."

I turned away, swallowing the lump in my throat.

*Time 1435.* Houston and Swish dove in and picked up a lone Brit, Thomas Carr. The rescue party hauled all three aboard. "Hey, captain, he's alive—just unconscious." Houston and Swish were bruised and bleeding from pounding and scraping against the side.

*Time 1444.* We picked up a couple of Aussies, Leo Cornelius and Robert Hampson. Hampson was hauled aboard on a line; Cornelius got a McNitt assist. The wind and waves steadily increased as the black clouds came. Our time was running out.

"Commodore, would you mind taking over while I drop below to check those last few men. Tuck has the watch."

"Not at all, Gene. I relieve you."

Arriving in the crew's mess, I gasped. "Doc, are you sure you're not running a mortuary? Look at that ugly ulcer on his leg. I could put my fist in it."

"No, sir, they're alive and I'm going to keep them that way." The nude forms on the mess tables were still. All that was left of the name "man" was a soul encased in some bones and impoverished flesh.

Our baker Elliman, now nurse, piped up, "Boy, have I got some feeding to do."

Now the production line was rolling. Donnelly grew a dozen additional hands. He zipped from one patient to another injecting, bandaging, administering, checking, and ordering nurses Ezra Davis, Joe Zamaria, and Julian Kosinski. "Don't forget. I said three tablespoons of sweetened water with chipped ice, repeated in short nips. Clean the oil from their eyes with boric acid solution, repeated every three hours. If eyes are bad, apply yellow oxide of mercury. Chief, the next two men you bring in— lay them on the deck. We're short of tables here."

Dashing to the bridge, I relieved Ed. Topside, the drama of life for the dying was unfolding: more rafts, bigger seas, danger. With the approach of black scud clouds, the sea began to lash the *Barb,* frothing up through the gratings, which made footing treacherous. With the wind howling, our ship fought up and down the deepening troughs. Two more survivors were seen.

*Time 1513.* We picked up Australian Neville Thams and a Brit, Augustus Fullar. Neville had taken his shorts off two days before because they were chafing him. When Fullar spied the sub, he used them as a flag and waved to us. Their hatch cover was also submerged. With the angry seas, they had figured another two hours before they'd have drowned. Hauled aboard by ropes, they were half dead. Only one survivor so far had been able to even sit up.

After passing through the assembly line rigamarole below, Neville was tucked in a bunk. With the rolling and shuddering of the *Barb* as she searched for survivors, he feared she might turn turtle.

*Time 1519.* We picked up a lone Aussie, Jim Campbell. Just an hour before, as the black clouds came, he had resigned himself to death. Praying, he bade farewell to his beloved parents, his sisters, his brothers, and, tearfully, to his fiancée, Jean. This done, he sank into his dying stupor. Someone called him. It was Jean: "Hang in there, Jim!" Soon the *Barb* appeared. He fainted. Houston went for him and brought him alongside unconscious.

Al Allbury's lone odyssey continued:

> The sea became even more boisterous. The raft stood up almost vertically, with the waves bursting around my head. I was home in London, helping my wife pick beans in our garden. Then I would splutter and spit and realize in despair my plight. This is the end, for I was making my spiritual farewell to everything I had longed for these last three years. Now devoid of emotion, I knew the calmness, the mental and physical negation of feeling, meant the last slip into unconsciousness and death.
>
> In a strange dream I fingered a rope that fell across my shoulders. In the dream I sat looking at it, wondering what it was. I heard shouts faintly through a mist of darkness. I put the rope over my head and felt myself being pulled backwards through the waves. There was a huge wall of steel, all shiny and wet. I was dragged up and splayed out on something that didn't heave and dip. A strange blur of faces, voices not Japanese, then I realized, as if it didn't matter, that I wasn't going to die, after all.
>
> A hand nudged under my neck and something trickled down my throat and exploded into fire in my chest. A deep drawling voice said softly, "Relax, fella. Everything's gonna be all right." I whispered, "Yanks," and fell headlong into a wonderful unconsciousness. Hours later I woke, wrapped in blankets in a bunk, and wondered if it was only a dream.

All of the survivors were 25 to 50 pounds underweight. Malaria, dysentery, pellagra, sores, ulcers, beriberi, ringworm, genital sores, skin abrasions, edema of the ankles and feet, dehydration, conjunctivitis, gingivitis, bashings, beatings, and starvation had left them human wrecks. These had been picked to go to Japan due to their physical fitness? God help the rest!

The *Barb*'s crew shall never forget the first dubious, then amazed, and finally hysterically thankful look on the survivors' faces from sighting to boarding. Their appreciation was unbounded. Even those who couldn't talk expressed themselves tearfully through their glazed, oil-soaked eyes.

"Tuck, I'll need the bullhorn. The wind is almost gale force. No more rafts in sight. Bob! Clear the deck as quickly as possible. Take all your gear, including the extra life jackets for the survivors in case a plane forces us to submerge and leave them on deck. We need to increase speed."

The bridge resembled a sardine can as the whole party madly wormed through the door from the forward gun platform.

"All rescue groups stand easy. At 1600 set the regular sea detail plus high periscope lookouts."

I had only one superlative for the performance of Bob's diving team: magnificent. I told him to get his men below and have Doc take care of their bleeding and scrapes before infections set in.

The black sky was ominous, about to explode. We cranked the *Barb* up to full speed. The oncoming watch topside was in foul-weather gear. As Dave relieved Tuck, I pulled my rain parka over my head. Dave yelled in my ear, "I've tried to pick them up by radar. With the gratings made of wood, it's no go unless the men have some metal on them."

"No luck there. The only metal they'll have is the fillings in their teeth." The weather was growing worse. "Dave," I said, "this whole patrol you've been hoping for a new experience, a typhoon. Well, you're going to get your wish, damn it! Your reaction will be interesting."

In the last shredded threads of daylight that sewed up our day, we saw more empty rafts—more ghastly floaters that brought tears to those topside. We kept reducing speed as the waves pooped us, breaking over the stern. Now we were down to four knots, barely enough to quarter the seas. The conning tower hatch was frequently slammed closed to avoid having water cascade below. The bridge became a bathtub as 20-foot waves swept across us. It filled with a rush. Looking down inside my parka, I saw water up to my armpits; bubbles burst out of my shirt. The water was warm. The bow came completely out of the water when we crested a gigantic wave. As we tipped over and plunged down, the propellers whirred when they broke the surface. I hoped that one of the floaters wouldn't land on someone, for fear the fright might cause him to jump overboard.

I was tempted to have the *Barb* dive. The calm of the depths offered a siren's call in a gale or typhoon and might have become a necessity for our survival. But not yet! Could one more life be saved? Lousy odds!

With darkness, the rains came. In torrents. I felt the heavens were being flushed along with me. Binoculars were of no use as the *Barb* rolled and pitched as if some giant hand had her in its grasp. I couldn't see the bow.

"ON SEARCHLIGHT! QUARTERMASTER, SWEEP FROM BEAM TO BEAM!"

The brilliant beam shot forth and came to a shimmering halt in a white wall of lashing rain at about 20 yards.

"Sir, all that's doing is ruining my night vision. Do you think it will attract the Japanese?"

"Dave, that's a chance we'll have to take. They'll more likely think we're Japanese. Our radar must warn us. There's just a tiny chance a survivor may see us and yell. All we can do is throw him a line from the bridge and haul him up. I'm going below to confer with the commodore."

It was too rough to cook, so people munched Spam-and-cheese sandwiches. Our survivors were all tucked in, with nurses standing close to reassure them. Doc was yawning, checking his remaining inventory of medicines and leafing through his *Medical Compend*. At age 26, he had never visualized anything like this challenge to his training. True blue, he was determined to bring in a live bag. My compliments, confidence, and support brought a tired smile.

Miller gave him a sandwich and a cup of joe. As he took time out to relax and eat, I told him of my own medical successes in my first sub, the *S-42*. Being an ensign fresh out of sub school and the junior officer on board, I was given my first job as ship's doctor and dentist. With it came a treasure: a *Medical Compend,* a *Dental Compend,* and two sterile packages. Curious, I opened them. One contained my surgical tools, the second my dental tools.

We were en route to the Virgin Islands from Panama when my first emergency case erupted, an abscessed tooth. Sitting the patient beside me on the wardroom bench, I assumed my best professional aspect (not knowing what in hell I was going to do). Confidently, I opened the book to where the index said, "Abscess." Then I put my finger in his mouth, pulled his cheek back, and compared the inflamed bubble on his gum with the photo in the book.

Wisely nodding my head I said, "Smitty, there's no doubt. This is it. Only one though—the rest are fine." He understood. So I opened the sterile pack and selected the molar extractor. I turned around briefly, and when I looked back, Smitty had disappeared. When I caught up with him, still clutching my extractor, he was hiding behind a torpedo. He assured me that it didn't hurt any more.

"Doc, you know I never lost a patient. Guess the word gets around—when you're good—that an ounce of prevention doesn't hurt at all."

Doc had fallen asleep. I slipped out.

In the wardroom Ed and Bob were huddled over the charts discussing the search probabilities. Bob had accurately put us into the survivor area. We waited for his recommendation. He had used what seemed to be magic in finding the survivors. He explained that he had had to consider a combination of wind and current. Wind-driven currents are deflected from the downwind direction by Coriolis effect caused by the rotation of the earth. Fortunately, he'd clipped from the U.S. Naval Institute's *Proceedings* magazine a "Professional Note" written by a Coast Guard officer, which gave him guidelines on the drift of life rafts. Through some complicated figuring of all the vector forces over the four days since the sinking, he had located the survivors.

"Bravo, Bob. Positively brilliant."

A radioman interrupted with a message from Loughlin. The *Queenfish* reported having picked up 18 survivors, two still unconscious. Her last one was retrieved at 1545, only six minutes after our last survivor; nothing since, only empty rafts and floaters. All of us, including Elliott, opted for continuing the search for 24 hours, in spite of the typhoon.

Ed's plan took us to Bob's maximum survivor position. At 0600 we would head eastward, a bit south of our entry search.

I had to get something off my chest about the unspoken decision after the previous night's combat as to whether to head for the survivors or reattack the convoy. Having seen the piteous plight of the 14 survivors we rescued, I could only say that I would forgo the pleasure of an attack on a Japanese task force to rescue any one of them. There is little room for sentiment in submarine warfare, but the measure of saving one Allied life against sinking a Japanese ship is one that leaves no question, once experienced.

"Well put," Ed said in response to my explanation.

No one slept, except some of the survivors strapped in their bunks. A nurse stood by each bunk, ever alert. Several times a wild scream rang out as delirious men dreamed of their ordeal. Yet they quickly quieted as a nurse reassured them that they were safe even in the typhoon. Rolling and plunging, the *Barb* groaned and battled the raging seas during the night. The wind velocity edged up over 100 knots during this night we would never forget.

# Chapter 12 Home Run

*18 September.* Dawn came. Young old men struggled to keep their binoculars to their bloodshot eyes. The searchlight was turned off and housed. No one said good morning; no one said anything. Dumbly, numbly, the search continued. The wind lessened, but whipped us with a velocity of over 60 knots. Skyscraper waves blotted out the horizon. Movement was pinned down to absolute crawling.

Men sickened, tossed their cookies, then dutifully cleaned up the fouling of their nest. Four of the rescued had diarrhea. The boat stunk. Taking the beating on the bridge was preferable to enduring the mental stupor below. Breakfast was forgone.

Throughout the day we floundered helplessly at a maximum speed of four knots. Now and then we passed rafts—all empty—and bodies, floating bizarrely face down, their legs and arms submerged. The beginnings of gassy bloat pushed their backs and rear ends high. Divine retribution was meted out to the officer who had wished to see a typhoon: he was too seasick to stand his watch.

The *Queenfish* notified Ed that one, rescued in a coma, died without regaining consciousness. At high noon his body was committed to the deep, shot out of a torpedo tube.

Toward evening the seas abated, but not the spindrift searing our eyeballs. Gradually, the exhausted wind blew itself out. Nature had eliminated any further chance of our rendering human aid. Finally, Ed ordered us out. Worn out, the *Barb* obediently turned her bloodied nose toward Luzon

Straits and Saipan. Never had she taken such a licking, not even from the Japanese.

Life seeped grudgingly through the ship. Repair groups formed and straightened out the chaos below decks. As soon as we were shipshape and ready to combat a human enemy, the men collapsed in dead sleep among the rags piled on the torpedo skids.

The commodore, Bob, and I stood in the control room charting the shortest course we dared take. Suddenly, a shrill scream pierced the quiet. We ran aft.

A nurse was huddled over one of the men.

"What's the matter?" I asked.

From an adjoining bunk came the answer: "Matey, he doesn't know where he is. He thinks he's back working on the Burma railroad."

The nurse gently patted the screamer's hand. "Now don't worry, Mac. You're in a submarine on your way home. We're Americans. In three weeks you'll be back in Australia with your family. Take it easy."

A weak voice chirped up, "Thank God I'm in safe hands at last."

Rubbing my stubbled beard I mumbled, "Let's all sleep. We're not fit to make any earth-shaking decisions."

*19 September.* Twelve hours later, Doc woke me. "Captain, I think they'll all live, but five or six need medical attention I just don't have. They're critical. Can we speed up?"

I clambered out of my bunk. "Doc, you do your best and the *Barb* will do hers. If you want advice by radio, I'm willing to gamble. If you can do without it and still keep them alive, we'll make faster time. The radio direction finders will plot us in anytime we transmit. You know the rest—planes, submergence, probable bombing."

"Sir, speed's more important than advice."

Topside, the tail end of the typhoon was still whacking us, now head-on. The center had arced to the east, to the Philippines. At eight knots the *Barb's* elevator antics barely kept solid water below bridge level. Dave was now on watch; his appearance, not green, not yellow, but a moldy mustard. I wouldn't chide him for missing the watch that Tuck stood in his place. He was fighting an internal battle, his eyes watering as he put his hand over his mouth to emit a lengthy belch. "Dave, Doc wants us to hurry. Try wringing out a few more turns, but no water over the hatch. We need all the air we can get below. I'm going down for a sandwich. May I do something for you?"

"Yes, sir. Don't mention food!"

The *Queenfish* notified Ed of a second death. The other chap that hadn't regained consciousness was committed to the deep.

In the afternoon, at standard speed, we were making only 10 knots.

Tomczyk's illegal diary entry for this period reads:

> Making all possible speed. What a job it is trying to write. I'm getting my brains beat out against the tubes. I think I'll screw this chair into the deck. Our guests after some sleep look a lot better. Three got up, were washed, patched up and look darn well. The others are nothing but skin and bones. They told us that a gunboat came to the scene and picked up Japs, but trained their guns on the Aussies and Limeys. Rotten! I can't see how they held out for six days. Boy are they happy. "On the ball" McNitt says we have 1800 miles to go, about 6 or 7 days. A message came in from ComSubPac saying someone sank an Aircraft Carrier in the South China Sea—that's us! That gives us 4 ships this time. Not bad—eh! OOPS! Forgot to mention the official number of depth charges for this patrol is 332. All wasted. The Japs probably reported about five subs sunk for that number. Also of 71 plane contacts, only 6 bombed us. Captain Gene says that class of Carrier has a full load of 48 planes. Guess that makes Barb some kind of ACE. Bet it's as much as any ship shot down in the Marianas' Turkey Shoot.

*20 September.* By dawn, seas had calmed to a chop. A quick trim dive for 15 minutes and we were back up making 15 knots.

We received an ULTRA intercept message on a convoy that would be crossing our path. Temptation! Before showing it to Ed, I drifted aft to have a private chat with Donnelly. "Doc, how are our lads doing?"

"Much better than I had expected. At Saipan we'll have only one stretcher case. All are off the critical list as of this morning."

"Doc, your efforts have been so outstanding that I'm recommending you for a medal plus a special immediate promotion to chief petty officer."

"Gosh, thanks, captain."

"Now, I have a problem. A convoy may be close. Will it affect their health if we lose about six hours to attack? We've still one torpedo forward and two aft."

"Not their physical health. Mentally, I don't know. Why don't you ask them?"

"Good idea."

Neville Thams, Jack Flynn, and Jim Campbell were in the crew's mess having a pot-belly-putting-on breakfast. They stood up when I sat down but I waved them back down. "Neville, Jack, Jim—I need your help. A convoy may be close. We have three torpedoes, but . . ."

"Go sink 'em, captain!"

"Wait a minute, I haven't explained . . ."

"You don't have to!" they answered in unison.

"Hold your horses and listen. We'll lose about six hours to attack and get away—if they are on schedule. They were spotted by our aircraft. Of course, there is danger and stress. You've been through too much already. Doc says all of you are off the critical list, physically. My question is, will it affect any of you mentally?"

Neville spoke. "Captain, I was in the hold of the *Rakuyo Maru*. We had resigned ourselves to death in Japan when the USA invaded. When those two torpedoes exploded, one aft and one forward, all of us cheered. We might have died then—or later—but something Japanese was being destroyed. That's how we feel." The others nodded.

"Okay, but some may have changed their minds, since Providence gave them another chance to live. Here's what I want you to do. Go talk to each and all, then return here. I'll wait for the consensus." They bounded out.

In 10 minutes, they were back, Neville acted as spokesman. "Captain, the vote is unanimous—attack!" With thanks, I breathed a sigh of contentment and went forward.

Ed was at breakfast. I handed over the message. He read it, then looked at me and shrugged. "Too bad, skipper. We've got to get to Saipan."

After carefully explaining what Doc and the others had come up with, I said, "Let us postulate, Ed. We know that Elliott is out of torpedoes. He can go on ahead. Six hours loss isn't hurting us. One, maybe two, ships sunk would move us another step toward winning this war."

"Gene, when you said 'postulate,' you meant exactly that. Think how long we've waited for convoys that were late or no-shows. Think how long we've been held down." He thought for a moment or two. "Tell you what, let's compromise. I'll let the *Barb* lag five hours behind the *Queenfish* on tonight's transit of Balintang Channel, but you must get through before dawn."

"Thanks, Ed." I turned to Bob. "Come on—we've got chartwork to do!"


The *Queenfish* reported sighting six planes that day and diving twice. She had been an hour ahead of us. We now had her in sight. As we dillydallied searching for the convoy, she drew away. At 2345 she passed Balintang Island abeam to port at 12 miles. Having no contact, we sped up along the convoy's track at 18 knots to save time.

NO SHOW. The rescued survivors were disappointed.

*21 September. Time 0440.* The *Barb* passed Balintang Island abeam to starboard at four miles at flank speed. I was tempted to wake Ed to inform him that we had carried out his orders to the minute.

At breakfast most of our recently acquired passengers were up and able to walk about to a limited, staggering extent. As they were carried or helped between the mess tables and their bunks, gales of laughter rang out. Slowly they grew stronger. Day and night between pills, vitamins, and mouthfuls of food, each of them entertained a circle of sailors.

Their saga of three years of prison life building the Burma-Thailand railroad hypnotized their audiences. The inhuman brutality they suffered, particularly from their Korean guards, would not be forgotten by those who heard their first bitter descriptions: bashings that broke jaws; shootings of men with diarrhea who went for the bushes without permission; the short guard who couldn't hit a big Aussie in the face, so stood on a stump, ordered him over, and broke his nose; the Dutch colonel-surgeon who amputated legs for gangrene and tropical ulcers without anesthesia; the patient, with a wet rag stuffed in his mouth, who was held down while the leg was removed with a hacksaw; Red Cross supplies that were stacked in warehouses but were not available for POWs. The normal day's work was 16 hours in G-strings, with the majority of the prisoners barefooted and on half rations. They bathed mostly in a polluted stream. No wonder over one-third died.

Flynn worked with two surgeons at different times when he was too ill to work on the railroad.

> The Dutch surgeon was provided cigars because he assisted the Japanese doctors who did not service POWs. Once his cigar went out while sawing away. He stopped, tapped the patient on his chest, and asked if he had a light.
>
> The other surgeon, Colonel Coats (later knighted), at the end was performing eight or nine amputations a day. He brewed a local anesthetic from plants. One bloke with gangrene had to have a leg off. Coats started on him just above the knee. The drug didn't work on Sam so we had to take him out and lay him down for a while. Also I am patting myself on the back for giving him my blanket. After he had rested a bit, four of us held him down and off came the leg.

Flynn was the first POW that we interrogated. His remarks on the effects of drinking sea water were surprising.

> While we were in the water, a sure sign that they were drinking salt water was their eyes. Pupils and all would become white, horrible, and they'd have hallucinations. They were seeing things. Not many of us left,

then that character from South Australia, Dick New, said, "Look at that dragon fly." Knowing that dragon flies are only found near fresh water creeks, I thought, Dick's a goner—crazy from salt water. Yet his eyes weren't white. Then I thought I saw it, and I knew I hadn't drunk a drop. I said to Dick, "Fair dinkum, then land can't be far away." He said, "No, about a mile." I reckoned I could swim that far, taking it easy, so I said, "Which way?" Laughing, he pointed with his thumb, "Straight down, Jack." I could have killed him.

You couldn't beat the Aussies even at that stage. Soon only three of us were left—Dick, Murray Thompson, and myself.

**23 September.** A message informed us that direction finders had located a Japanese weather ship whose position was close to our track. Since the POWs were in good shape, Ed agreed to a gun or torpedo attack as necessary. We diverted to the location. All we sighted was a patrol plane that forced us to dive; it had probably been a weather plane.

Late in the morning the cooks set out our long-overdue celebration cake, topped with the carrier and tanker plunging into the frosting. Even a few rafts with survivors showed. Our guests were ecstatic.

"Splice the main brace!" Cheers!

Sitting in the wardroom with the officers who were stingily sipping their libation, Ed complimented us on our handling the *Barb* in the heavy seas and during the typhoon. I told him that Tuck did most of the conning during the POW rescue. We had experienced sudden 20-foot waves in the Okhotsk. Wave height during the typhoon was over 35 feet. With slow speed and quartering, the officers did well.

"I've never been in a typhoon or a hurricane. Is this one the worst you've been in at sea?"

"No, Ed. When I was a midshipman on a summer cruise in the old battleship *Wyoming,* we ran into a lollapalooza out of Halifax. Seventy-foot waves!"

"My God, what happened?"

"Even quartering the seas at slow speed it seemed as if we were in a submarine. Sliding down a mountainous wave we kept on going down into the trough until we had green water over the high turret, heavy spray smashing against the bridge and over and down the stacks. The skipper was worried the boiler fires would be extinguished and we'd be a goner once sideways to the seas. Inside, lockers and equipment broke loose from bulkheads and had to be tied down. Three days of fighting the sea, and one night of calm in the eye of the hurricane."

"What was that like?"

"Seemed like heaven. I learned some novel navigation. The *Wyoming*'s navigator hadn't been able to take star or sun sights and was using 'dead reckoning.' No one knew where we were. At that time my classmate, Skinny Ennis, and I were linked together in the navigation detail. The hurricane eye was cloudless, the sea a choppy calm in brilliant moonlight. It was dark when we entered the eye. About 2200 they started a movie. When it finished at 2330, Skinny and I noticed that the horizon was quite sharp and decided to try some star sights. When we plotted them, we had a fix with a triangle less than five miles across. We compared it with the *Wyoming*'s dead reckoning position and found a difference of 205 miles.

"At dawn we were in the hurricane again with a black sky. No sights again. In the after barbette below the twin 12-inch gun turret, our navigation classroom, we proudly showed our fix to Lieutenant Callaghan, saying the navigator was 205 miles out of position. What a rebuff! He said that no one could take star sights at night. We had made it up. This we denied. The next morning the sky was clear for star sights, and everyone was shocked to find the *Wyoming*'s position was in error 220 miles. Our prof made no comment. Ed, it takes a war to wake people up."

"Concur."

The togetherness of mortal combat had now made Ed and me fast friends for life. Likewise with Elliott, for we all worked well together, passed contacts and information quickly, and were mutually supportive. The submarine force was undergoing a major transition. Consequently, Ed and I had many thought-stirring discussions concerning the new trends and weapons coming into being.

At the time of the Pearl Harbor attack, the average age of sub skippers was 42. Ten percent produced good results; 90 percent were too cautious, probably due to restricted training in peacetime—which prohibited attacks on merchant ships—and faulty torpedoes. The difficulty for Admiral Lockwood was that he couldn't tell who would blossom and who would fade. The age group as a whole didn't cut the mustard, so maximum age was lowered to 35, with entry as low as 30.*

Another subject we discussed at length was the effectiveness of the new Mark 18 electric torpedoes we had used. In fact, I had been waiting for them since leaving the Naval Academy in 1935. My ordnance and gunnery professor was that same lieutenant from the *Wyoming*. We studied the old steam torpedoes, which always left a wake and showed the exact firing position of the submarine. Talking with the prof, I asked about new

---

* The British had identical trouble and lowered their maximum age to 36. Kretchmar, the top surviving German submariner, told me that the *Kriegsmarine* had the same problem and dropped the maximum age to 33.

improvements to eliminate the enormous disadvantage of the wake. He said it was impossible to eliminate.

That evening I drew up a sketch of a torpedo, replacing the motive power plant—air flask, water tank, alcohol tank, and steam turbine—with a set of batteries and a motor. After the next class, I presented the prof with my idea. He took one look and threw it in the wastebasket, calling it asinine. He said it would sink because the batteries were too heavy. Meekly, I suggested buoyancy could be worked out, but he told me, "Forget it."

"Ed," I said, "I knew its day had to come."

Ed agreed. He said that judging from our patrol, they certainly provided undetected firing against merchant ships during daylight submerged attack. Their slower speed of 1000 yards per minute versus the 1500 of the Mark 14 steam torpedoes required more skillful handling of the periscope to avoid being sighted.

"Right," I said, "but at night it's a toss-up in surface attacks. In high phosphorescence, the electric wake is not seen; in low, the speed of the steam torpedo wins. Against escorts with good sonar, I favor the steams in daylight submerged for speed—or electrics for firing at closer range. The torpedo must be on him before he can outmaneuver it. Yet there may be another problem with the new electrics. I suspect they may be running deeper than set. Remember that frigate we fired at on the night when we misled Elliott into the hunter-killer trap by our message of two small merchantmen. The three fish were set at three feet. The range was 2000 yards, which is good enough to remain undetected in a night surface attack. He changed neither course nor speed until after the torpedoes crossed his track. That frigate should have been sunk. Our solution was perfect."

Ed added, "God knows we had a helluva lot of trouble in 1942 with the steams not exploding on contact, until correction of the faulty detonators. The electrics' depth keeping should be rechecked."

On a new subject, Ed was concerned—as I was—about Bob McNitt's leaving the *Barb* in Saipan to attend a postgraduate course in ordnance. No one is irreplaceable, but Bob came close to it. He deserved to move on after all his patrols to his own command or school. His replacement, Naval Academy graduate Chic, came to the *Barb* from PT boats, very senior but not yet ready to be qualified in submarines. Chic could be my exec, yet was he technologically capable of saving and fighting the *Barb* if I were killed on the bridge? I thought not. Ed agreed.

"So what do you do?" Ed asked.

"I've talked it over with Chic. He knows he needs another patrol to qualify. At the Saipan tender, I'll check their latest lists of execs of subs to find one who is senior to Chic. Meanwhile, I'll make Jim Lanier my exec, even though he's a reserve officer and junior to Chic. Still, Chic doesn't want to be transferred and has offered to accept his position as junior to Jim's. This is agreeable to me; it's what a person can do that counts."

Ed told me I couldn't do it. "Gene, it's against Navy regulations; it's never been done before. The authorities won't permit it."

"If this were peacetime, no problem," I replied. "If I died, the ship and crew wouldn't be endangered. Yet off Takao, with midget subs, planes, hunter-killer groups, and decoys—who minds the shop if I'm gone? The *Barb* and her crew are paramount. Her skipper must be experienced."

"I'll back you up on that when I get to Pearl, Gene."

It was time for us to continue our preliminary interrogation of our guests, now that all but two were ambulatory. The horror stories started anew. Punishment for minor infractions had consisted of standing at attention holding a log over one's head. An American from the sunken cruiser *Houston* was ordered to stand at attention for 72 hours. The man was beaten every time he moved until he lost consciousness. A battalion commander was put at attention for 48 hours and beaten every time the watch was relieved. The one exception in their three tortured years was shown by a Lieutenent Yamada, who publicly whipped a Korean guard for an inexcusable assault. He also provided some recreational gear.

Intelligence material on Singapore—the number of ships in the harbor and the frequency of their movement—was important, so we sent it off to Admiral Lockwood by dispatch right away. Everyone interrogated was offered a beer; only one refused. Then it was back to the mess for homemade ice cream.

Chief Saunders approached us, his hat held upside down, crammed with bills and change. "For our new friends, sir. Thought we'd give each man a small stake to begin life once more." Automatically, everyone went for his wallet and turned it upside down.

"Swish, how are you making out?"

"Pretty good. I've collected every cent on board."

"Wonderful!"

**25 September.** The *Barb* moored alongside the submarine tender *Fulton* and received an excellent welcome to Saipan. The survivors were all well and happy, proudly dressed in khaki and a sailor's white hat. Each had a ditty bag with razor, toothbrush, comb, skivvies, spare shirt, money, and a secret supply of sandwiches. When I queried Swish about that he said, "Captain, I told them they'd be fed, but they raided the galley anyhow."

Hundreds of Navy men and Marines, including senior officers, came

Ten days later and 25 lbs. heavier, our guests line up for a photo. Pictured are: (*bottom row*) Harold New, Murray Thompson, Jack Flynn, Cecil Hutchinson, Ross Smith; (*top row*) Jimmy Johnson (Brit.), Leo Cornelius, Thomas Carr (Brit.), Lloyd Monro, Al Allbury (Brit.), Robert Hampson, Neville Thams, Augustus Fuller (Brit.), Jim Campbell. Those not identified as British are Australian.

out to greet our Australian and British friends. As the mooring lines were passed over, all were topside except the two stretcher cases. Everyone was embarrassed as each survivor thanked us over and over again. Saying farewell to these gaunt bodies, men tried to joke and make conversation.

Stretcher bearers went below and brought the last two up. Each one insisted on stopping his bearers and clasping my hand as he passed by. Big tears rolled down their faces. I choked up. People all around were looking at me. Hell, a captain is not supposed to be emotional, so I headed for the bridge.

Alone, on the bridge, weeping privately, I noted two "Barbarians" below me blowing their noses. Said one, "I've never felt like such a great man in all my life. What makes us so great anyway?"

The other replied with garden-variety simplicity, "Well, Syd, maybe it's this. We always try to do our job—all the way."

The *Barb* could never receive a greater tribute.

When the tumult and the shouting died and the captains and admirals had departed, I got ahold of Saunders. "Swish, we still have four cases of beer that were put on board for a picnic ration. I'll take one and a group of officers and find a place on terra firma where we can relax and wiggle our toes in some real live sand. I suggest you do likewise with those off watch—take the last three cases."

"Would it be okay to swim over the side? This berth is practically an open roadstead."

"Sure, it's clean."

Six officers quickly volunteered to picnic. In true democratic fashion, we voted to let the one with the longest beard carry the case of beer. Max won, being the only one who wasn't cleanshaven. Joyfully we ambled down a coastal road singing "Waltzing Matilda," so loved by our Aussies.

"Halt! Who are you? Where are you going?"

Startled, we turned around to face a burly Marine in camouflage battle dress, face streaked black, holding a Tommy gun on us. I explained.

"Captain," he said, "you may be killed. No one leaves the base unescorted. We're still mopping up the remaining Japs in their burrows and caves. They're armed. Sorry, you must turn back and be careful. I'm the last outpost on this side."

Thus ended our picnic. We swam off the *Barb* instead.

*Time 1915.* Sirens screamed: AIR RAID ALERT! "Stations for emergency getting under way on the batteries." This was a new one for Ed, Bob, and me. On the bridge, dusk had ended; Tanapag Harbor was blacked out completely, the night dark. None of us could make out the buoys or the reef. The mooring lines were singled up for a quick getaway, but to where? Without a searchlight the natural dangers appeared more formidable than the enemy. We awaited orders.

*Time 1930.* Secured from air raid alert. They had been Japanese scouting planes.

Jim Lanier relieved Bob McNitt as executive officer. They were big shoes to fill, for Bob was the perfect naval officer, in every respect outstanding. He regretted leaving, yet if he missed this opportunity in postgraduate education to broaden his horizons, there would be no other. The crew idolized him. His classes for officers had been of great value.

Many skippers required two watch officers on the bridge, as had my predecessor. I noted how tired his officers were. Eight hours of watch, plus four or more hours running a particular department, plus hours of training, plus meals completely filled a day. All were exhausted before they got to tracking or battle stations. I changed the schedule to one well-rested and alert watch officer, four hours on and 16 off. This gave them enough time to be tutored.

*26 September.* At dawn we were under way for Majuro Island in the Marshalls. We would pass between Wake and Marcus islands en route. I mentioned to Ed that Wake was a good bombardment target for his two subs. Out came the charts.

After two days' study, Ed settled on Peale Island, which adjoins Wake on the north side. The *Queenfish* agreed. A plan was developed to conduct the joint bombardment just before moonset on the morning of 30 September. Ed notified Admiral Lockwood of the plan. As we approached, the Japanese were in evidence. Twice we were forced to dive to avoid their patrol planes. With everything in readiness a little before supper on the 29th, in came a message from the submarine force ordering us to refrain. What a disappointment!

Tomczyk recorded some of that letdown in his diary:

> Peaceful cruising—on the Bridge getting my suntan. Captain Gene says it's good for blackheads like me as a source of vitamin A, or is it D, but not for freckling redheads. He told me that he won a freckle contest when he was six years old. I believe it. He could win another one after that day picking up the POWs. Making preparations to shell Wake Is. Do not ask me why, but we are going to give it hell or receive it. This ship is sure a fighting Son of a Gun. Boy oh Boy! Ready to go, when we're told not to. Why? October 1. Received radio news that Halsey's planes bombed Wake Is. yesterday. We would have beat them to the punch. Guess that's why they didn't want us to shell it.

> *3 October.* Well we tied up to the Tender Gilmore at Majuro. It's good to feel that old dirt under my feet—I mean sand. The place is beautiful, never have I seen anything like it. But that's about all. The rest of the setup stinks like H——. If you're lucky you get two beers a day. I'll be transferred to school, a new sub, and be home for Christmas with my little Honey. The mighty Barb did it again and still floats very proudly. Hope to God she will always be the same Forever.

On 16 October 1944 I received the admiral's "off the record," private, personal reactions and comments after he had studied our patrol report. These were "eyes only" sent to the skipper.

> An extremely active, well-conducted patrol. An excellent report. Data sheets well written and fully detailed. An excellent percentage of hits.

Fine bag. I hope the indicated sinking of the aircraft carrier will be confirmed [it was].

The spirit, determination and coolness displayed are in keeping with the splendid record already established by the *Barb*. You are to be congratulated on a most "successful" patrol resulting in extreme damage to enemy and further enhanced by the rescue of Allied P.O.W.s.

The *Barb* is welcomed to Majuro Island by the band from the submarine tender USS *Gilmore*.

Letters and messages of appreciation were also received from Prime Minister Winston Churchill, the Army Council, the British Joint Staff, the Australian Prime Minister, and the Department of External Affairs. The general tenor of these follows:

> The incident has touched our people profoundly and our Government would like the American Government to know of the deep gratitude which it feels for the kindnesses shown, which go far beyond the requirements of duty, and contrast so strongly with the barbarous treatment from which the prisoners have been recovered.

> The prisoners who had been rescued felt that they could never praise

too highly the treatment which they had received aboard the submarines after they were rescued. The officers and crews acted as mothers and sisters to them. As they came off watch, they spent their entire rest periods cooking for, feeding, shaving, and washing the survivors, and gave them complete outfits of clothing. Emaciated, diseased, they were nursed back to health.

I fully realize that the commanders of these submarines endangered the safety of their own craft and personnel in going to the rescue of our men. Please accept our high appreciation for your splendid performance.

Thus ended the *Barb*'s ninth war patrol, my second in command.

# Part III The Tenth War Patrol of the USS *Barb* in the East China Sea, Adjacent Kyushu, Japan, 27 October–25 November 1944

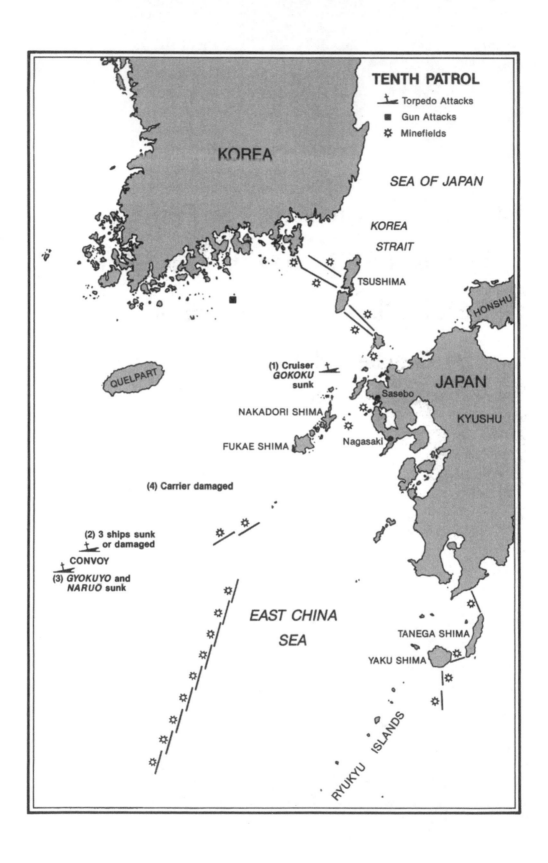

TENTH PATROL
⊥ Torpedo Attacks
■ Gun Attacks
✿ Minefields

KOREA

SEA OF JAPAN

KOREA STRAIT

TSUSHIMA

HONSHU

QUELPART

(1) Cruiser
GOKOKU
sunk

JAPAN

Sasebo

NAKADORI SHIMA

KYUSHU

FUKAE SHIMA

Nagasaki

(4) Carrier damaged

(2) 3 ships sunk
or damaged
CONVOY
(3) GYOKUYO and
NARUO sunk

EAST CHINA
SEA

TANEGA SHIMA

YAKU SHIMA

RYUKYU ISLANDS

# Chapter 13 Hey! Let Me In!

Our three-week stay at Majuro Atoll was far different from that at the Royal Hawaiian. The best thing that happened there was getting mail from home, and, for me, the next best was receiving my promotion to commander.

Majuro appeared to be a paradise of swaying palms, balmy breezes, and two weeks in which to enjoy complete relaxation before a week of training at sea. But on an all-male island there was little to do except write letters— no natives, no athletic fields, no movies, little reading material, and poor swimming, for the beach was steep and loaded with sandflies. We slept in Quonset huts with canvas flaps for doors and unscreened windows. The sun broiled us in the daytime, and we slept in our undershorts through the hot nights. Things crawled over our backs, and food came from cans.

To break the dull routine I asked air operations to let me go with the pilots on a dive-bombing mission of the bypassed islands still held by the Japanese. Instead, they permitted Ed and me to qualify as waist gunners in a dumbo flying boat used for rescuing ditched fliers. At 0400 we took off for a dawn strike on Jaluit. The airstrip was cratered there so the remaining enemy planes couldn't take off. We flew low outside of the range of their manned antiaircraft guns. Even this mission turned into a disappointing farce. Unbeknown to the pilot, Ed, and me, the strike was canceled the previous evening, and we flew around fruitlessly for several hours looking for the bombers. Someone goofed.

Two days later, Ed took a flight to Pearl with Chic, who was being transferred to the *Swordfish* because she had an executive officer senior to him. With them were two of our best: top electrician Ezra Davis, and our spirited, peerless top torpedoman, Charles Tomczyk. Both would be missed—as friends, shipmates, and combateers. Davis was up for a Silver Star Medal; Tomczyk, a Bronze Star Medal.

During the training period at sea, Jim Lanier (now executive officer and navigator) amazed the new officers by showing them some of the navigation magic that McNitt had originated and passed on to him. One was the "Annie Oakley," or over-the-shoulder, sight taken with a marine sextant when there was a poor horizon under a celestial body but the opposite horizon was clear. The other was taking low-altitude sights through a periscope while submerged.

Sea exercises went well and we returned to Majuro. Patrol orders came for our next run. Again in a wolfpack, we were given a great area—the East China Sea off Kyushu, the southwestern island of Japan proper. Kyushu meant proximity to the mammoth naval bases of Sasebo and Nagasaki and the heavily mined Tsushima Strait. With Elliott in the *Queenfish* commanding the pack—the *Barb* and *Picuda* (under my class-mate Ty Shepard)—we were eager to shove off.

All three subs were ready early to cut the umbilical cord from our mother ship, the submarine tender *Gilmore,* as the scorching sun rose on 27 October. We waited for the escorting destroyer that was scouring the area off the reef entrance to Majuro to signal "All clear." With only Eniwetok, Kwajalein, and Majuro recaptured, the rest of the Marshall Islands were still being supplied by Japanese subs.

Topside, a mail clerk made a final round for the last letters. I let Tuck censor three more love letters to Marjorie, then dropped them in the sack. I wondered when she would find out that we had been in combat. My lies to keep her from being worried were getting weaker and weaker: engine failures, a new battery to be installed, and so on. I had a burning urge to bring this cruel war to an end as fast as I could. What little I could do with the *Barb* might only be a drop in the bucket, but with enough drops, some day that bucket would well over and we'd be sailing home.

As we waited for the destroyer to clear our departure, Tuck said, "Captain, you look like you're all charged up."

"I am Tuck—just been thinking."

"How many are we sinking this patrol?"

"Bet your boots, it'll be as many as we can find."

I dropped down to the wardroom to size up our new prospective commanding officer, Lieutenant Commander Tex Lander, who had just come aboard. As I sat down, the *Barb*'s two new officers—Tom King and Ensign Jack Sheffield, a publisher—were busy censoring the last-minute

The entire crew of the *Barb* at the start of the tenth war patrol, with the battle flag.

mail. Meanwhile, Jim was briefing Lander on the operation order for the coming patrol. Tex came to the *Barb* for temporary duty, expecting to take command of the *Ronquil*. He would be an observer of our modus operandi.

Jack's squawks of laughter drew our attention. "Here's one that needs no censorship. Listen to this cock-and-bull story. Tuck Weaver is writing to his uncle":

> I must tell you of what is, for me, one of the unforgettable moments of the war. On our last outing we were making an approach submerged on a freighter with two escorts. When we arrived at our shooting position, the Captain raised his attack scope for the final bearing. Then lowered it and raised it again, turning it. Obviously he couldn't see. Time was running out, for the target was passing.
> My mind raced back to when I was sunk, but depth control was perfect. He was in a real jam with the scope problem. Then this red-head laughed—

this broke the tension. Calling for the camera, he had the other scope raised, turned it toward the attack scope and took a picture. Then turned to the freighter which was passing by, fired three torpedoes for three hits, and six of us watched her sink. This proved to be a luxury, for the escort rushed in to drop his calling cards.

The culprit had been a bird draping his tail over the periscope. I looked at the Captain after we evaded and wondered; how many men are there in the world, who in such a tense, critical situation could laugh and then take a few seconds to calmly photograph a bird before sinking an armed, escorted enemy ship?

Tex and Tom joined Jack in his laughter. Jim said nothing, got up, and went to his cabin. Returning, he laid a picture of the bird on the periscope down on the table. He said to the newcomers, "The first thing you'd better learn in the *Barb* is not to jump to conclusions until you know what you're talking about."

At last the destroyer signaled all clear. Shortly thereafter, Elliott led his wolfpack single file through the channel en route to Saipan for a fuel stop. Already, salty humor had immortalized us as the Queerfish, the Boob, and the Peculiar.

*6 November 1944.* At night "Loughlin's Loopers" entered the assigned patrol area through the four-mile-wide Tokara Strait, carefully avoiding the patrol boats on guard. Then we separated to our respective areas to cover the west coast of Kyushu. For any convoy contact we would gather the pack, if feasible. Areas would rotate every five days. The *Barb* started in a beautiful area for major warships, patrolling the south entrees to the major naval bases of Nagasaki and Sasebo.

*7 November.* According to Japanese records, Captain Takayoshi Mizuno commanded the light cruiser *Gokoku* (CL-11) berthed at the naval base of Kirun (Kiryu), near Taipeh, Formosa. The cruiser was scheduled to leave on this date for the naval base at Kure, in the inland Sea of Honshu. This voyage was not without its problems.

The *Gokoku Maru* (10,438 tons) had been built in 1941. After Pearl Harbor, the Imperial Japanese Navy converted it to a light cruiser. Fore and aft were 5.5-inch gun turrets; 3-inch and 2-inch double purpose guns were mounted along both sides, plus sonar and two aircraft. A cruiser-raider, she served well until she caught a bomb on 20 September 1944 in the Formosa Straits, which damaged her port shaft. When the ship limped into Kirun, north Formosa, Commander Fujii—dockyard staff officer for naval technology—recommended a complete overhaul, but Captain Mizuno objected due to enemy raids. He insisted on temporary hull repairs only, with the port shaft repair at Kure,

Japan, as the quickest way to bring his *Gokoku* back to fighting status. Due to his rank and the fact that the south Formosa dockyard at Takao— overflowing with damaged ships—sent many north, the Navy concurred with Mizuno.

Commander Kokichi, his executive officer, had loaded on board some 400 Formosan volunteers for the Navy, 2 soldiers, and 3 civil employees for Sasebo. Also on board was the entire cargo of a freighter that had been sunk: 600 tons of sugar and 300 tons of aluminum nuggets. In the mailroom were 81 urns of the bones of the dead, plus mail.

Normally, the *Gokoku* traveled with four destroyers, but such were not available. The Japanese Admiralty had arranged for three antisubmarine vessels to join up en route to Sasebo, and they sent the destroyer *Hibiki* from the Baku naval base in the Pescadores Islands to depart with the *Gokoku* until the other destroyers arrived. The *Hibiki* was coming to Kiryu for repairs; her ability was affected by lack of crew and illness on board. Only the starboard propeller shaft of the light cruiser was operable. This reduced her speed from 22 to 15 knots, with a cruising speed of 12 knots. The *Gokoku* and *Hibiki* got under way at 0700.

*8 November.* Patrolling submerged two miles off a major air base in the afternoon, the *Barb* ran into a heavy rain squall. We surfaced for radar coverage to keep ships from slipping by. The squall lifted suddenly, exposing our position just off a lookout post. As no planes were up, I decided to use our certain sighting as a ruse. The area was devoid of traffic, so we headed south at slow speed on the surface. After dark we would sweep west and north around the islands to cover the north entrance. I hoped the Japanese would divert all ships away from where we had been sighted.

*Time 2030.* Moving at flank speed on four engines, the *Barb* hustled northward, not wanting to lose a moment of coverage.

"Contact report from the *Queenfish* some 100 miles southeast. She's attacking a convoy of two ships and three escorts heading east."

"Tex, what should we do?"

"Follow your present plan, Gene. They'd be in port before we arrived."

"Concur."

Fifteen minutes later, Elliott fired four torpedoes at the *Hakko Maru* and two at the *Keijo Maru*. The first blew up with two hits; the second took one hit. As the *Queenfish* turned, the lead escort opened fire and forced her to dive. Unnoticed by *Queenfish* personnel, a torpedo that had missed the *Hakko* hit the patrol craft escort *Ryusei* and blew her to pieces, dumping depth charges around with a shotgun effect heard below. The *Keijo Maru* upended and sank. One escort made a brief search while the other picked up survivors. Within an hour both departed. The *Queenfish* surfaced at 2217, having drawn the first blood.

Commander Kokichi handed the captain a message from the *Hibiki*. "Dysentery is now epidemic. Have only one watch section and no ability to support you. Request permission to proceed Sasebo." Before a reply could be sent—which depended on when another escort could arrive—Captain Nakakazu sent a personal message to Mizuno. "Buried one officer at sea. My four other officers sick. I am the sole watch officer. May we leave you now at fastest speed?"

The *Gokoku* replied: "Granted."

**9 November.** The *Queenfish* was not to rest long. At 0110 Elliott was rousted out of his bunk. "Radar contact. Convoy of three ships and five escorts heading east."

A second *Queenfish* solo was under way, with the *Picuda* off Quelpart Island, south of Korea, and the *Barb* off the Tsushima minefields, covering the northern entrance to Sasebo.

With a quarter moon and rough seas, Elliott ended around, dived ahead, and came in at periscope depth by sound between the two columns. He let fly three torpedoes at the lead ship and three at the second ship. While setting up for the third ship, all three hit the first and two hit the second. An unseen escort rained depth charges, preventing Elliott from observing what was happening to his two targets above.

The naval gunboat *Chojusan* absorbed three indigestible *Queenfish* torpedoes; 20 minutes later she exploded, scattering debris over the whole area. Remnants hit the *Queenfish*. The other merchant ship successfully swallowed her two torpedoes and shoved off for dry dock. At 0400 the last escort bade farewell, with a final depth charge, taking the survivors with her.

*Time 1200.* The *Gokoku* received special warnings in a navigation message: "From North to South, patrol plane spotted submarine [*Picuda*] between Quelpart Island and Korea at 0748. Submarine [*Barb*] sighted surfaced yesterday afternoon twice; first by Fukue Shima Air Base lookout post and later by O Shima lighthouse keeper, patrolling the southern routes to the naval bases. Last night two convoys south of Koshiki Islands were attacked by subs [*Queenfish*]. Escorts sank one submarine."

Her captain then changed the *Gokoku's* course to go through the slot between Saishu To [Quelpart Island] to the north and Fukue Shima to the south to enter the northwestern entrance to Sasebo. The three escort frigates had not yet arrived.

At 1400 frigate *168* arrived as the seas and winds increased. She could only make nine knots heading into the rising seas, which caused the *Gokoku* to slow down. Two more frigates were due at 1800.

Since dawn the *Barb* had been patrolling on the surface along the northwestern approaches to Sasebo, a mammoth naval base. Only one ship had been sighted. We dived and conducted an approach. She turned out to be a large patrol trawler with radio antennae. Her armament consisted of a 2-inch cannon forward, a 20-mm. gun mounted on top of her bridge, and depth-charge racks fully loaded aft. The officers discussed the possibility of a quick battle surface sinking. All agreed that it was too rough for accurate shooting, nor did we want to disclose our position by an extended gun battle, so we secured our approach.

During the day we dived three more times to avoid pairs of patrol planes. So far undetected and with decreasing patience, we looked forward to the appearance of a large convoy.

*Time 1800.* On the dot, frigate *87* hove into sight ahead, wallowing as she hurried down sea to join the task force. Commander Kokichi assigned her to protect the starboard quarter of the cruiser; frigate *168* was positioned on the port quarter.

The *Gokoku* queried as to the whereabouts of the third frigate. The reply was that orders for frigate *112* were changed to assist the fleet convoy HI-86 coming up along the China coast.

At dark the *Gokoku*—endangered by the slow speed of the frigates as they pounded into the waves from the east—increased speed to 12 knots.

At 2200, the frigates had dropped out of sight astern. Now the *Gokoku*—sheltered from the eastern winds and seas by the Goto Retto Island chain—speeded up to 15 knots. The Sasebo Channel was only 90 miles away.

The *Barb* had a busy day patrolling parallel to the Tsushima minefields in the rough seas. Sighting masts, she dived and made an approach, only to find a large armed trawler with depth-charge racks patrolling. It was not worth exposure by an attack. Surfacing when clear, the *Barb* dived three times to avoid planes. After dark we watched the searchlight drill simulating a bombing raid on Sasebo. With the entire coast blacked out afterwards, it appeared that we would have a restful night. In the wardroom with the rest of the officers, I wrote out my night orders, which all of the officers read and initialed.

"Patrol east-west line across entrance channel to Sasebo at one-third speed, zigzagging. Keep unlighted Kashiki Light in the channel center as your center reference point. Do not approach the light closer than three miles nor be more distant than five miles. Immediately evade all small craft and aircraft. Call me for all contacts or anything unusual. Call me if in doubt. Call me if you doubt you are in doubt. Taut watch—this is a hot spot."

*Time 2354.* "Captain! Koshiki Lighthouse has been turned on. Its characteristics are normal as noted in the *Coastal Pilot*. No other lights. No radar contacts."

"Aye, Max. Do you think you are silhouetted?"

"No, sir."

"It must be lighted for something. Be careful. Be alert. Do you want another officer?"

"No, sir."

"Okay. I'll sleep with my clothes on. Don't hesitate to call me."

*10 November. Time 0245.* "Radar contact 22,000 yards! STATION THE TRACKING PARTY! RIGHT FULL RUDDER! ALL AHEAD FULL! They just turned on the channel buoy lights, captain."

"Well done, Max. How's the moon, the sea, and phosphorescence?"

"Quarter moon, visibility is excellent, sea is calming, and we do have phosphorescence."

"Tuck, relieve Max and let me know the second you have the target visually. We have a single large ship speeding along the coast. We must catch her before she turns into the channel. Jim, man battle stations torpedoes! Tex, get below and take a gander at the coastal chart as to safe depths. Too bad, we'll have to make the attack at periscope depth—maybe radar depth. We'll only have one crack at her before she's safe in the channel. I don't want to bounce off the bottom. Jim, take charge here; I'll be on the bridge."

I asked Tuck what was happening.

"Possibly a lump on the horizon—could be an island."

"Jim, get a single ping sounding for Tex, just one. This baby may have sonar. Keep the ranges coming every thousand yards. What's Max's best estimate of her speed?"

"Aye. Range 17,500 yards. She's zigzagging. Plot has no base course yet. Max has her at 15 knots."

"Have plot assume a base course parallel to the coast. We'll shoot on a 60° port track to keep us well off the beach at a range of 2000-plus yards submerged. Use tubes 4, 5, and 6. If we miss, that will give us a second shot on a 120° port track. If she crosses our tail, we'll use the stern tubes. Make ready all tubes."

"Captain, I've got her! She's big. Looks like a battleship from head-on."

"Range?"

"Seventeen thousand yards."

"Tuck, you've got the eyes. You lookouts—Bluth, Lego—we've picked up the target. But keep sweeping. Could be midget subs around."

"Range 12,000 yards. Target appears to be slowing."

"Probably slowing before she turns into the channel, Jim. What does Dave's plot show?"

"Dave says he slowed two minutes ago and ceased zigzagging. Now on a steady course of 074, speed 12 knots."

"Good. That'll help submerged."

Signalman Sever clutched my arm and whispered, "I think there's a floating mine broad on the port bow, about a thousand yards."

"Could be. The nearby Tsushima minefield is 60 miles long. Some of the chains are bound to rust out and break live mines loose. Good work. Look for others."

*Time 0327.* "Captain, we better dive before she sees us. I can see a big gun turret near the bow."

"Okay. Tuck, take her down."

"CLEAR THE BRIDGE! DIVE! DIVE!" A-oo-ga! A-oo-ga!

"Paul, radar depth, on the deep side, about 47 feet. Waves are small, quarter moonlight; we don't want to be seen.

"Captain, range is 4500 yards."

"Set torpedo depth eight feet. Open the outer doors, all tubes. We're coming in nicely. Down scope. Tex, this is when I like these electric fish. Otherwise the enemy would spot their wakes. We're almost in perfect position for a 60° port track."

*Time 0334.* "Up scope. Damn! she's turning away. Max, angle on the bow is 90° port increasing. Passenger superstructure, turrets fore and aft, other guns all along the side. Range?"

"Twenty-three hundred yards."

"She's steadied—angle on bow, 110° port. Spread aft forward, 50 yards between. Final bearing—mark."

"Set."

"FIRE 4! FIRE 5! FIRE 6! JP sonar, stay on those fish."

"All hot, straight, and normal, sir!"

The *Gokoku's* after-action report for this day recorded that she had made good time on a steady course at the highest speed possible on the starboard shaft. Now she had to slow before entering the channel. In addition, currents had moved the proud "Guardian God" a bit too far offshore. She came right 60° to close the beach to get back on track.

*Time 0336.* "Captain Mizuno! Sonar reports torpedoes approaching port side!"

"FULL HARD PORT RUDDER! BATTLE STATIONS, FIGHTING POSITIONS! ENGINES MAXIMUM SPEED FORWARD!"

"Port lookout reports four torpedoes in sight, port 070. Range 1000 meters!"
"Stand by forward and port cannons!"

"Damn it, Jim, she's zigged 40° toward us. We may have to fire again, but this time we'll be closer and it will be for keeps. Get ready for a quick setup." WHAM!

"Wow! One hit smackeroo just aft of the funnel . . . and a second a bit forward of her bridge. Stand by to surface. She'll sink."

*Time 0339.* We surfaced at 1400 yards and turned away at flank speed to clear the sinking ship's guns while we watched her go down.

According to the *Gokoku* battle report, the violence of the first torpedo explosion aft blew some of the 18 lookouts overboard. The port engine room was demolished, watertight doors were blown open, generators flooded, and all engines stopped. Damage control was hampered by no electricity, no lights, no phones, and an inability to counter flood quickly. The *Gokoku* rolled over 30° to port, coasting at a slowing speed with a swing to the right. Captain Mizuno gave the order to send the emergency message.

"Impossible, sir; the antennae are down and we have no electricity. But all hands are in good spirits, though repairs are very difficult with the ship leaning."

"Then order the head signalman and the head telegrapher to seal the secret books and codes in the confidential storeroom."

I lowered my binoculars. "Tuck, she's not sinking. She's under way at about two knots and slowly turning right. The captain's going to beach her. All ahead full speed. Jim, I'll give you a setup for a *coup de grâce*. We're going back in. Tuck will keep on the firing bearing for a surface shot. You won't need plot."

*Time 0349.* The *Gokoku* slowed and swung left for the finale, coming in on her high starboard side. Her big guns were now useless: they couldn't depress enough to hit us, with the 30° list. "Put four cases of beer in the cooler. Setup! Angle on the bow 40° starboard. Use speed 1 and ¾ knots. Open the outer doors tubes 1 and 2. All stop."

*Time 0353.* "Range 970 yards."

"Tuck's on. Final bearing—mark. FIRE 1! QC sonar, get on that torpedo. It took a jog to the left on firing."

"Sonar reports torpedo circling!"

"ALL BACK EMERGENCY! LEFT FULL RUDDER!" We didn't breathe.

At about 400 yards the torpedo broached, turned right, and went off into the night. "All stop."

"Sonar reports erratic torpedo has now returned to its normal course."

*Time 0354.* "All ahead two-thirds. Jim, angle on the bow 20° starboard.

Final bearing—mark." FIRE 2! QC sonar, watch this torpedo, it took a jog to the left!"

"Sonar reports torpedo now hot, straight, and normal." We saw its phosphorescence passing down the side of the ship. Another erratic. I couldn't tighten the noose. The situation was becoming tense, exciting, and dangerous for all. Now at 500 yards, Max expressed our proximity perfectly by wanting to "throw spuds"—using oranges for tracers—to sink her. Lego, the battle lookout, suggested we put our nose against the side and roll her over. "Damn the torpedoes! All stop!"

I was becoming irritated at this grotesque picture of the target lurching drunkenly on. A blinker gun from her bridge flashing AA—the international call code for "Who are you?"—made me laugh and dispelled my anger. One never thinks well in anger. I watched the gunners valiantly trying to fire some weapons at us. "Tuck, get Lego, Bluth, and Sever below."

"Surfaced submarine starboard bow!" Captains Mizuno and Kokichi scurried up to the railing and watched the two torpedoes pass alongside. "Kokichi, have Lieutenant Hayashi disarm the four depth bombs on the launchers and throw the other 16 depth bombs over the side. All hands put on life jackets. Lower all life boats. Prepare to throw over the life rafts on the forecastle."

"Captain, Hayashi and the gunners have been trying to fight back, but that submarine is tricky. Either it disappears or stays on the high side where we cannot shoot."

"Jim, they're dismounting a machine gun which they can't depress enough to shoot us. All ahead two-thirds. We'll be diving momentarily to radar depth. Setup for 90° track. Open the outer door on tube 7. Speed is zero, angle on the bow 90° starboard. Use radar range. Time to submerge, Tuck; everybody else is down. Let's get below."

Taking a look on both sides of the bridge, I pressed the diving alarm twice and we dropped down. The ballast tank's vents opened with a roar as the *Barb* started down with Tuck on the lanyard pulling the hatch shut.

"Hey! Let me in!" Everyone gasped, horrorstruck.

Sever, having his hand on the hatch wheel ready to secure the lugs, pushed the hatch back open. "Do you want to come in too, Mister Teeters?" Dave dropped into the conning tower without touching a step. Water poured over all of them as Tuck yanked the lanyard, shutting the sea out. Dave was too breathless to talk as I turned back to the problem of sinking this ship.

"Tex, here's a chance for you to prove that you're ready to command.

Take over the periscope. I'll let you sink this ship. Now put that torpedo in the middle of the target."

"Thanks, Gene. Up periscope." I turned back to Dave, but he disappeared below to his battle station in plot.

"Gene, take a look. They're lowering their lifeboats." I did; they were.

*Top:* The auxiliary raider cruiser *Gokoku* (XCL-11) after its conversion from the 10,438-gross-ton *Gokoku Maru*. *Bottom:* Japanese artist Ichirou Ohkubo's painting of the sinking of the *Gokoku* on 10 November 1944.

Bumping and tumbling down the high side, some sailors were falling out into the water. Then I swung the scope over toward the Koshiki Light. An antisubmarine craft or patrol boat was coming up the channel.

"Here, Tex, help's coming up the channel; we can't wait."

"Range is 1400 yards. Final bearing—mark. FIRE 7!"

Donnelly pushed the firing plunger. "Tex, sonar?"

"JP, get on the torpedo."

"Hot, straight, and normal, sir."

*Time 0410.* WHAM! Smack in the middle. Tex qualified.

"TORPEDO STARBOARD BEAM!" Captain Mizuno, on the compass deck, saw it and yelled, "BANZAI! ABANDON SHIP ALL CREW AND PASSENGERS!" The torpedo explosion eliminated many. The forward life rafts were tossed overboard by crew members. With a swelling chorus of BANZAI! nearly everyone jumped overboard as the *Gokoku* sank rapidly by the stern. Captain Mizuno bravely tried to control the abandonment from the top deck. The light cruiser rolled over, then stood on end, her bow reaching for the dawning sky. Hesitating, she slid down so fast her eddy sucked one lifeboat and many down with her. Some of the crew saw the captain swimming and talked to him, yet both he and Kokichi vanished.

*Time 0414.* We surfaced and cleared the area on all four engines, with the patrol boat passing Koshiki Light. I wanted an hour's run before daylight, when I expected all hell to break loose. Choosing a course paralleling the Tsushima field, we headed for Korea, too exhausted to splice the main brace.

Securing from battle stations once we had determined we hadn't been sighted, all officers piled into the wardroom to hear about Dave's adventure. Unafraid and smiling he let go.

"Captain, I must apologize for violating a ship's order. After the first two hits, when you headed back in, you said to secure plot. Everything became super exciting. I have never seen battle action from the bridge. With nothing to do, I crept up on the bridge while you and Tuck were busy as a one-armed paperhanger avoiding the erratic torpedo. I didn't ask permission to come up, which is required, because I thought you would negate. Then I climbed up to the high lookout stand and was fully enjoying the spectacle. That is, until I heard the vents open and looked down on the empty bridge with the hatch closing. You know the rest."

"Well, Dave, I must congratulate you for setting the world's record for clearing the bridge from the high lookout platform to the conning tower. I think we've all learned a lesson, myself included. At times, probably because I'm the boss, I forget to ask or ignore asking permission. Let's

Dave Teeters, the zealous, top-notch officer who uttered the famous, mind-boggling cry "Hey! Let me in!"

follow the ship's order. Now let's all hit the hay and rest up for the next fray. We've 18 fish left. One down, four to go."

The patrol boat that the *Barb* had picked up was an antisubmarine vessel sent out to escort the cruiser *Gokoku* to Sasebo. Not knowing that Captain Mizuno had speeded up and was two hours ahead of schedule, she went to investigate the loud explosion. There she found Lieutenant Hayashi in a lead lifeboat sailing for help.

As no attack message had been received, Lieutenant Hayashi radioed for emergency rescue and medical assistance. Frigate *200* was dispatched along with hunter-killer coverage by planes and ships.

*Time 0607.* Distant depth charging and bombing commenced—infrequent and sporadic—until 1100, when the increasing crescendo made sleep impossible. Everyone was up, though the distant drumming was pleasant. Some heavy salvos of 10 to 16 depth charges in one pattern were as annoying as a Tin Pan band in a small room, so an announcement was cheered: "Splice the main brace!"

Along with the beer, stories poured forth, aided by the certainty that our sinking must have been pretty valuable to warrant this plastering. For a while we recorded each explosion until they averaged two per minute; we gave up at a count of 300 plus.

Dave sipped his beer in the wardroom, his face becoming whiter and whiter, drawn and tense. "Dave, what's wrong?"

"Well, good God, think what would have happened to me if I'd been left up there, or caught."

At 1344 that day, Lieutenant Hayashi reported the survivor search ended. He tallied the missing list sadly: 17 officers, 92 crew, 2 soldiers, 212 Formosan volunteers, 3 civil employees. Their proud and mighty "Guardian God" met the foreign devil she had enraged at Pearl Harbor and was vanquished.

# Chapter 14 Piggybacking

Finally, in the late afternoon, quiet settled in and the *Barb* surfaced and headed for her lifeguard station. A large group of B-29 bombers from China was scheduled to strike Kyushu after dawn on 11 November. Loughlin's Loopers were ordered to cover their approach and retirement, rescuing any downed aviators. Elliott assigned us positions about 40 miles apart on the transit corridor.

At 0500 the *Barb* was surfaced on station, ready for rescue. Two hours later, Japanese patrol planes were picked up at seven miles and we submerged. The strike force knew our locations. Should it become necessary to ditch some aircraft for mechanical failure or enemy action, the pilot would head for the nearest lifeguard submarine. The sub was required to stay on the surface as much as possible. Consequently, as soon as a plane disappeared, the *Barb* would surface, if submerged, to be seen by friends and to receive any distress messages.

As it was now close to strike time, I stayed on the bridge with the regular watch. As long as the patrol planes stayed about 10 miles distant, we faced them to reduce our silhouette. If they headed at us, we dived, stayed down for 10 minutes, then popped up again, yo-yoing.

John Arthur, a great lookout who could spot a flyspeck, queried, "Captain, have the Japs been tipped off, with so many patrols flying?"

"Doubtful, John. The B-29s are from deep inside China. The Japanese radars on that coastline could pick them up and pass the info up the line. Obviously, the Japanese patrol planes are not searching for subs. Our boys will probably get a warm reception. I pray that our rescue effort won't be needed."

*Time 0528.* The *Queenfish* snuggled into her lifeguard station south of Fukue Shima and commenced a trim dive. On her way to 100 feet, the radar operator was taking a final sweep before the SJ radar went under. "Convoy!"

Startled, Elliott responded, "Surface, surface, surface. Four main engines. Come on, Jack, we've just a few minutes of darkness remaining to get ahead of them."

Surfacing amongst a school of fishing sampans who were exercising their squatters' rights on the lifeguard station, the *Queenfish* surged ahead and the race was on. While Elliott was broken-field running, Jack Bennett had been ogling targets on the radar scope.

"Captain, the convoy is south southwest at seven miles. I've counted 12 or 13 ships with at least 6 escorts. Plot figures two hours for the end around. Being on the far side of the strike transit corridor, possibly we won't be forced down."

"Good, Jack. That'll give us enough time to attack once and return to our station before the strike is over."

The Japanese convoy commander of convoy Mo-Ma 07 had frigate *28* for his flagship. Mo-Ma 07 was the abbreviation for a convoy departing from Moji for Manila. Moji was just inside the Inland Sea, close to the heavily mined Tsushima Strait between Kyushu and Korea. Its sheltered bay was a major assembly point for convoys to China and the southern war front. This convoy consisted of six merchant ships—the *Naruo, Tamatsu, Ninyo, Fukuyo, Tatsusho,* and *Miho*—and five escorts—frigates *8, 9, 28, 54,* plus subchaser *24.* Prior to getting under way at noon the day before, they had received the daily submarine warning message covering attacks and sightings. The convoy commander noted the sinking of the cruiser *Gokoku* to the north, the two sinkings by a sub to the southeast, and submarine sightings west of Okinawa. The route west toward Shanghai, and down the impenetrable, shallow shelf of the China Coast to the Formosa Straits, appeared clear.

With five escorts, the merchant ships were well protected. That is, they were until dawn that morning, when three ships—the *Hakigawa, Jinyo,* and *Shinfuku,* with a single subchaser for an escort—requested permission from the convoy commander to piggyback on his convoy. They would fall in astern of his columns and follow the maneuvers of his ships ahead, without him being responsible for their protection. Permission was granted.

Within the next half hour, three more ships—the number 1 *Konan Maru, Tatsuaki Maru,* and *Gyokuyo Maru*—sailing independently slipped in astern of his lengthening columns. Only the latter, which had departed from Miike on the tenth, asked permission, which was granted.

Farther astern the commander noticed another independent tailing along with his convoy's zig plan without announcing herself. In response to his signalman's query, she gave her call sign JEMT, which he could not identify from the shipping list.

**11 November. Time 0726.** Arriving at a position some 10 miles ahead of the convoy, the *Queenfish* spotted a covey of patrol planes passing over it.

"Take her down, Jack. The convoy's air cover has arrived. Battle stations torpedoes! We're going in."

The zigzagging convoy slowly lumbered over the horizon with all its escorts echo ranging in the near-calm sea. Elliott moved in slowly to avoid being picked up by the escorts' active sonar as they tried to bounce an echo off a sub to reveal its presence.

The *Queenfish* neatly evaded the flagship by going deep as she passed overhead. Elliott's plan was to shoot the lead ship in the center column with the bow tubes, then swing hard right and shoot the stern tubes at the second ship.

**Time 0900.** Frigate *54* on the port bow of the convoy made sonar contact. "Sub sighted" signals indicated a submarine inside the convoy.

"Emergency right turn!" was ordered as frigate *54* turned in to depth charge.

**Time 0902.** "Okay, Jack, this will be final bearing . . ."

"Captain, sonar reports convoy is changing course. High-speed screws approaching. His sonar has shifted to short scale for an attack."

"Up scope! Shift target to second ship center column. Angle on the bow 30 port, range 1500 yards."

"Set."

"FIRE 1! FIRE 2! FIRE 3! FIRE 4! Right full rudder. Oh, oh—here comes an escort. Take her deep. Rig for depth charge. Close all tube outer doors."

"Target's screws appear to be passing ahead."

WHAM! "One hit. Stand by! He has started dropping."

The *Queenfish* jumped as the first pattern bracketed her. Elliott spiraled downward, happy that she was a new thick-skinned sub with a test depth of 438 feet instead of the *Barb's* 312. Still, this first pattern blew light bulbs and knocked men around. Cork dust flew. These charges were close. Added depth helped complicate the frigate's problem. Reaching water-density layers that would affect the accuracy of the attacking sonar was an assist.

"Sonar reports a second escort has arrived on our beam."

Elliott shook his head. "He'll echo range to provide our position while the other does the depth charging. Experts! I can see we'll be celebrating Armistice Day in reverse."

The convoy commander, maritime transportation commander #5, ordered frigate *8* alongside to shift his flag. This accomplished, he ordered frigate *28* (which had the best German sonar) and frigate *54* to stay and sink the sub and render necessary assistance to the damaged *Miho Maru.* The rest of the convoy sailed on.

The *Gyokuyo Maru* reported to her home base that a submarine had attacked the convoy. Though she had no damage, the commander thought she had and ordered the same frigates to assist her. She stayed with the convoy, however, while the frigates kept hounding the *Queenfish.* The unknown tailing ship left the convoy after the attack.

The *Barb* knew nothing of the *Queenfish's* attack. Yet at that time, we sighted a group of black smoke clouds over the horizon, followed by a series of explosions. The strike had commenced. At 0930 we dived to avoid a group of Japanese fighters. When we surfaced at 1000, our aircraft radar scope was like a pin cushion, with planes everywhere above the overcast.

At 1051 Jack turned my way. "What a beautiful sight. Here come the B-29s, in the lower edge of the clouds, returning to their base. Permission to pass the word to the crew?"

"Sure, Jack."

This being his first war patrol, he was elated and proud to see something American, beside ourselves, close in on Japan.

After the B-29s passed overhead, another squadron of Japanese fighters were picked up by radar. These, too, were searching above the clouds down the retirement alley that the *Barb* was monitoring after bombing was completed. At 1500, lifeguard duty was completed with a final round-up message after all the bombers had checked in. One B-29 was down 170 miles to the southwest.

Though Jack had been a publisher in civilian life, he quickly qualified as a watch officer under Tuck's sage tutelage. Alert, he responded. "All ahead full. Left full rudder. Helmsman Petrasunas, steady on southwest. Captain, permission to put four engines on line?"

"Jack, normally we would, but not with all these planes around; we'll be leaving a wide, foaming swath for them to sight. Put three on and keep one on battery charge."

"Captain, control room. Tuck's decoding a message from *Queenfish.*

She attacked a convoy at 0900 close to her lifeguard station, damaged one ship, and took 54 depth charges from a pair of expert frigates. She evaded and just surfaced, though the escorts are still searching for her. Convoy base course was 260, zigzagging at nine knots."

"Perfect, Jim. That's roughly the same direction as the ditched B-29. We'll combine the mission of mercy with a mission of murder. Neither will have an accurate position. Have Dave estimate what time we may contact each mission."

*Time 1539.* "Plane coming out of clouds, astern four miles!"

"CLEAR THE BRIDGE! DIVE! DIVE!" A-oo-ga! A-oo-ga!

"Left full rudder. Paul, 200 feet. Take a sounding. This area can be shallow. Rig for depth charge!"

*Time 1541.* One bomb, very close. The *Barb* shook as the reverberations slammed against her side.

"Indicator shows depth bomb to starboard and slightly above. Lucky we made a swing to port. All stations report damage."

"Jim, we can't afford to stay down long. Paul, your sounding?"

"At 200 feet we've five fathoms under the keel."

"No major damage, sir. Elliman got a bad cut on his forearm from a flying knife and scalded one hand when his pots and pans jumped off the stove. Doc Donnelly's stitching him up. Otherwise, shattered bulbs, cork, and a broken tube in the SJ radar, replaced by Maher."

"Good work. Now let's get up and get going. Periscope depth, Paul."

A quick sweep now showed our bomber had called in a friend for the kill. "Max, these fliers won't go home. Go aft and have Syd pull the torpedo from tube 7. En route tell machinists Russel Custer and McKee to take an empty five-gallon tin from the galley, fill it with diesel oil for Syd to fire. We'll give them their kill so we can get out of here." Max beamed.

"Tube 7 is loaded, captain."

"Fast work, Max. Open the outer doors and fire locally. Hold your torpedo reload until we level off deep and speed up."

Seconds later came the swoosh as tube 7 fired the oil. "All ahead flank. Make your course southwest, 100 feet. Jim, roughly when do we make contact?"

"With another hour down, a bit before midnight."

BOOM! Another bomb well astern.

*Time 1647.* "Up periscope. Three planes circling our oil slick. Radar depth. Give me a range on the planes."

"Average 10 miles for the group."

"Secure from depth charge. Surface. Four engines." Away the *Barb* went leaving both sides happy.

Tom King had the watch, so I stayed with him until dark to avoid our diving unless mandatory. We soon encountered a sampan close to our path. At Tom's query, I explained that the Japanese would know a submarine was in this vicinity if the sampan radioed. All we had to do was avoid his probable gun range, should he have a machine gun. We must not dive. Tom passed him abeam at 4000 yards.

As the *Barb* plowed deeper into the vast open spaces of the East China Sea, we lost the protection of the land masses. With the falling darkness, the waves rose to 15 feet, pushed by northerly winds. Going below to have a bite of leftover supper, I cautioned Swish and Jim to have everyone off watch catch bunk time now. With contact around midnight, the high seas would bedevil our attacks, surfaced or submerged. We should expect a long, tough, exhausting night.

With the *Barb* seesawing, the crew, though excited, obliged.

*Time 1800.* The Mo-Ma 07 commander, now clear of the attack area, changed the base course of his convoy to south southwest. The headquarters of Surface Escort Division One informed him of the attack and probable destruction of the sub by two patrol planes. He then ordered frigates *28* and *54* and the *Miho Maru* to rejoin the convoy. Two hours later frigate *28* radioed that they had completed the attack on the *Queenfish* and would rejoin the convoy by 1200 on 12 November.

*Time 2255.* "Radar contact; convoy 18,000 yards. Station the tracking party. Captain, he's farther south than we estimated. I've come left 15° to 210°. We may have to slow a bit to avoid being pooped. Visibility is poor."

"Well done, Dick. As soon as Tuck relieves you, assist Max on the data computer. We must have his speed and course on the nose in this lousy weather. It's going to be wild. I'll stop by plot and the radar on my way to the bridge."

Red-goggled and wearing my parka, I stopped by the control room. Dave and Chief Machinist Thomas Noll were laying the convoy out from the radar bearings and ranges. Tex and Jim were checking the layout before taking their stations in the conning tower.

"How does it look on your plot, Dave?"

"Not enough for course and speed yet. But we're definitely short two escorts and two ships from *Queenfish*'s original contact. She reported hitting one ship. Assuming it is the same convoy, there are just 11 big ships, 4 escorts, and no air cover. Station keeping in columns and between columns is ragged. Starboard column has 3 ships, the center column 4 ships, and the port 4 ships, the last of which straggles between center and port columns. One escort is well ahead, one on the port bow, one on the starboard flank, and a smaller one astern."

"Well lads, on the bright side, we have targets and no air cover. Now what's the best way to attack? Visibility is poor, so we go in surfaced. Ask Tuck what the wind is doing."

"Captain, wind is from west northwest about 40 knots."

"Gad, those ships must be rolling their guts out. Those 15-foot waves are smack on their beam. That's a partial advantage—I hope they're seasick. Here's our plan. We'll attack on the starboard bow, aft of the lead escort and ahead of the starboard flank escort. All going well, we'll fire two bow tubes at each of three ships, then swing right with the seas and fire two stern tubes at each of two ships. Anyone with a better idea, please don't hesitate to speak up. We still have a couple of hours before we catch up to them."

A study of the radar scope confirmed plot. "Tuck, may I come up?"

"Permission granted, sir," chuckled Tuck. "As long as I live, I'll never forget 'Hey! Let me in!' "

"No ship blobs yet?"

"Nothing. We're closing them about 8000 yards an hour. Lots of spindrift. Moonrise at 0315."

"We must attack surfaced before then. Depth control will be very erratic. Darn it, the spindrift is fogging my binoculars. When we slow down and put the wind astern, they will clear."

*12 November.* "Mr. Weaver, I have the convoy in sight 20° on the port bow."

"Well done, Novak. I have him. Keep sweeping. There may be a straggler or a sleeper escort."

"Radar, what's the range?"

"Ninety-eight hundred yards, sir."

"Captain, in another hour and a half we'll be in position, so we'll beat the moonrise."

"That'll help. I've been thinking about what other advantages we can count on. These mountainous seas have such a stultifying and stupefying effect on the brain. Studies should have been made on the lowered efficiency of men in really rough weather. Not just the violently seasick, who would just as soon die, but the others in responsible positions, after they've been bashed and banged around for hours or days. It's something akin to accelerated exhaustion."

"What effect do you anticipate in tonight's attack?"

"I wish that I could be sure. In this David-and-Goliath game for keeps, all of us must be on our toes, mentally alert, and ready for the unexpected. With the natural forces working on all ships tonight, desired control by rudder and engines could be found wanting. I wonder, will their watches be alert? How will they respond to the first torpedo hit? Will the sheep obey the shepherd and let his dogs fight with us, or will we find ourselves in a cage of wild animals unable to position for a kill. Torpedoes just aren't an instant weapon, and enemy gunfire will be erratic. Their best tactic might be to ram us. Depth charges in these shallow waters are a certainty with our poor depth control due to the waves. What other surprises? We can only wait and see."

"Captain, I'll be transferred to new construction after this patrol. Perhaps my relief should join us for this attack as a 'makey-learn.'"

"Smart idea. Send Jack up as Tuck's understudy for attack training." Jack was delighted to look, listen, and learn.

*Time 0120.* "BATTLE STATIONS TORPEDOES!" The gongs resounded as the *Barb* became fully alive.

"Men, our night surface approach has started on a fat convoy of 11 ships and 4 escorts. The weather is bloody awful, with high winds and 15-foot waves; the night is dark. We'll be attacking with both bow and stern tubes. If the attack goes as planned, we'll probably reload while surfaced. The torpedo reload crews may need help and must be careful to avoid any personnel casualties. Let me know if our wallowing makes such too risky. The enemy must be groggy. If you feel groggy, shake it off. We must be more alert than ever. You're the best crew alive, so heads up. We're going in now. Make ready all tubes. Left full rudder. Put four cases of beer in the cooler."

*Time 0130.* The *Barb* commenced the attack, surfboarding into the convoy and leaving a foaming path abreast 100 yards wide to each side. Their formation keeping was so poor, the targets were designated and redesignated twice before I was able to say, "Tuck, pick out any suitable target ahead and stay on her. Conning tower, use the target bearing Tuck has on the transmitter. Angle on the bow is 60° starboard. Open the outer doors all tubes."

"Sir, the starboard flank escort has increased speed, easing up along the formation, putting him on a collision course with us."

"Nuts, he's forcing us to shoot early."

*Time 0140.* "Range 2600 yards; torpedo depth set eight feet."

"Final bearing—mark! FIRE 5! FIRE 6! QB sonar, get on those torpedoes. Tuck, shift to the ship astern."

*Time 0141.* "Tuck's on the second target. Angle on the bow 70° starboard. Final bearing—mark! FIRE 3! FIRE 4! Tuck, take that ship just forward in the center column."

"I'm steady on her."

*Time 0142.* "Tuck's on. Angle on the bow 50° starboard. Final bearing— mark! FIRE 1! FIRE 2! Right full rudder."

"Sonar reports all torpedoes hot, straight, and normal."

"Dick, how much time?"

"Ten seconds, captain."

WHAM! Lego, the lookout, hollered, "Beautiful hit forward in the large freighter." The rising geyser was unmistakable, but the second torpedo missed.

"Conning tower, set up for the stern tubes."

WHAM! Jack yelped, "What a sight, right in the bow of the freighter aft of the other ship.

Ten more seconds went by. I said, "Another miss."

WHAM! All eyes switched to the center column freighter as she caught this torpedo amidships, and another miss.

The convoy now became a violent free-for-all. The large freighter we hit had her bow nearly under with her stern high. Our second target, the medium freighter, had a 30° dive angle, heading for the bottom. The last target could not be separated from the other ships, and the lead escort had moved back alongside the large freighter leading the second column. This freighter zigged sharply right, almost colliding with the escort.

"Left full rudder. All ahead full. Jim, we've got to clear the escort's path. I'm hooking over to the port side of the formation, passing close ahead to give her a stern shot when she stops turning."

"Radar range 700 yards."

*Time 0153.* "Max, she's steadying down. Open the outer doors on tubes 7 and 8. Angle on the bow is 80° port. Tuck's steady on the after bridge target bearing transmitter. Let me know when you're set."

"Set."

"FIRE 7! FIRE 8! Sonar, report on the torpedoes."

"Hot, straight, and normal, sir."

Turning to the quartermaster, I said, "Broocks, keep your eye on that ship. Tuck and I will look around for a suitable target for our last two torpedoes." Utter confusion reigned in the convoy.

WHAM! "Captain, hit in the after hold!" All eyes followed the geyser of water shooting skyward. Driven by the wind, it arced over and down as if someone were trying to fill the ocean from a giant bucket.

Tex broke our reverie. "Gene, it's high time the remainder of this convoy falls out and falls in again. Jim and I are watching three pips disappearing on the radar scope. How's it look up there?"

"Tex, the only sure thing up here is that the weather and the convoy are unsettled. There's no use shooting more torpedoes. I'm hauling out on their port bow for reload and observation. Try to find out what's causing us to miss with every other torpedo. I can't believe our target speed error could be as much as half a knot. So that's no answer."

"Gene, our master gyro follow-up system is out, so the input to the system has to be set in manually. This could be the cause."

"Concur. In these seas, maybe a degree or more."

Watching the last large freighter hit was tantalizing. She slowed, but did not sink. The whole convoy slowed as they started to reform. Our whereabouts were known. Two escorts edged over toward us. Needing more space and undisturbed time for the reload, we moved out to 5000 yards and quartered the sea.

"Stand easy on stations. Commence reloading torpedoes fore and aft. All hands that can be spared, lend a hand. Tuck, since we can't see all the ships, I'll drop down to the conning tower to make a count on the radar scope."

Dave, Tex, Jim, and I huddled over the scope. Definitely, only eight ships and four escorts were left. Three ships had been sunk. We waited for the difficult reload to be completed.

Radioman Hinson stuck his head up the hatch. "Message from the *Peto*, sir. She received the Loopers' contact report on this convoy yesterday at 2230 and ours at 0055 today, but can't find anything. Her position is 17 miles north of ours." ("Underwood's Urchins," the neighboring wolfpack, consisted of the *Spadefish, Sunfish,* and *Peto*.)

"Captain, you realize that we have now penetrated more than 100 miles into their area. This is against all the rules. I believe you'll hear about this from Admiral Lockwood."

"Jim, don't forget there was a ditched plane in this direction. If I know Uncle Charlie, he'll cheer this night's sinkings. After all, we haven't interfered with any of the Urchins' attacks, nor endangered them so far. Besides, *Peto*'s skipper is a close friend of mine from prewar Panama days in the S-boats at Coco Solo." Tex concurred.

"Send Hugh Caldwell a new contact report—position, base course, speed, eight ships, four escorts. Then add this personal: 'Hugh, to make sure you make contact, home in on our radar, because positions may be in error. We'll stay with convoy until you arrive, then shove off to our area. If possible, we will make one more attack to slow them down for you, perhaps even lighting up the horizon to help you. Lucky Fluckey.' "

"Captain, per usual, you get your teeth into something and never let go. Do you think he'll believe it?"

"Jim, git!"

"Gene, radar shows nothing on any stragglers or cripples left behind. Scratch three ships, one damaged. The convoy is reforming in two ragged columns of four ships each, with the four escorts milling around."

"Okay Tex. I'm going up to the bridge."

*Time 0305.* "Bridge, reload completed."

"Great. Safely?"

"Yes, sir."

"Well done. Stand easy at battle stations for the next onslaught."

"Captain, moonrise in about 10 minutes. Stuffing us with oatmeal cookies, carrots, and vitamin pills to improve our night vision is having an effect. I honestly believe that I can see a perceptible improvement in the zodiacal light as the horizon approaches the moon."

I found Tuck's summary unusual, but technically correct. The use of "rise" for heavenly bodies probably started when the earth was considered to be flat. "Use your super vision, then, to give me a visual count of the remaining convoy before we submerge."

*Time 0317.* Whether the moon came up or stood still, it was less than a quarter on the wane. The clear sky overhead was rimmed with dark clouds to the south. We were easing into a position at 0200 on the starboard bow of the convoy, preparatory for a submerged attack.

"I have a count—eight ships, four escorts certain."

"Perfect, Tuck. I concur. Two columns with escorts spread as before. Conning tower, visual count of convoy agrees with radar. Make a note of this. It's essential that we have their speed accurate to a gnat's eyebrow."

"Speed nine knots, base course 270, zigs 30°—plot and torpedo data computer check exactly."

"Good. Since we will be unable to control the *Barb* at radar depth, I plan to attack at periscope depth. Use my periscope bearings, estimated angle on the bow, and estimated range to shoot with near-zero gyro angles. We'll submerge shortly."

"Radar reports SJ interference. *Peto* is near."

*Time 0354.* "Take her down, Tuck. *Peto* will home on us."

Depth control was next to impossible. At two-thirds speed the *Barb* oscillated violently between 55 and 80 feet. Visibility—which had been fair on the surface after moonrise—was now exceedingly poor due to spoondrift. Would that it were only the windblown froth of spindrift rather than the wind-driven cloudburst of water lashing the periscope exit eyepiece. This was a surprise feature I hadn't counted on.

"Jim, we'll never sink anything using the periscope this way. Raising and lowering it, the water takes so long to drain off, I can't see. The only solution is to leave it raised until after we shoot. How much water do I have to play around in?"

"Single ping soundings are steady—200 feet."

*Time 0400.* "Okay, leaving the scope up I can see quite well. Even at 75 feet I can see part of the time in the wave troughs. We're commencing an attack on a large freighter with a 5° starboard angle on the bow. Convoy is still on course 270. Range is about 2500 yards. I'm cutting across the convoy from starboard to port for a stern-tube shot on a port track. Make ready all tubes."

*Time 0409.* "Open the outer doors on tubes 8, 9, 10. Angle on the bow . . . she's zigging to her right. No sign of slackening her zig. Let's try another target. Here's a small freighter in the port column. All ahead full speed. Angle on the bow 10° starboard; range 400 yards. I'll cross ahead. She hasn't yet zigged. Max, this is going to be a very close shot, but I'm certain we can make it."

*Time 0411.* "Whew! Just crossed ahead at 200 yards or less. We're on a 120° port track. She must zig. Angle on the bow 20° port; range is less than 100 yards. She's zigging right."

In high power on the scope I could not see anything except the funnel. In low power I could not see the whole target, so I commenced swinging the scope back and forth to obtain a retention-of-vision picture of the angle on the bow. In doing so I noticed a medium freighter overlapping ahead. Estimated range was 150 yards (actually 137 yards as measured by plot), which was too close to be inserted in the data computer without running it up against its stops. Thus I didn't want to alarm or confuse the fire control party. Target zigged 30°.

"Max, set range at 500 yards. We're on a 150° port track, so the torpedoes will have at least the 300-yard run they need to arm and explode. Jim, keep calling the scope bearings; scope's steady on the funnel. Coming on! Sonar, get me a ping range! Spread fish from aft forward."

I could hear Max muttering, "She must be a lot closer." A whale of a lot closer, but we controlled the situation. An old headline flashed through my mind of a sub firing at the suicide range of 2000 yards. What a joke!

Things were happening very fast now. Torpedo gyros raced toward zero.

The sonar operator said we were too close to get a range. We couldn't use it anyway; we couldn't miss.

*Time 0412.* "FIRE 8! (10 seconds) FIRE 9! (10 seconds.) FIRE 10! Sonar on the torpedo . . ." WHAM!

The first torpedo hit in the forward hold and the target blew up in my face, literally disintegrating. My mouth hung open. I was flabbergasted.

The engine room, not realizing that a torpedo could hit within four seconds of firing the third torpedo, immediately reported, "Depth charge has blown off our afterdeck."

"Sonar reports other torpedoes normal. Target is so close he can hear water rushing into her hull."

Parts of the target commenced falling on top of us, drumming on the superstructure. The concussion had forced the *Barb* sideways and down. The periscope was now ducked.

*Time 0413.* WHAM! Another timed hit was heard by all, undoubtedly a lucky one in the overlapping freighter ahead. This we knew because the torpedoes were spread from aft forward and the first torpedo hit the forward hold.

"Sonar reports high-speed screws approaching."

"Paul, take her down easy to 180 feet. The bottom is at 200 feet. Left full rudder. Rig ship for depth charge."

Breaking-up noises could now be heard throughout from the close target. These were followed within a minute by a crunching thud. "She's hit bottom!"

"Captain, do you want to take the rudder off?"

"No, Jim. Circling this sinker may stop our enemy from depth charging us, for he'll blow up the survivors."

"Tex, for your info, on our last patrol we picked up 14 British and Australian survivors from the *Rakuyo Maru,* a prison ship unknowingly sunk by the *Sea Lion.* Interrogating them, we learned that a number were injured or killed when the escorts were depth charging after the attack. None depth charged in the vicinity of Japanese survivors. So, instead of putting the rudder amidships, we are circling with the hope that Japanese survivors will be our overhead protection."

*Time 0415.* Breaking-up noises from the medium freighter commenced that were definitely separate and distinct. She was probably on her way down. I wished we could have a visual check and the two pinging escorts close by would shove off. Rigged for silent running, we were at slow speed.

Paul, the diving officer, had a real problem. The trim was heavy; the *Barb* was settling slowly at 185 feet with a 3° up angle. An increase would cause our screws to hit bottom. Both choices were noisy: speed up or blow ballast. Either would bring the escorts.

"Captain, we must blow."

"Granted. Rig in the sound heads and speed swordarm to avoid their striking the bottom." Paul caught her at 190 feet.

The escorts also caught us. The screws of one could be heard through the hull. A hush descended on all hands. The escort was shifting to short-scale pinging preparatory to attack.

"Right full rudder. All ahead flank. One hundred seventy-five feet. Lower the sound head. Darn it, we're sandwiched. Pings are raining off our sides."

"Sonar reports he's started his run."

"Left full rudder—stand by. He's passing to starboard, judging by the flap, flap, flap of his screws."

Click—Boom! First depth charge close. We could hear the small detonator explode. Depth set estimated 125 feet and to starboard. Another charge hit the water above. Click—Boom! And another charge at our depth—175 feet—close abeam to starboard.

"Tex, he's trying to bracket us with his depth-charge pattern vertically as well as horizontally. Smart guy."

A series of splashes were now heard as he dropped his pattern. The conning tower was as silent as the proverbial nun, breathless with adoration, waiting. Waiting, waiting—I couldn't hear anyone breathe, myself included.

"Sonar has a group of thuds broad on the starboard bow."

"By golly, he screwed up. The charges hit bottom without exploding. We're saved by the water being too shallow for his bracket." Happy smiles appeared all around.

"Shankles, steady as you go. Course 315. We'll evade upwind and up sea until those waves are putting green water over their bridges. All ahead full speed. Paul, make your depth 185 feet. We're going to scurry away as close to the bottom as I dare. Hopefully, their pings reflecting off the bottom and the *Barb* may intermingle."

"Sonar reports both escorts have stopped."

"What now?"

We heard two side throwers go off, followed by two splashes—5, 10, 15 seconds. Click—Boom! Click—Boom! Close astern.

"Sonar has escorts shifting to long-scale pinging."

Evidently, they had lost us. Screws passed overhead like a freight train. "Listen!" We heard a pitter-patter of splashes on the surface. "Stand by!"

BANG! A close, small explosion. Hedgehog? Sono bomb?* Only one? Others too deep?

Another patter of splashes were followed by three close explosions, then a most-welcome silence. With this final display of "rile and bile," the escorts gave up and shoved off.

The convoy commander, in frigate *8*, sent a message regarding the ships for which he was directly responsible. He reported only the last attack at 0413 and the sinking from among Mo-Ma 7 of the heavily loaded *Naruo Maru* (4823 tons), which exploded. He was responsible for her; the other, the *Gyokuyo Maru* (5396 tons), had sent her own message before she sank. His two subchasers had probably sunk the submarine.

The *Peto*, using the contact reports from the *Barb*, homed in on her radar interference. She arrived in the convoy vicinity without making contact as the *Barb* submerged. At 0413 Tokyo time, she saw the large flash and heard the explosion of the *Naruo Maru*. The *Barb* had connected. A burning ship farther away (the *Gyokuyo Maru*) was hull down, though clearly visible.

In a few minutes the *Peto* had radar contact—32,000 yards. Moving in on the remainder of the convoy, her crew counted a loose formation of 8 pips and 2 escorts for a total of 10 ships after the *Barb*'s sinkings.

With the sky now overcast, shutting off the flow of silvery moonlight, Hugh attacked on the surface at 0530. The *Barb* was still submerged. Two hits resulted from the four torpedoes fired at 3400 yards. The damaged ship slowed and kept going. The *Peto* tried to hurriedly reload to attack again before the break of day, but heavy rolls prevented it.

*Time 0540.* Escorts gone, the *Barb* surfaced 17,000 yards astern of the convoy. A radar count showed a total of 10 pips, including ships and escorts. We found one grating blown out of the deck aft. Radar interference from the *Peto* showed she was close to the convoy. Half an hour later she attacked. We witnessed a beautiful tower of explosion from one ship and felt one other hit.

As the sky brightened, we submerged. Everyone was too tired to either splice the main brace or eat. Drifting through the boat on a cloud of satisfaction for the sterling performance of every shipmate, I set 1130 as our target to celebrate.

---

* Small explosive rocket weapons simultaneously fired in groups of 10 or more. One hit could hole a submarine.

# Chapter 15  Rest and Be Thankful?

Exhausted and elated, the crew was putting the *Barb* to bed as well as themselves. I left the conning tower to the regular watch. We were lazing along at two knots to save battery power. Swish, the section diving officer, kept her at 130 feet while the after torpedo room was completing its reload.

"Swish, you did a superb job with the planes, keeping us 15 feet off the bottom when we were evading the two escorts up sea at full speed. Well done."

"Kinda tricky, sir. We've never practiced running like that. Sure glad they set some of those depth charges too deep. Down here we could hear some thudding against the bottom without exploding."

"Well, we must remember our *Barb* sweetheart is a thin-skinned gal whose test depth is 312 feet. She won't bounce off the bottom or withstand dimpling by depth charges as well as the new, thick-hulled *Tang* class. Their test depth of 438 feet provides superior tactical and protective capabilities."

I ambled through the ship in a daze, congratulating the crew on their most-recent superb performance. As I walked into the forward torpedo room, I was bashed in the head with a flying ball of rags; the relief from tension had led to some horseplay. "Don," I said to the culprit, "do you know the punishment for striking the captain in wartime?"

"Dare I ask, sir?"

"Well, normally, keelhauling, but in your case perhaps I may make an exception and exile you back to the after torpedo room from whence you came."

"Mercy, sir. Anything but that. After two patrols aft, I finally achieved seniority to move forward. I enjoy the extra room here and the fresh air."

Since I was the judge and jury, I took cognizance of the 50-percent torpedo hits achieved forward and the arduous reload on the surface in heavy seas. "Rest and be thankful! I'm going aft where there's less danger."

In the other areas of the ship the men were quietly going about tasks such as reloading. In the crew's quarters the darkness was filled with the cacophony of exhausted snoring, which caused me to whisper, "Rest and be thankful. God bless." I finally wound up in the crew's mess, where, after accepting a cup of coffee from Chief Williams, I fell asleep, my arms and head flat on the mess table. Heeding the advice of the cooks, I shuffled off to bed.

"Splice the main brace!" I awoke with a start. No doubt, it was 1130, and I felt as if I had been dragged through a knothole. The crew would be waiting for the cake cutting. Turning on the light, I shuddered at my reflection in the mirror. The crew would have to wait for their captain to shape up. A quick pass with the electric razor, a hair combing, and a splash of cold water worked a miracle. Adding a smile, I bounded belatedly into the control room like a young impala, ready to greet the mob.

On the bench in front of the bow planes control, Jim sat with Bob Phillips, the apprentice ship's cook. Their backs rested against the diving wheel. In front was Phillips' masterpiece—the cake, ornately decorated with four ships sinking, one listing drunkenly, undecided as to whether she should sink or not, and the last with a tower of icing as a torpedo struck her side. Bella! Bella!

"Congrats! Bob. You've outdone yourself." Looking about I saw no one but the few on watch. "Where is everyone?"

"Captain, when I passed the word, I guess they only rolled over."

"A fine celebration, Jim. Like a wedding without the bride. Too bad we don't have a boatswain's mate."

Just then our two new and most junior officers stumbled in. "Where's the party?"

"Tom, Jack, let's take a cue from the British tradition of olden days. At reveille, their bo's'ns mates would cry out:

Paul hammocks! Paul hammocks!
Rise and shine! Rise and shine!
Show a leg! Show a leg!
The sun's shining her bleeding eyes out!

Then they'd pass through the compartment and swat those still asleep on the bottom with a billy club, unless they showed a leg."

"What does it mean?"

"Paul is 'up all.' The leg meant a lady's leg, which gave you permission to sleep in. Now, Tom go forward, Jack aft, and let hard experience crown your efforts with success."

They were game and it worked. The control room soon filled.

While the photographing of the cake with Bob proudly sitting beside it progressed, the beer ration was passed out. From the stodgy quiet as they assembled, the crew's beer cheer now ballooned the decibel level to the point that it would drown out the tintinnabulation of ten thousand silver bells.

I called for quiet. "Men, time flew so fast that the galley hasn't had the opportunity to bake the celebration cake for the cruiser *Barb* sank only two long days ago. I know it seems like two weeks. Our master artiste Bob, however, had nothing to do earlier on the galley watch, since no one cared much for breakfast. So he whipped up a cake and got carried away with the ships sunk early today—thus the embellishment. It's more than we've ever had on one cake. Let's have two cheers for Bob and the *Barb*. Hip-Hip-Hooray! Hip-Hip-Hooray!"

"We'll save the cruiser cake for a less-happy occasion. Our new exec now has the honor of cutting the cake."

All smiles, Jim picked up the long bread knife and started cutting. Gracefully, he ended around one of the sinking ships, slicing through the waves. His knife stopped. He tried sawing. He pressed down with both hands as he stood up to get more leverage and pressure. Then he looked at Bob.

All talking stopped. Bob clapped his left hand over his mouth and muttered, "I wonder if I baked it at too low a temperature for too long a time." Agonizing, I bit my lip.

Swish stepped forward, "Mr. Lanier, as a Boy Scout, I've been half-baked and double baked. May I have the knife?"

"Gladly. You try cutting this Gordian knot."

Swish grasped it firmly, upended it, raised it on high, and plunged it into the heart of the cake with a force that would have done credit to Saint George slaying the dragon. Cheers erupted. Even Bob smiled.

Crumbling slices pasted together with icing were passed around. The texture wasn't jawbreaking. The taste? After a mouthful Dick opined, "Most unusual; a sweet-and-sour cake."

Testily, I tasted it and gazed at Bob. "Salt?"

He nodded. "Not too much, I hope. Navy recipe says a pinch per pound.

I had butter on my hands when I picked up the box of salt. It flipped into the mix when I gave it a tap, but I scooped most of it out."

"Bob, we're all old salts. The photos will be great. If you put egg in your beer, don't cook it first."

*Time 1400.* We surfaced and commenced our search for a downed aviator. As the other wolfpack was searching to the west, we chose the southeastern area.

Four hours later we sighted the *Peto* surfacing and closed. Hugh thanked us for bringing him in on our radar, otherwise the *Peto* would have missed the convoy due to differences in navigational positions. At 16 miles he saw a flash like a ship exploding, then another ship burning east of the flash. Our respective times were just a few minutes apart for our final attack. This agreed with our estimates. He also confirmed our ship count prior to his attack. The *Peto* sank two and damaged one before the escorts worked her over. The Japanese reported that the *Barb* sank the *Naruo Maru* and the *Gyokuyo Maru.*

*Time 2300.* We secured from the vain search and joined the *Queenfish.* Elliott had divided the Loopers' area west of Kyushu into three parts: Elliott had the southeast, my sub the northeast Tsushima Straits area, and Shep a 50-mile strip west of these. The East China Sea beyond belonged to the Urchins' wolfpack. Every four days we rotated areas counterclockwise. It was a highly intelligent plan that covered our 40,000-square-mile area. Practically any convoy contacted by one sub could be passed to a backup sub, if the backup was too far away when reported. If close, the two would make a coordinated wolfpack attack.

The convoy that the *Barb* had attacked yesterday had been passed to her from the *Queenfish,* attacking well south of Nagasaki near her lifeguard station. In hot pursuit we were allowed to enter and attack in the Urchins' area until they attacked. The *Peto* was most appreciative to the *Barb* for her contact reports and for homing her in. Otherwise, she would not have made contact. In turn, she passed the remnants on to Underwood, her pack commander in the *Spadefish,* and to Ed Shelby in the *Sunfish.*

As the *Queenfish* headed for Tsushima Straits, the *Barb* moseyed around the area east of Quelpart Island, spoiling for combat. Regretfully, we had inadequate weapons to fight our nemesis—aircraft. So, we yo-yoed to evade.

Tex avidly questioned me on strategy and tactics. I explained that some skippers stayed undetected, submerged all day, using their periscope to search a circle with a radius of five miles. Our *Barb* system was to stay on the surface searching, with high periscope up, covering a circle with a radius of more than 10 miles. The difference: 70 square miles versus 350. Surfaced, we also augmented our coverage by using as much speed as our diesel oil supply would permit.

"I'll buy that," Tex said, "but you take more chances of being seen by planes—and convoys may avoid the area."

"And we have a greater chance of being bombed, granted. The choice is to patrol either the coast and open seas submerged or the open seas surfaced. I believe that more worthwhile targets come with escorted convoys and naval task groups that ply the high seas at some point in their movement. Thus we stay out unless watching a major port. No doubt, attrition of everything has merit. The small ships and spitkits of less than 1000 tons, which haul local freight along the coast, aren't worthy of more than one torpedo. Most unescorted coasters should possibly be gunned to save torpedoes. You'll have to choose. Remember, though, that each sub must produce, or the admiral puts in a new skipper. He might accept an empty bag once, but never twice."

After lunch I broke out a drawer full of patrol reports that I considered worthy of Tex's study, including some from all the most productive subs. At Submarine Base Pearl there is a skippers' room that houses a library of war patrol reports. After the *Barb*'s eighth patrol, I spent hours there studying patrols in the South China Sea, which was to be our next area. I was provided with copies of whatever I wanted to carry on patrol to read. I now spread these out for Tex's perusal.

"Where do you start?"

"Tex, Jim, I'm a torpedo bug, with more experience with the fish than most. I was torpedo officer on an old four-piper destroyer for two years, and again in the *S-42*. I have studied and worked on the design of both torpedoes and depth charges in my free time."

"Depth charges?"

"Absolutely. In our practices in the destroyer *McCormick*, I watched the ashcans roll down the racks and drop. As soon as they hit the water, they tumbled over and over while sinking. Who knew how long they took to sink to the depth we set to explode? No one. There was a general rule of thumb, but to judge a score for the exercise, the observers clocked the time the charge touched the water surface. A terrible system. Yet when I went to design engineering postgraduate school in 1942, the ashcans had not been improved since World War I. So I went to the head of the Ordnance Engineering Section to find out what thought had been given to a more accurate, fast-sinking depth charge.

"What for? Like what?"

"I explained it was from my own observations and my experience. A tear-drop design to fit current racks and side-throwers could be a deadly improvement. He became curious and excited. I had three precious weeks of leave before my classes started. Perhaps he could telephone the Bureau of Ordnance to find out if work along this line was in any program. If not, I could draw up a tear-drop casing with the depth-setting pistol detonator insert in the stabilizing-vanes end. Dimensions used would be exactly equal to the present charges.

" 'I'll do better than telephone,' he said. 'I'll go to Washington tomorrow.'

"I was elated. By noon the next day he was back and reported that there was nothing going on in this area. But they were interested. 'Get cracking. I'll take your design over, personally.'

"Three weeks later, my precious leave gone, I presented my design. He concurred, took it to the Bureau of Ordnance, and returned, boiling.

"He said, 'I was told to thank you for your effort, but they have had such on the drawing boards for two weeks. Frankly, it's theft. The 'not-invented-here' syndrome. At least you started something. I am sorry.' "

That was that. Jim, Tex, and I went back to the reports.

"I start with the endorsements," I said. "They give a precis of the report. Then I turn to the enclosure section that shows the statistics on the torpedo attacks. Study these for type of torpedo, hits, misses, erratics. The last may determine your modus operandi. At the skippers' room I was lucky to encounter Sam Dealey, the skipper of the *Harder*, studying alone. Sam had flown up from Australia. I had read all of his reports and considered him tops, though we differed on one tactic. So we had it out."

"Which tactic?"

"Jim, Sam is fearless, and he's famous for his down-the-throat shot against escorts. I questioned him at length. His tactic was to head for the escort at about 3000 yards, then leave his periscope up to be seen. He sank two destroyers with this tactic, firing three torpedoes when the range closed to 1500 yards with a spread one-half degree left bow and one-half degree right. If the destroyer, seeing the wake, turned to evade, he would catch one of the wing fish."

"Sam said, 'The other destroyers were sunk with *Harder* undetected, using track angles. I believe that if you are detected, down the throat is the answer.'

"I concurred, providing there was no other recourse. My differing was due to erratic torpedoes and torpedoes running deeper than set. If the hitting torpedo missed, a good escort would cremate the sub. Yes, he had had erratics, but not against the escorts. Believe me, Sam is my hero."

Jim shook his head. "Captain, you mean was. *Harder* is overdue by two months and presumed to be lost."

"I know that, but feel that the loss may not have been due to a down-the-throat shot. The info in Majuro was that his other wolfpack mate evaded a lone escort that had sonar contact on both of them and *Harder* did not."

"Gene, is that why you're so eager to have one sonar immediately follow the fish?"

"Positively. I watched erratics and circular runs in the calm seas in which we exercised before the war, to avoid losing any. Think of the seas we fire in now. These torpedoes were never tested under these conditions."

"*Barb* has had erratics?"

"Certainly. I don't think you will find a sub that hasn't. On *Barb's* eighth patrol we had zero, probably three that ran deep. On her ninth, one, plus one circular run we had to avoid. The present patrol—firing at the light cruiser—we had one, followed by one that jogged left before it settled down. Then you fired the one that polished her off. Take a quick look at a few reports to see if we are average."

Jim picked up the reports of the *Tang's* third and fourth war patrols while Tex perused the *Tautog's* and *Harder's*.

"Captain, *Tang's* third had three erratics. Her fourth had four. She's on her fifth now—Formosa Straits."

"Tex, you must know instantaneously what action to order to evade any erratic. Circular runs make perfect circles. Action varies as to whether they are fired from forward tubes or stern tubes and whether they are circling right or left. All torpedoes turn using full rudder. You know the speed of the different types and the turning radius. Now calculate the length in yards of the circumference. Divide this by the speed in yards per minute and you have the minimum time to avoid your own departure from this world. Your direction of motion must turn toward. Got to go now."

Just then I was notified of a plane report and went up to have a look. Taking the periscope from Jack, I watched the plane circle then disappear to the north.

"Nothing but a routine patrol plane, Jack. Go ahead and surface to expand our search. I see no evidence of traffic coming through this area. Bend on four engines and head to the northeast of Quelpart Island toward Tsushima. I note from the log that the barometer is slowly dropping. Let's take advantage of this piece of fair weather to cover as much as we can."

Main ballasts blown, Jack and I scurried up the ladder, accepting the partial drenching as the *Barb* shed herself like a Labrador fresh from a romp. With the clap of the exhaust and the engine air-induction valves opening, the engines roared and we took off for more fruitful waters.

*14 November.* Dawn was aborted by a heavy overcast and drizzle. The morning seemed ideal for a small ship gun shoot. Our "total war" orders were to sweep the seas clean of any craft that supported Japan. Jim, Max, and I studied the chart for the shortest and most likely route between Korea and Japan. Using his innate ingenuity, Max hypothesized that they wouldn't even zigzag to save fuel and shorten the crossing. Smart—we bought his plan.

*Time 0701.* Paul Chapman, the port lookout, sang out, "Two schooners, port beam, opposite course! They're just barely visible, sir. The decks are almost awash. Fully loaded."

"Well done, Chapman. Don't take your binoculars off of them until we pick them up. Dave, I relieve you. Drop below and try to get a radar range. Take an air sweep, too. Right full rudder. Chapman, we're reversing course to end around. I don't want them to see us until we're ready to shoot. Control, send up another port lookout on the double, and also Mr. Weaver."

*Time 0741.* "MAN BATTLE STATIONS GUNS." At the sounding of the gongs men streamed up the ladder from the conning tower to man the 40-mm and 20-mm guns. The 4-inch gun crew emerged from the gun access trunk under the forward part of the bridge deck.

With Max in charge of the guns and Tuck the watch, we headed down the reverse of their course, planning to shoot to starboard when they could be seen through the sights. With simultaneous sighting a few minutes later, the best-laid plans of mice and men were dumped at 1100 yards.

"The leading schooner's turning toward to ram and she's speeding up, sir."

"Tuck, speed up and slip in between the two. Max, put your 40-mm on the target to starboard and 4-inch on the target to port. His courageous maneuver gives us a golden opportunity to make a double attack, firing both starboard and port broadsides."

Tuck put her in the slot while Max rearranged his targets. "Fire when ready, Max."

Max bellowed "COMMENCE FIRING!" so loud that he didn't need the telephones. At a range of about 1000 yards, a majority of the shots hit, and the schooners sank in a few minutes. Dick, the spotter, used the high periscope to call "overs and shorts" and reported that their guns aft appeared to be wooden.

We secured from battle stations in a hurry because we were only 7000 yards from the Kanjo Gan Lighthouse. Certainly we'd been heard, if not seen. No sooner had we secured than another schooner was in sight. Again battle stations, and again the target was sunk.

Later, a plane was picked up in the overcast at three miles. Diving, I heard six distant explosions as I left the conning tower heading for my bunk. Swish, the diving officer, heard me mutter, "Sometimes I think they're actually testing our nerves."

Smirking, he added, "Not with distant thunder below, sir."

Lying in my bunk reading war patrol reports was my favorite pastime at sea, and it was educational. Life is not long enough to personally garner sufficient experience for anything. Without blood, sweat, tears, responsibility, or danger, one can absorb vicariously and harvest the experience of others. Otherwise, their history of errors is bound to be repeated.

I soon became immersed in the reports of the *Greenling*. Her skipper in 1942, Chester Bruton, was a quiet, self-effacing officer. One of the most intelligent in the Navy, he was not the bull-in-the-china-closet type that specialized in public relations. Few realize that in 1942—a year when skippers were hamstrung by unrealistic peacetime practices and defective torpedoes—he sank more ships than anyone else. His 10 ships would be equal to 30 in 1943 or 1944. Many of his trials and tribulations had since been overcome, but it inspired me to see that submarining had risen to its full, mature potential. Where dare I take my ship, and with what, to shorten the war and to get back home to my wife and our young daughter?

My musing was jolted by a quiet knock on the panel beside my doorway. It was Doc. "Sir, Mr. Lanier has had a heart attack."

"WHAT?"

"Yes, sir, I said heart attack. There is no doubt."

"Is he all right? Is he in further danger?"

"He's comfortable now. I have him in his bunk, treating him with nitroglycerin tablets. Definitely angina pectoris."

"How do you know?"

"Typical symptoms I've seen in the New London Hospital: tightness in the chest, shortness of breath, pain in the neck and shoulder when he came down from the bridge after taking star sights."

"Doc, shall we secure from patrol and head for Saipan?"

"Not yet, captain. He could recover quickly, if it's minor. If not, I don't believe he will become critical as long as he rests. I have plenty of nitro tablets."

"Do you want advice by radio?"

"Not now, sir."

"Donnelly, what can I do to help?"

"Keep him in his bunk for the present. Later he will probably be able to go to the head and take some meals in the wardroom. He must confine his activity to officer's country. No climbing through hatches or up ladders, no smoking, no tea or coffee, and above all he must not go to battle stations. Please order Ragland and Jones to keep him in his bunk whenever an alarm goes. I'll give Ragland a low-fat diet to feed Jim. He's not an easy patient. He asked me to promise not to tell you. I said N-O!"

"Doc, well done. Thank God I have you on board. You fully deserved the special promotion to chief after our last patrol. Bringing those Brits and Aussies back alive merited it. Now, another challenge. May I talk with Jim?"

"Certainly, sir, but not at length."

I first beckoned the officers together and explained the situation and restrictions. Max would take care of general exec business and Tex would be navigator. Tuck would handle the periscope and bearings for me in submerged attacks. Tex would coordinate tracking and fire control.

Having spread out Jim's jobs among the officers, I then joined our patient. "Jim, Doc told me about your problem. So take in the slack for a few days and let's get you back to battery, okay?"

"Damn it, captain, there goes my heart right in the middle of the patrol. I've let you down." Tears came.

"Stop it, Jim! We're not in any foreseeable action. We only have seven pickles left. You follow orders and get well or I'll put you in irons!" I explained who would handle his duties while he was *hors de combat*. They would come to him for advice and keep him informed of essentials, but the object of the exercise at the moment was for him to relax down to his fingernails. "Now, rest easy on your oars."

*15 November.* The planes came with the dawn as we patrolled southeast of Quelpart Island. After six ups and downs we gave up, convinced that a convoy must be coming through. We stayed submerged shortly before noon to have a peaceful lunch.

Our first slurp of soup was accompanied by distant thunder below as we counted, in unison, three explosions followed by 15 depth charges. Then silence. A prayer was said for the *Queenfish* when sonar reported that the emanation was from her direction. Tom reported planes slowly working toward us. In mid-afternoon 20 distant depth charges and numerous distant explosions elicited more prayers for the *Queenfish*. When the planes disappeared we surfaced, hoping for a contact report, since the convoy should be close unless it had been rerouted.

After supper Elliott's report arrived, showing that the convoy was headed directly into our arms. The *Queenfish* had sunk a 10,000-ton carrier and had been worked over. With that we headed west southwest at full speed to catch the convoy. The *Peto,* choosing a southern reroute too, also chased southwest. The weather turned sour as the seas churned up. A black night fell. The radar took this inopportune time to conk out. What a mess! Instead of looking for an elephant in a haystack we were reduced to the proverbial needle. After an hour of feverish activity, Dave, Maher, and Lehman booted it up so our night eyes were opened again.

*Time 2300.* "Radar contact 19,000 yards, 245°."

Only 5° away from our search course! We had outguessed the convoy and were running up its tail. Jackpot!

"Station the tracking party!" Tuck came up in a flash.

"What good luck, Tuck."

All smiles, "Right, sir."

"Bridge, range is closing rapidly!"

"Come again?"

"Correct. Range now 17,800 yards."

For a second Tuck and I looked at each other in disbelief, then lightning struck.

"Right full rudder, all ahead flank. MAN BATTLE STATIONS TORPEDOES! Tex, this is not our convoy. Check the scope. What have we caught? Shankles, make your course 300. Max, I'm going to cross ahead and attack down sea from the port bow."

"Gene, the pips look like a huge ship with four escorts. Max says she's clocking 21 knots, zigzagging."

"Tex, have Jack send out a contact report on a carrier naval task group to all subs East China Sea. Make ready all tubes; we'll shoot our last seven fish. Dave, Noll, give me their zig plan as soon as you have anything. Max, what's our distance to her track?"

"Twelve thousand yards, now making approximately 19 knots."

"All ahead standard. Tex, I've got to slow; we're shooting columns of spray high in the air. If we get sighted by the near escorts, we won't get in a shot. Dave, I need that zigzag plan bad—give me any indication."

"Recommend coming right to 330° true. Rough zig plan 000, 030, 000, 050, legs five to seven minutes."

"Course 330."

Tuck said, "Barely make her out—their biggest—*Shokaku* class!"

"Jackpot! Can you see her clear enough on the target bearing transmitter?"

"Not good enough. The close escort interferes with the long hull."

"Tex, use radar bearings. Helm right to 340; all ahead two-thirds. Open outer doors. Set depth 10 feet. Dick, divergent spread, 50 yards between."

"Target has zigged right to 055. Range is 2680 yards. We can't get any closer with his speed. Angle on the bow is 90 starboard. Set!"

"Identification, new carrier *Katsuragi*."

*Time 2323.* "FIRE 1! FIRE 2! FIRE 3! FIRE 4! FIRE 5! Get sonar on those torpedoes. Left full rudder! Get me a range to near escort and cross your fingers. We have almost a three-minute torpedo run with these electric fish."

"Sonar reports all torpedoes hot, straight, and normal. Range to near escort 1350 yards, opening. Fifty seconds to hitting time—Captain! Target is zigging to his left, damn it!"

"Oh no! No! No! Keep your times coming."

"Twenty seconds—15—10—5." WHAM! With Tuck's and my binoculars riveted on the near escort, Lego called out, "Hit aft! Near his stern." An undeadly quiet ensued, leaden with disappointment. The other four torpedoes passed ahead. Mechanically, quietly, lost in choking back my own tears, I gave orders to trail at emergency speed and send out contact reports every 15 minutes. Tex closed the outer doors aft, harboring our last two torpedoes.

*Time 2328.* The starboard destroyer turned out and commenced dropping depth charges.

"Bridge, radio reports *Queenfish, Peto,* and *Sunfish* are zeroing in on the *Katsuragi* from all directions and urge us to keep trailing. They need our radar to home in due to navigational position variations."

*Time 2332.* "Gene, *Katsuragi* has slowed to 17.5 knots. She may be in trouble. Rub that rabbit's foot."

*16 November.* Midnight fading, along with hope, firehose blasts knocking us around the bridge, the *Barb* buried her head in her run for the roses. Minute by minute we gained a yard.

"Bridge, SJ radar interference from two different points forward of our port beam." I was listless, shaking my head.

Tuck replied, "Captain has the word," then muttered to me, "those subs will be unable to attack. They're not far enough ahead. If only someone could smack her again to slow her down, all four of us could devour her."

"Tuck, you sound awfully bloodthirsty, but how right you are. Those two must be the *Peto* and *Sunfish* coming from the west. That leaves only the *Queenfish.*"

"Not bloodthirsty, just disappointed." He voiced his frustration. "The most important target the *Barb* ever contacted is getting away due to four broken links that let her through. Lousy weather with such a high-speed target was the first. Chasing a convoy going the other way, so our certainty on contact cost us critical minutes and a closer firing position was number two. Third, the initial radar range of 19,000 yards was poor, due to the radar being down. Fourth, we took the final wallop with the short zig less than three minutes after firing."

"Good analysis. Let's bet on the *Queenfish.*"

*Time 0105.* "Bridge, *Katsuragi* has slowed to 12.5 knots. We're now closing her at 270 yards per minute."

"Terrific! Send out a revised report." My palms closed prayerfully in front of my lips as hope arose again.

"Tex, the destroyers are all ahead and on her beams since her speed prevents any attack from astern. We only have the two fish aft. The plan now as *Barb* catches up to her is to slip in between the carrier and starboard beam escort. We'll shoot at 500 yards broad on her bow and dive."

"Understood."

Tuck was rubbing his soaked, salt-stung hands gleefully.

*Time 0113.* "Bridge, *Katsuragi* speeded up to 19 knots. That ruins our plan. Three more depth charges. Contact report going out . . . Wait one . . . radio reports *Peto* made radar contact at 43,000 yards. She now has two big targets close together. One is at 42,000 yards, the other at 44,000 yards."

Tex added, "Gene, she also says that she is in trouble trying to catch this task force because she was convinced our reported contact was the same as that of the *Queenfish.* So she's been chasing the contact south instead of northeast for some minutes, the same as we did. We didn't include a course in our original contact report because we didn't have one. Perhaps she missed our follow-up report which gave the course and speed 15 minutes later."

"Okay, Tex, keep those reports going out every 15 minutes. It's essential that someone slows her down."

*Time 0200.* "Bridge, carrier has changed base course to 060 heading for

Tsushima. *Peto* and *Sunfish* are falling back. *Sunfish* gave up on this mission impossible."

"Tex, have everyone secure from battle stations except the tracking party."

*Time 0315.* "*Peto* reports she's unable to attack. Wait one . . . *Queenfish* reports her closest range was 16,000 yards. She's secured."

"Send '*Barb* has ceased trailing. Sorry.' "

Thinking out loud, Tuck said, "I know what went wrong."

I shrugged.

"Captain, you forgot to put four cases of beer in the cooler."

The *Barb*, her eight-hour full-power run completed, headed back to her west area with her tail between her legs, worn out. *Katsuragi* went home. The parting was such sad sorrow—I hoped we would meet again.

At dawn we submerged for the morning to rest on our defoliated laurels. Jim was snoring away, a good omen that indicated his shortness of breath had disappeared. My head hit the pillow, and the next thing I knew, Ragland was shaking me to announce soup and sandwiches.

After lunch the *Barb* headed for her new area. As we had only two stern torpedoes left, we decided to take a crack at the small coastal ships. Noma Misaki, a southern cape of Kyushu, seemed to be a likely focal point.

At dawn the next day we submerged when we could see homes on the beach through the morning mist. As it lifted, we saw smoke going away up the coast and a local plane.

*Time 1140.* "BATTLE STATIONS TORPEDOES!" We sighted two small ships with two small patrol boats and one plane coming down the coast. As they came closer, we saw that one was a mast-funnel-mast freighter and the other an engines-aft freighter. Against the background of beach homes, both looked big. Not finding them quickly in the recognition manual, I estimated the funnel height that we used for solving the torpedo triangle.

*Time 1213.* The *Barb* turned tail and fired one torpedo at each ship at 1200 yards; both missed. One exploded on the beach.

Sonar reported both running normally and passing ahead of targets. No thunder, no depth charges, no bombs resulted. The ships, which had been coasting at 1000 yards off the beach, now moved coastward to 500 yards. We commenced clearing the scene, homeward bound.

Once outside the strait I thanked the Loopers for their teamwork, with the hope that we could wolfpack together on our next patrol. Elliott was the perfect wolfpack commander—faultless, keenly intelligent, dependable. The *Barb* then sent her zero-torpedoes departure report, requesting a base.

Commander Submarine Pacific replied that Pearl Harbor was chock-a-block. We were to go to Midway Island for refit.

Jim was sitting quietly in the wardroom twiddling his thumbs. To be left out of everything is not good therapy.

"Jim, like Mohammed, if you can't go to your work, have your work brought to you. Unless, of course, you don't feel up to it. I don't want you to overdo it. We need your brain, okay?"

"Okay, I'm bored. What can I do?"

"To start, have Higgins bring charts to the wardroom and lay out our route to Midway. Let him take your star sights. His may not be up to your expertise, yet it's good training for him. You can check your positions together. Where your cabin may be too small for some of your normal work, use the wardroom. I'm sure Lego has a lot of paperwork for you. Actually, nothing of consequence has occurred since you've been *hors de combat*. Ready to start cracking?"

"Yes, sir!"


*18 November. Time 0819.* After a brief and uneventful skirmish we had with a plane, I wandered through the boat. Men were busy writing to their parents, sweethearts, wives, friends, siblings, uncles, aunts, et cetera.

I pondered the stack of letters addressed to me from many of their families. They would ask me not to inform their son or nephew that they had written, because he would be angry. Still, they wanted me to understand their feelings: So-and-so was very young and had a bright future ahead; would I please bring him back alive and unharmed? Submarining, understandably, involved risks, but perhaps they could be avoided by staying down deep. Did I have any educational courses he could take on his weekends off? He was missing schooling. Don't let him stay up late; he's grouchy when he doesn't get about nine hours of sleep.

Deep love abounded in these letters, as did the necessity to cut the umbilical cord. They had to understand that they had given the Navy a lad, and it was up to me to return a fully responsible man. One who might be ready to cuddle up to a mate. (The forward bases were devoid of female companionship. As the war progressed, *Barb* men returned to Hawaii every 200 days. There the ratio of unattached females to unattached males was about one to a hundred. The odds in combat were better for sonny.)

Pulling out a stack of letters, I settled at my desk. There came a gentle knock on the aisle panel. It was one of my younger crew members, Ted Lord.*

---

* The sailor's true name has been changed.

"It's a very private, personal matter, sir. I need your advice. I don't know what to do."

Drawing my curtain, I patted my bunk. "Sit down, Ted. Relax and unburden yourself. What gives?"

"It's this letter, sir, from my mother. Please read it. She has never written to me before in my entire life."

It was several pages long. She talked of her abiding love, how she longed to see him, his married sister's family, and how business was slow. She had talked with a Navy chaplain. He said Ted could allot half his pay to his mother in her dire poverty. It was his filial duty, being unmarried.

"So, what's the problem?"

"Sir, she's a prostitute. I don't know what she looks like. From what my sister has told me, we were bad for her business. When Sis was six and I was a little less than two, mother would take us along to the bars and saloons. Her system was to pander her all to support us. It had worked since I was a babe in arms until the customers got fed up."

"And then?"

"Then she dumped both of us in an orphanage and never visited us nor sent us anything, says Sis. Sis was adopted at age nine. No one adopted me. As I grew up, they treated me well. I went to school and helped around the place before and after school. Yet it wasn't home. I couldn't bring the other kids around, and so had few friends outside."

"Ted, tell me, when did you join the Navy?"

"I saw the poster 'Join the Navy and See the World.' That was for me, but the chief said age 14 was too young. Come back when you are 16. The next year—1939—with the limited emergency, the posters read, 'OUR NAVY NEEDS YOU!' Down I went to the recruiting office. The previous chief was off to the Murmansk run. I was interviewed by a real old chief from World War I who had been called back to active duty. He balked at my age—15. I explained that my home had always been at that orphanage. He looked at me, took off his specs, cleaned them, wiped his eyes, then ran my application up in his typewriter."

"And?"

"He kinda winked at me. 'Son, these old eyes act like a bay fog. I could have sworn your birth certificate was dated September 9, 1924, instead of 1923. You just got under the wire. Put that certificate away. You're 16. Now go down to the far door, where they'll give you a physical and a new home. As a boot apprentice seaman, when you get aboard ship, remember this. If it moves, salute it; if it doesn't, paint it with red lead. Keep your nose clean and serve your country well. It's a great Navy. You'll love it.' "

I picked up the letter again. "Ted, are you sure this is your mother?"

"My sister is sure and wrote that my mother had been inquiring about me, wanting money and my address. She gave her neither."

"Do you sense any love for your mother?"

"None, yet not knowing a mother, it's very upsetting. She never cared one scrap for me until I started making some money. Am I wrong? The Navy is the only home I've ever known. I do have a girlfriend in New London I may marry after the war. Mother would demean everything."

"Ted, I know you are right. I suggest that you not answer this letter. Would you mind if I answered it for you? She may take it better from me."

"Not at all, captain. Thanks."

"Many people have skeletons in their closets known only to themselves. Leave them there forever and let them die with you without casting a shadow over other lives. You are a top-notch sailor with nothing to be ashamed of—don't you forget it."

Holding Mrs. Lord's letter, I pulled out a sheet of paper and wrote:

Dear Madam,

Ted, your son, father unknown, has shown me your letter, practically demanding half his salary to feather your shoddy nest. Forgetting the greed, I sense a deep undertone in your writing of something you want to hide. Though you hate to admit it to yourself, buried inside your turtle-shell hide is a flame that will not die. Something that men don't have and makes them such bastards as you well know. This is the true, instinctive, eternal, maternal love. Feel it, let it out.

What you have done to ruin and make a shambles of Ted's life is unforgivable. Yet, out of this inherent morass of being not wanted, shunned by the friendless outside world, he has arisen. Mentally stable, morally straight, he has found a new mother, the United States Navy, and a new home within her hull. Above all, hope for a future.

I'll bet your bottom dollar that he will succeed inside or outside the Navy. He has a nice girlfriend from a good family, whom he plans to marry after the war. Being decent, your letter upsets him. If he acknowledges you in any way, it will never stop. Delving, he is not sure what he owes you. I say nothing, but you owe him something. I beseech you, don't snuff out his bright candle again. He can never explain you to anyone. Yet you can be a heroine to yourself. Atone for what you have done to him. Let him alone. Don't try to communicate. Leave him and his-to-be in peace, a peace I'm sure you've never breathed. This may be your everlasting moment of glory.

Good luck and God bless.

After signing this letter as the captain of the *Barb,* I thought of an old Scottish prayer: "O would some gift, the giftie gie us to see ourselves as others see us." I prayed that this depth charge I was sending off would strike home. (Neither Ted nor I heard from Mrs. Lord again.)

The skies clear, the *Barb* surfaced. To take a break from answering letters, I nosed into the wardroom. As only Jim and I had 12 x 18–inch desks, the other officers used our fixed mess table for all their reports and the never-ending censorship of letters. Dick, Tuck, and Tom were busy at that tedious task.

Flashing my letter at Dick—"I'm writing to the Lord"—I then tucked it in an envelope and sealed it.

"Hallelujah! I hope you put in a good word for me," remarked Dick as he pressed down his censor's stamp.

As lunchtime grew nigh, sea stories commenced. "Jim, what brought you into submarines?"

"Being in the armed guard in a merchant ship with a union problem. We had a 3-inch antiaircraft gun, a Navy gun crew, and a ready box of ammunition on the Murmansk run. A magazine below had plenty of ammo, but the crew refused to form an ammunition train unless they got extra pay. The captain had no authority, in spite of knowing that the ready box would be used up in one minute of combat. The union leader on board said 'Nix.' When the German Stukas attacked, the armed guard sat down. The crew screamed to fire, but there was no point with only the ready box. The ship ahead was sunk, the ship abeam afire. Lucky. The crew was no problem after that."

"It's too bad," I said, "that a few bum union leaders damn the good merchant ship crews. At Guadalcanal last year during the heavy fighting, Marine officer Stan Adams told me that one crew refused to unload a ship because it was Sunday. The Marines had a solution. They went aboard, threw the crew over the side, and unloaded the ship."

*24 November.* The "Nightly News" message from ComSubPac bore the sad news that we lost five submarines in October. The *Darter* had her compass knocked out and ran aground. The *Shark II* was sunk by Japanese escorts. The *Seawolf* was probably sunk by our own forces after a Japanese sub sank the USS *Shelton*. The *Tang* was sunk by her own torpedo on an errant circular run. The *Escolar* was probably sunk by mines. She had been ordered to contact us Loopers east of Tokara Strait to pass on information when we took over her East China Sea area. A no-show, she may have been forced down in the strait where the Japanese had laid 780 mines during the summer.

On the bright side, the *Queenfish* sank four ships, including a carrier, and the *Picuda* sank three ships.

In honor of the lost and the Loopers, four cases of beer went into the cooler, and we spliced the main brace quietly.

*25 November.* We arrived at Midway Island with our battle flag billowing and pennants for the ships we'd sunk on this patrol flying from the antennae. We received a tumultuous welcome with the band playing.

Presentation of medals for the *Barb*'s eighth war patrol, 6 December 1944, by Vice Admiral Lockwood to Commander Fluckey, Lieutenant Monroe, Chief Gunner's Mate Saunders, Chief Motor Machinist Mate Starks, Chief Pharmacist Mate Donnelly, and Gunner's Mate 2d Class Murphy.

*8 December.* I received the admiral's private, personal reactions and comments after he had studied our patrol report.

> *10 Nov.* — Attack [against the light cruiser *Gokoku*] well planned and executed. Excellent reasoning while delivering the "coup de grâce."

> *12 Nov.* — This attack against a large convoy of eleven ships and four escorts demonstrates in a most brilliant manner how night radar and periscope attacks should be made. A determined and beautiful battle. Surely "Fortune favors the brave!"

> *13 Nov.* — Another inspiring battle [on the three large schooners], and excellent gun practice.

*15 Nov.* — Hard luck on the Carrier. You deserved better. Please accept and express to your officers and crew my admiration for you and them.

Charles A. Lockwood, Jr.

P.S. Congratulations on all three of your swell jobs, Gene. The Presidential Unit Citation is already underway for patrols 8, 9, and 10.

Thus ended the *Barb's* tenth war patrol, my third in command.

The *Barb's* finest relaxing at Midway Island

# Chapter 16 Gooney Birds and
## Gooney People

A doctor at Midway came aboard and confirmed Doc Donnelly's diagnosis of Jim: it was angina pectoris. "Mr. Lanier is doing as well as can be expected," the doctor said. "I plan to Medevac him on the first plane to the Aiea Hospital on Oahu. He requires first-class treatment."

"I'm sorry, Jim. You've done a great job. I've recommended you for a Silver Star Medal. Be a good patient. Max will relieve you. Good luck."

I returned to the wardroom where Submarine Division 61 Commander Jack Broach, Commodore Edmonds, and I had been engaged in the customary ritual of going over our voluminous report that covered the *Barb's* tenth war patrol and considering major repairs to be accomplished in the next two weeks. I explained that my exec had had a heart attack near the end of the patrol, but I had wanted the *Barb* to continue as long as we had torpedoes. They were surprised and shocked yet pleased that Donnelly's diagnosis had been correct. A new exec would come from the pool of senior lieutenants at Pearl, though I was satisfied with Max Duncan. Tuck Weaver was ordered to new construction subs in Manitowoc, Wisconsin, and his replacement, Bill Walker, was present.

The old wooden Pan American Clipper Hotel had been taken over for officers' quarters. Of typical tropical construction, it had screens and storm shutters for windows, and twin beds and a shower stall between each pair of rooms were the luxuries. Bare boards with cracks and a broom made cleaning this one-story shelter simple. In one corner of the building a recreation room with a bar, refrigerator, and three or four card tables

213

served as a gathering point for all officers off the refitting subs. Beer, soft drinks, and liquor were available, self-served on an honor system.

Letting off steam, comparing strategems, tactics, and problems, and, above all, horseplay abounded until dark. After dark, blackout was in effect, and the necessary curtains cut off the cooling breeze.

The crews had barracks and a clubhouse built during the war, well separated from the officers. Softball was the most popular sport on the submarine base. Ships' parties and cookouts brought each sub together once a week. Swimming, fishing, volleyball, and horseshoes were other attractions.

As the brass debarked, both praising the *Barb*'s tenth patrol, I was asked to dine with them that evening. Then the relief crew took over responsibility for the *Barb* and her repairs.

The officers billeted in Gooneyville Lodge (the Pan Am hotel) assembled in my small suite to have a farewell drink with Tuck, who was flying to Pearl that afternoon. All were there except Jim, who had left with the doctor. Gibson, held up by relief crew work, was to be along in a few minutes.

The refrigerated beer was ice-cold, yet, to toast Tuck properly, we needed glasses. Jack Sheffield hustled to the bar to rustle some up. He returned with champagne glasses, which lent more dignity to the occasion. The glasses were filled with the sparkling, fizzy brew, and each officer offered a toast to Tuck. Then it was my turn to wrap up the farewell.

"Tuck, your outstanding service to *Barb* will never be forgotten. Fearless in combat, you've been my right arm on the bridge in surface combat. Your dry, Will Rogers humor has been a godsend in relieving stress when *Barb* was on the receiving end. You are the only shipmate who has had the experience of being sunk and lived to tell about it. All of us will miss you. Good luck and Godspeed. Gentlemen, by Act of Congress, raise your glasses to *Barb*'s indomitable Tuck Weaver!"

Dick had just arrived. "Well, what d'ya know—champagne!"

"Dick, I'll get you a glass," Jack said with a twinkle. "The bottle's in the toilet."

"The toilet?"

"It's the only bowl deep enough to cool it with ice."

We all maintained poker faces. Jack hurried back with a cold, bubbling glass.

"Dick, we're giving a farewell toast—bottoms up—to Tuck."

After he knocked it back Dick said, "Golly, it's good to have champagne again. Sure beats that beer we've been guzzling. Do we have a refill?" Jack obliged.

Sipping it slowly, Dick smacked his lips in appreciation, then burped. He looked at the rest of us cautiously. "Funny, this champagne has an aftertaste similar to beer."

The poker faces melted into gales of laughter as Dick flushed at his naiveté. "Guess I've been at sea too long. Anyhow, word just came down that Tuck has been bumped from his flight and has been rescheduled for 28 November. Also, I brought up all your mail." The session broke up as each of us was doled his personal packet and headed for his bunk to debunk the harshness of war.

Separating Marjorie's letters, putting them in sequence by postmarked dates, I tore open the first to lapse into the sweet oblivion of everlasting, ever-longing love.

> Dearest Gene,
> I feel like a perfect fool. How could you do this to me? I have never been so embarrassed in my life. Your deceit . . .

Boy, is she ticked off, I thought. I got up and went down to the bar for another cold beer to better concentrate on this unexpected problem. Something must have really upset her.

Settling down at a table this time, I carefully read every word. Tears moistened my eyes till the type blurred as I chuckled inwardly. My secret was out, the bubble burst. I could envision the wonderment on Marjorie's face when it broke: her denial, the resultant heehaws of the assemblage, her crimson blushes when it sank in that she had been had.

My beloved Marjorie had attended the monthly meeting of the Naval Academy Garden Club. In charge of the chapel flower arranging, she sat at the head table. Chitchat had ceased as they got down to business. Anne Brindupke burst in.

"Girls, have I got news for you, fresh from the Navy Department. Guess what? Gene Fluckey's just been approved for a Navy Cross for the *Barb*'s eighth war patrol."

Marjorie stood up. "Anne, that's impossible. He hasn't even been out yet. I get a dated letter from him at the most every three days or more often."

"And furthermore, *Barb* nine was an even greater patrol. They sank a carrier and lots more. The admiral says he's a shoe-in for a second Navy Cross."

"You must be mistaken, Anne; they've had all kinds of engine and battery breakdowns."

"Oh, my. Girls, sympathize with her. Why is it always the wife that is the last one to know? Honey, you've been had!"

With congratulations being showered on her, Marjorie was so embarrassed at her predicament that she left and went home crying.

After she exacted a promise from me to cease and desist putting phoney dates on letters, the rest of her letters were back to normal. My strawberry-blonde daughter was also ticked off about her dated letters. It was two against me.

After writing a promissory note to each, I went out to watch the gooney birds, a positively fascinating pastime.

The gooney bird—an albatross the height of a turkey, with a yellow-hooked beak—spends one year on Midway Island then disappears for seven years. Many are banded by the Forest Service, but no one knew where they go. Their population on Midway never changes perceptibly, however. Returnees from their long pilgrimage use the macadam streets as a runway, probably thinking they're water. They light with their webbed feet stretched out in front. As they hit, one can almost see smoke rise from their feet like from the tires of an aircraft. Then they tumble headfirst, over and over. After a few rolls, they stand up, groggily shake their head, and amble off into the sand and underbrush.

Before mating, they go into a sexy dance, squealing and squawking. Though they have the dark wings and white chassis of a seagull, their goslings are the size of a goose and pure cottony white. Walking as if drunken, gooney birds have no fear of man, but they scream if anyone picks up their offspring. For reasons unfathomable, the males or females often stand around in groups of six or seven discussing the war.

Dinner with Shorty Edmonds was all war talk. We had lost three more subs this month—the *Albacore, Growler,* and *Scamp.* Good friends, all. So far, 44 subs (with over half my submarine school classmates) were on their eternal patrol. Yet, we knew their deaths were not in vain. The attrition of shipping was definitely strangling the Japanese Empire.

Being a bit overwhelmed by the kudos for the *Barb,* and the barrage of questioning regarding our tactics and strategy, new to submarines, I begged off at 2200. Then I stretched my legs, walking back to the Gooneyville Lodge in the full-moon light.

Tired, I brushed my teeth by moonlight, undressed except for my shorts, and thrust my feet into the full-sized luxurious twin bed. HORRORS! Something alive was in there! To hell with the blackout! Turning on the light, I ripped back the top sheet and counterpane, only to guffaw, "Those wonderful bastards, always ready for a frolic or a fray." I hauled out two very comfortable, wide-eyed goslings embedded happily between the sheets. Opening a screen, I dumped the tender gooney birds out onto the sand. Honestly, what was I going to do with a crew like that!

Broach dropped by at 0730 after reading the morning messages. Shorty Edmonds had been relieved of his command and would depart that day. Captain John "Dutch" Will, replacing him, would arrive the next day. Jack would be at sea on a sub undergoing prepatrol training, so, as senior skipper, I should meet Dutch's plane at 1600. Also, two days hence ComSubPac wanted the *Barb*'s commanding officer to catch the plane to Pearl for a few days of staff discussion. I could see my recuperation period was going to be busy.

I spent the rest of the morning at the cable station, discussing with its boss his equipment and its worth. The incoming lines from the United States divided at Midway and then spread to various parts of Asia. The value of the equipment was well over two million dollars. Naturally, he was curious as to my interest in his classified operation. I explained my problem in trying to knock out the Japanese cable station in Kunashiri Strait at Kushibetsu, Etorofu Island (the Kuriles). Twice I had been unable to bombard it because it sat in a fog pocket.

The manager was cleared for ULTRA intercept information. I informed him that it was a rarity for the subs patrolling the Okhotsk Sea to receive ship movement information. Thus I felt sure their movements were being passed by cable. With that station bombed out, more ships would be sunk. He concurred and stated it would take months to manufacture the necessary material to place it back in service. Some day, I said to myself, I would get that station.

Jim Webster, the *Barb*'s new executive officer, arrived just after lunch. He was a naval reserve lieutenant from Fountain City, Tennessee—husky, affable, quiet. I put him to work as soon as he unpacked. Walker, replacing Tuck, also reported in.

Bill was from Columbia, South Carolina, Max from Forest City, North Carolina, and Tom King from Nashville, Tennessee. Dave and Paul came from the West Coast, while Dick and I hailed from around the nation's capital. Jack from Buffalo and departing Tuck from Illinois rounded out the demography. In volume, however, the confederates permeated the *Barb*. Our heavies were Jim, Max, and Bill.

Taking the newcomers around to meet the crew, I informed them all to do a bit of sprucing up in the morning. A new commodore would arrive on the morrow and probably inspect their quarters. The following day, I would go to Pearl for staff discussions concerning the three men who had put the gooney birds in my bed. Swish, Houston, Hatfield, and Broocks burst out laughing and tried to hide—I had found the culprits. There would be no punishment because I needed men of such astuteness to form a boarding party under the command of Tom King.

The following morning at 0700 these men would report to Captain Stan Adams, who headed the Marine contingent, to commence their training with hand and rifle grenades as well as hand-to-hand combat. Cheers resounded as men offered to buy the selectees' places on the boarding party. As we left I realized how unsuspecting I had been. Now I'd have to check my bed nightly for gooney birds placed by others trying to join the party.

Marine Captain Adams had set the number at 15 so we would have replacements for expected attrition. In conference with the other officers and leading men, Max made up a list, which I accepted.

Borrowing the division commander's jeep, I met flinty Dutch Will. Handing his luggage to the steward, he wanted first to see Gooneyville Lodge and the recuperating crews' quarters. The moment we entered the submarine side of the base, we parked. He wanted a walking tour after six hours in the metal bucket seat of the cargo-passenger plane.

I was spruced up with a cap, long-sleeve shirt and tie, and long khaki trousers; Dutch had on a short-sleeve shirt and khaki field hat. Chatting as we walked alongside the lodge, we were brought to a shuddering halt. A body holding a sloshing beer bottle came hurtling out a window, carrying the screen with it. It was Dick Gibson, dressed only in khaki shorts and sandals. Picking himself up, he brushed off the sand and grinned at us. "Those bastards really threw me out the window, didn't they? I didn't think they were serious." Then he went to the front door, carrying the screen, and went back in. We were struck dumb for a moment.

"Skipper, does this happen here all the time?"

"It's the first time I've ever seen anything like that. It must be a little horseplay or letting off steam."

We entered and went directly to the room, knocking on the door. It opened a wee crack and Tom peeked out, then flung the door open wide as the officers jumped up to a rigid, frozen attention. Dutch looked at them sternly then snapped, "Well, isn't anyone going to offer me a cold beer?"

That broke the ice. Jack pulled out a metal waste basket he had shoved in the closet and uncapped a libation. After a swig, Dutch remarked that he had never seen anyone thrown out of a window before. He didn't think we should make a practice of it. Someone might be injured and miss a patrol. Then he launched into sub stories of going up to China from Subic Bay in the springtime, until he finished his beer.

Away we walked to the crews' quarters. Knocking on the door, we could hear things being pushed under bunks. Swish opened the door and shouted, "Attention!" The men stood stiffly, facing Dutch. Grimly, he looked from one to the other in the hushed barracks. "Well, isn't anyone going to offer me a cold beer?" Five appeared in as many seconds. His prewar days performance was repeated, springtime at Tsingtao and on the Whangpoo River at Shanghai. He made a hit with everyone.

As Swish said, "Underneath that gruff exterior is a happy submariner just like us."

Tuck and I took the same plane to Pearl. How I hated to lose this rare gem, as great an officer as any I've known. Yet nine combat patrols merits a blow, and he'd be home for Christmas. This thought brought bits of envy. Jack, his replacement, was his equal only in enthusiasm, and hadn't been tested under fire. We said our farewells. I had the driver take me to Aiea Hospital. Jim Lanier was coming along, but would be physically retired, the doctor thought.

After a night's sleep in the Bachelor Officers' Quarters, I went to work with various staff members. Dick Voge, a wizard in operations, said that the admiral wanted to keep the Loopers together as long as possible. The *Queenfish* and *Picuda* had been sent to Guam for refit. When the *Barb* was ready, she would join them there to start on the next patrol.

"Gene, do you still want a hot-spot area?"

"Absolutely."

"Well, I've got one that's hotter than it's ever been—the Formosa Straits and the southern East China Sea, right where the *Tang* was sunk. It's tougher now: the Japanese know we're preparing to land somewhere in the Philippines. They'll make a major effort to reinforce their positions. The Loopers' job will be to bottle up the straits and stop supplies and warships from getting through."

"Perfect."

"Okay. Now for your future. When you finish this patrol, you'll have finished the four patrols the psychiatrists claim is maximum. How about a job with me here in operations?"

"Thanks, Dick. I'm complimented, but there is much I still want to do in command. Let's see how the war is going."

"Gene, give up the thought of a fifth patrol. At your red-hot pace no one will let you stay on. The Board of Awards has some good news for you, so take them next. Good hunting."

At the Board of Awards the reception was ecstatic, even though the *Barb* had given them more work for three patrols than any other sub. Admiral Lockwood wanted to speed awards up, as Admiral Nimitz was pressing for them.

I laughed remembering how I met President Franklin D. Roosevelt—

with Admiral Nimitz and General MacArthur—in the back seat of his limousine at the Royal Hawaiian after the *Barb's* eighth patrol. Admiral Lockwood had given FDR our war patrol report the previous day for his bedtime reading. At our meeting, the President told Nimitz to send every *Barb* report to him from then on. Next, he wanted the *Barb* to reenact her return to Pearl with battle flags flying as he filmed some home movies.

The next morning Jim Lanier, Paul Monroe, and I—with Chief Brendle and a few key men to help the relief crew—had gone back to the submarine base to prepare the *Barb*. She had been moved from the refit piers to the dock behind a loaded ammunition ship in Magazine Loch. This cleared the *Barb* of the general background clutter per Roosevelt's instructions. With battle flag flying and five pennants on the antenna for the ships sunk, the *Barb* was ready. Admiral Lockwood was on hand, acting as messenger boy for the President. He would signal us when to come alongside. The *Barb* backed out in the stream as the President arrived, shifted to his wheelchair, and checked his movie camera.

On signal, Jim brought us alongside with our men at quarters. After a brief confab, Admiral Lockwood hustled over to us. "The President wants more action; bring her in faster." Again Jim brought us in at two-thirds speed. Another confab. "Gene, the President wants you to come in faster, so your flags are billowing out."

"Admiral, that's pretty dangerous. If we lose power, we will ram that ammunition ship and damage our bow."

"Never mind. I'll be responsible." The *Barb* backed clear.

"Jim, I'll take the conn, now." I was angry. Damage might delay our next patrol. The President dropped his hand for a "take."

"All ahead full! Look lively with those lines." Admiral Lockwood was tense, but our battle flags were billowing. As the *Barb* arrived, I shouted, "All back emergency! Left full rudder!" The sub shook and shuddered as her rear end foamed. The admiral's face was white. The *Barb* stopped alongside, 10 yards aft of the ship.

"All stop! Throw over all lines—double up and secure." The President was clapping his hands as he passed the camera to his chauffeur. "Skipper! Absolutely perfect."

As I crossed the brow to greet the President, Admiral Lockwood muttered, "Don't you ever do that to me again."

"Admiral, then don't ask me to endanger my ship again."

A short chat and everyone was happy again as Roosevelt shoved off for his vital strategy meeting with Admiral Nimitz and General MacArthur.

Ed Swinburne, the senior member of the Board of Awards, banished my reverie. "Gene, you look like you're on cloud nine. Come down to earth, and I'll clue you in." To my surprise, the board was already at work on my recommendations for the recent tenth war patrol; they would be approved. For the three patrols, the awards approved by Admiral Lockwood for the *Barb* were the following: 4 Navy Crosses (one for the wolfpack commander on board), 9 Silver Stars, 12 Bronze Stars, 4 Navy-Marine Corps medals, 26 Letters of Commendation with Ribbon, and 4 Navy-Marine Corps medals and 8 Letters of Commendation for the British-Australian prisoner of war rescue.

Furthermore, the board was working up the Presidential Unit Citation recommendation for the *Barb*'s eighth, ninth, and tenth war patrols.

This *was* good news because all of the men would get the Ribbon with a Gold Star. I had tried so hard to spread the medals and letters around to all sections, without repeats, except for the executive officer and the chief of the boat.

Everything was coming up roses. I went for my physical, and there, too, all was normal, except I had lost 12 pounds and was told to eat more. I assured the doctor that I would gain them back during refit, and that on patrol we consumed a tremendous amount of cake.

The session with the admiral was private and of great interest. Admiral Nimitz and he would be moving to Guam in January with their principal operators, leaving their staffs at Pearl. Echoing Captain Voge, he said he wanted me on his staff after our coming patrol. He could see I was not happy about this. He reasoned four patrols in command was enough psychologically for any skipper. After that he either became overcautious, and did little, or over-cocky, and his boat would be lost.

"Admiral, as I start out on each patrol, I feel as if it's my first. I'm healthy, eager, full of new ideas and plans to shorten this war. With my apprenticeship in command so far, it's obvious I can do more in *Barb* as I develop. Look at the changes you've made in submarining since our fallacious peacetime exercises as the war has progressed. From token sinking for prestige to wolfpack teamwork for strangulation, the war develops further submarine concepts. For instance, it may be better to put ships out of commission by damaging rather than sinking."

"Explain that."

"Example. Forty-four of our subs have been sunk. What would happen to your tenders and repair yards if you had them all back, holed and partially flooded? Consternation all around."

"True, but . . ."

"Admiral, please. Let's make an agreement based on the results of the coming patrol."

"Has Voge mentioned the Loopers job?"

"Yes, sir; bottling up Formosa Straits to stop supplies and warships supporting the Philippines. Now, provided *Barb* satisfies you in this, I hope you will permit me to have a 'graduation' patrol. *Barb* is due for a two-month overhaul at Mare Island. I'll have time to develop harassment."

"Gene, would you like that patrol from Australia?"

"No, sir. You've offered me 'down under' patrols before. I don't like the way they operate, far too much communicating and directing movements constantly from the home office. Each patrol area has its own ambient of weather, opposition, and possibilities. With study of what my predecessors did in my assigned area, and full intelligence support from home, I believe your system of freedom of movement is far superior."

"Where do you want this graduation patrol?"

"I'll need to talk to Jasper Holmes when the time comes to find out where the remnants of the Japanese fleet are holed up. How far we have progressed in advancing our bases for the final onslaught on Japan will be a factor. I have some ideas that haven't quite jelled—to strike and strike hard, forcing a diversion and occupation of enemy forces far out of proportion to the effort or means involved. It will be a new angle to test in submarining as shipping thins out. I need your approval."

"You have it, if you're in good shape. I've mentioned to Admiral Nimitz *Barb*'s Presidential Unit Citation for patrols 8, 9, and 10, now that we have confirmed the sinking of the carrier *Unyo*. He concurs, but keep it under your hat. Good luck and good hunting."

"Thanks, admiral. Aloha."

*10 December.* Back on Midway the two-week refit ended, and testing and loading commenced. Only Sunday afternoon offered a holiday routine. I sent my new exec with Doc Donnelly to draw the medicinal whiskey or brandy from the hospital for our depth-charge ration—and I learned how the Confederates raided the stores behind the Union lines in the Civil War. Jim came back with our normal ration and three cases of champagne. "Sir, a new supply officer mistakenly sent a shipment of champagne to Midway instead of the stores in Pearl Harbor. I bummed enough for half a fifth per man to celebrate our first sinking. Special services is sending down our picnic ration of beer."

"Jim, you have the makings of a great exec. Stow away!"

After four days of underway training, the *Barb* was ready. Last mail came in and went out. I received packages from a few families so I could play Santa Claus on Christmas Day. A bundle of letters went to Marjorie and Barbara.

*19 December. Time 0800.* With the entire crew at quarters, I spoke to the men. "We're off to Guam. The Loopers will ride herd again together under Captain Loughlin, area unknown but important. *Barb* will be in there pitching. Now, Lego, hand me *The Mast Book.*" Holding it on high, I said, "New shipmates, we have had no punishments since I took command. Sense our pride in *Barb!* One respects her or leaves. *The Mast Book* is no longer needed." I threw the book overboard. "We're under way at 1430. Dismissed." All eyes watched the book slowly sink.

Part **IV** The Eleventh War Patrol

of the USS *Barb* in the East China

Sea, Formosa Straits, along

the China Coast to Wenzhou,

20 December 1944–15 February 1945

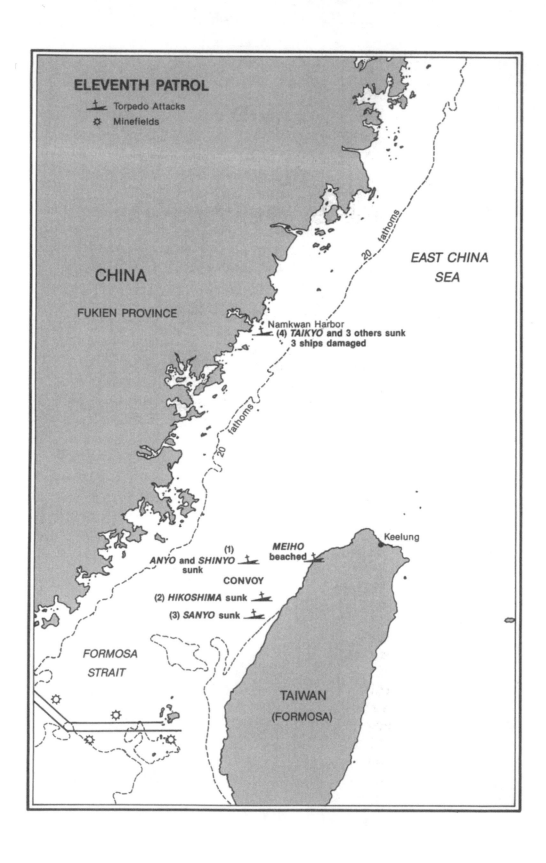

ELEVENTH PATROL

⊥ Torpedo Attacks
✿ Minefields

CHINA

FUKIEN PROVINCE

EAST CHINA SEA

20 fathoms

Namkwan Harbor
(4) *TAIKYO* and 3 others sunk
3 ships damaged

20 fathoms

*MEIHO* beached ⊥
Keelung

(1)
*ANYO* and *SHINYO* sunk ⊥

CONVOY

(2) *HIKOSHIMA* sunk ⊥

(3) *SANYO* sunk ⊥

FORMOSA STRAIT

TAIWAN (FORMOSA)

# Chapter 17 A Jewel for Christmas

*19 December. Time 1430.* All aboard, the *Barb* took off through a hailstorm of coins thrown by our own gooney crewmen for a safe return. Once clear of the Midway channel, I called down for my shorts and sandals and stripped off my formal attire. Awaiting my shorts, bare assed, I watched Midway recede. I cocked my head when I heard Novak, the port lookout, stage whisper to Arthur on the starboard side. "Psst—the Old Man's a real ginger, isn't he?"

"Knock it off, Novak! You're up there to look for periscopes." Snickers. "And no further remarks about that either." Such characters made me feel like Gypsy Rose Lee.

As I sat on the upper-bridge platform with one foot on the bridge rail, the *Barb* swayed on the swells following the course clock. This was a new gadget. Setting base course southwest, we could program either a zig plan or a constant helm plan into the clock. The helmsman thus stayed on the mark southwest, or 225°, yet the ship wandered back and forth across this course on a timed sequence. It was most convenient, for the navigator would know precisely what actual speed was being made along the base course line of advance, dependent upon the zig plan input.

The twentieth of December was a dead loss. Navigator Jim and his assistant, Quartermaster Higgins, simply wiped it off the calendar as we crossed the International Date Line. Because we would follow a holiday routine on Christmas, we had only five days to hone the *Barb* to maximum combat sharpness before we arrived at Guam to sortie with the Loopers.

Daily dives, battle surfaces for gun shoots, tracking and fire control party drills, anticipated casualties of every conceivable nature consumed our waking moments as well as our sleeping ones. (That surprised our eight new unqualified men, who imagined the night was made for sleeping.) The gongs, a-oo-gas, drills for chlorine gas and electrical and oil fires, and rigging for depth charge, silent running, and collision startled them awake. Practices even included evading enemy torpedoes, or our own circular runs, and being illuminated or shot at with guns. If a man in an essential position was eliminated, his substitute was ready to step in. No one was indispensable except the *Barb* herself. Tom King's boarding party had been well trained by the Marines at Midway and was ready.

During the various stages of training, I wandered around from room to room, observing the verve and seriousness with which instruction was being given. Each man realized his one mistake could sink us all. Better ways to accomplish things and good ideas could come from bottom to top. No enmity, no professional jealousy, no incompatibility existed; the *Barb* melded all into one proud team. Even after a second advanced base refit, and in spite of a long siege of shallow-water combat operations, this team was captivating. They knew that this patrol would be followed by a period of recuperation during navy yard overhaul in California. Yet, all were going into it with a determination and an inspired zeal for action that sent tingles of pride running up and down my spine. I knew I would have to cudgel myself to keep pace with them. I vowed I would not let them down. Drive yourself and lead others was my philosophy.

At supper time on Christmas Eve, training paused. Someone had drawn a Christmas tree on the back of a chart and hung it on the bulkhead in the crew's mess. Others joined in and drew ornaments. A spirit of longing cheerfulness pervaded. Would that we could have Peace on Earth, Good Will toward Mankind.

In my cabin, I unwrapped a number of packages for the men sent to me by their sweethearts and families as a surprise, asking me to slip the Christmas present under our Christmas tree. The more I thought about our tree, the more it came alive. Its spirit grew until it became as precious as any tree in the whole world. I stopped, realizing the young child in me still wanted to believe in Santa Claus.

I opened a letter from Augusta, Kansas, dated 14 November 1944:

> Dear Captain of the Barb,
>
> Although I've never had the opportunity to meet you, you do seem so near to me. I feel you mean so much to our son, Glenn Lawrence Koester, and all those that are with you. I wanted to send this record as his little sister, Eva Jewel, sang the songs for him. We thought it would be sort of

a surprise. Would you play it on the phonograph? The surprise is for him, for you, and for all the Buddies in the Barb.

She hopes you'll all get some enjoyment from knowing she thinks and prays for the whole crew each and every day. We feel like you Boys are the grandest of the sea. I know you are one grand Commanding Officer. Our son says you are one of the swellest guys that's to be found. I know he means it.

We look forward to the day when Glenn and each one of you can come home. We hope it won't be too long.

Thank you so much for this favor. God bless you and your Boys.

Yours sincerely,
Mr. & Mrs. Otto Koester
and Eva Jewel

Holding the record, I reread the letter. This time I read the love and yearning that lay between the lines. Thank you Mr. and Mrs. Koester and Eva Jewel for your precious gift. The *Barb* now had a Jewel for Christmas.

Up on the bridge I gave my night orders to Max and left them in the conning tower for the oncoming watch to read. The night was dark, cloudless, hushed, awaiting moonrise. There was no conversation. The watch mechanically carried out their exacting duties protecting the *Barb* from her enemies, while inwardly imagining Christmas Eve half a world away. The Big Dipper spilled its brew, Orion brandished his sword at the twins Castor and Pollux. The brightest stars were Sirius and Cappela. I wondered what would have happened if Pontius Pilate had fined Jesus instead of nailing him to a cross. Would the mushroom that started in Bethlehem and refused to die have spread faster or not? The *Barb* buried her nose in a swell and came up spuming, soaking the watch. I turned in.

*25 December. Time 0700.* The crew's mess was filled with jovial breakfasters. Entering quietly like one of the wise men with my jewel, I was stopped by Phillips, the cook. "Captain, try one of my hot cross buns."

Price, an electrician, waved a letter in the air. "Don't worry sir, it's Madeline's recipe."

Tasting it, I told him, "It *is* good; bravo Phillips!"

As I munched away, Koester entered with a polite "G'morning, sir," gave his order—"scramble two, soft, and ham"—and sat down with his back toward me. Someone passed him a cup of joe and the tin of evaporated milk. Eva Jewel's hour had come.

Glancing at the Christmas tree, I slipped the record onto the record player. Glenn sipped his coffee. Soon the mess was saturated with the angelic, soprano voice of a child. "It came upon a midnight clear, that glorious song of old. . . ." Glenn put his cup down and cocked his head in disbelief. "From angels bending near the earth, to touch their harps of gold." Swinging around he shouted, "It's Eva Jewel!—my little sister— singing." In the silent mess, I nodded. Glenn listened, unashamed of the tears moistening his cheeks.

To take the attention directed at him away, I read his mother's letter. This precious jewel was for all in the *Barb*. They were touched. The first carol finished, such thunderous applause erupted that the control room chief stuck his head in questioning the commotion. "Silent Night" stilled the mess as I went forward to write the whole Koester family, for I had also had letters from his uncles and aunts.

At 1100 we started celebrating Christmas with carols and festivities, including our picnic ration. Various tenors tried out over the public-address system. Don Miller was in good voice until the bridge command switch cut him off.

"Captain, floating mine dead ahead, 300 yards."

"Circle it, Dick. Away the boarding party, weapons only. Let's get some live practice."

After 10 minutes of shooting, the mine, though hit, was still afloat. I called Buel up to turn the twin 20-mm guns on it. Without exploding, the mine sank.

Returning to the picnic ration, the cooks—Bentley and Phillips— brought out their beautifully roasted turkeys with all the trimmings. Baker Elliman followed, surprising everybody with strawberry shortcake with real whipped cream.

At night, in my cabin, I unwrapped my present from Marjorie and Barbara, a silver Dunhill lighter inscribed "Love." Tenderly, I held it in my hand and felt the warmth of everlasting Christmas, the basic desire of humanity for world peace.

*27 December.* The *Barb* eased into Guam; the Loopers were united once more and went into a huddle to plan the coming patrol. The submarine tender *Sperry* and Commander Submarine Division 101 were gung-ho in their welcome and eager to provide us with voyage repairs and supplies. Two days later Loughlin's Loopers eagerly left Guam. Outside, we formed a scouting line 10 miles apart, headed for the Tokara-Strait entry into the East China Sea near Kyushu.

*1 January 1945.* Many stayed up to see the New Year in. The champagne was in the cooler awaiting our first sinking. Ceremoniously, I opened the new ship's logbook and wrote, "0000 Item (Tokyo time). Celebrated the advent of what we hope is the final year of the war."

Of those awake, most were motor machinists working hard. Three of our four main engines were out of commission. They badly needed the forthcoming overhaul. Chief Williams—with Whitt, Howard Peterson, Wearsch, Houston, Jesse Penna, McKee, Zamaria, and Custer—had been working around the clock for 30 hours. Our supply of spare parts was being used up fast. We were now three hours behind our scouting-line position.

By 0400 two engines were on line, but we needed one to charge batteries. By 0600 we had three engines working. Though four hours behind, we ceased losing ground. Breakfast rolled around with all four engines rolling. At full speed, it would take us 12 hours to catch up; at flank speed, once the battery was charged, 8 hours. I stayed on the bridge to avoid unnecessary dives.

*Time 0847.* "Bridge, radio. *Queenfish* reports contact on a patrol boat. She invites all Loopers to rendezvous for a gun shoot."

"Send her, '*Barb* regrets. Sixty miles behind. All engines now in commission. Thanks.' " What a letdown.

The *Picuda* joined the *Queenfish*. Another message said, "At 1030 commenced coordinated attack." Disappointment. The *Barb* was now hustling along, with her skirts pulled up, at flank speed. At lunchtime, there was no further word; we assumed the patrol boat was sunk. Our wolfpack members had lost at least two hours, so they should be only 25 miles ahead.

*Time 1233.* "Captain, we've picked up a thin column of white smoke dead ahead." Ten minutes later, Quartermaster Bluth on the high periscope had the patrol boat in sight. All fires were out.

"Jack, ask *Queenfish* for information."

Her reply: "Sink target at discretion." Unbeknown to us, the two subs had set the target on fire, then were forced by a plane to dive. Much later they surfaced and cleared the area.

*Time 1305.* "MAN BATTLE STATIONS GUNS! Four-inch and twin 20-mm guns only. Boarding party stand by with full equipment." Looking the ship over carefully, I judged her to be about 300 tons. She had been hit in the bow with 4-inch shells. Automatic weapons had sprayed her with the usual lack of effect. One fire had been started forward and another aft. Having put the fire out, the boat's personnel were hiding. Her hull, bridge, and engine room were unaffected, since no flooding had taken place. I decided to board.

"Dick, take her alongside. Send Walker and Epps up with grapnels. Max, keep your automatic weapons on all exits."

*Time 1318.* "Away the boarding party!"

Tom King and his swashbuckling group of modern pirates jumped aboard. The ship had two decks. Watching the party captured my entire attention. They were doing a magnificent job of tearing things apart and stripping the ship of everything worthwhile. There was a darkened crew's bunkroom aft. Not having flashlights, the party closed the door without entering. They filled bags with charts, professional books, codes, binoculars, a radio transmitter and receiver, rifles, compasses, barometers, signal flags, the Japanese rising sun flag, and other loot. Everything was passed over to our gunners.

From my bridge position I could monitor their work in progress on both the upper and lower decks. Jim Richard, stripping something with a crowbar down below, heard footsteps up above him. Thinking it was a Jap, he came out with his Tommy gun. "Shivers" Houston above heard footsteps below him. Thinking it was a Jap, he snuck out to get him. I hooted and hollered "DON'T SHOOT!" as they poked their guns in each other's faces.

*Time 1328.* "Boarding party, two-minute call!"

After bringing the rest of the loot on board, we noted that the patrol boat was armed with machine guns. A quick look at the bags of loot proved her to be a naval weather ship.

Since we had sighted this ship almost an hour ago, it was possible that her crew had radioed for help on seeing the *Barb*. It was also certain they had on seeing the other subs earlier. Thus I decided against sending the party back to pick up a prisoner. If any had surrendered we would have taken them, but as they were probably armed, the individual danger was not worth the risk.

"Let go the grapnels! All back two-thirds. Max, at 40 yards start shooting 4-inch gun only."

One hit lighted off the engine room fuel tanks, turning the midships section into a blazing inferno. Thirteen shots for 13 hits and she sank stern first as we cleared the area. We took Kodachrome movies of the event.

*Time 1930.* "Splice the main brace!" The cake was ready, a combination of several recipes from home. Elliman outdid his previous masterpieces. They had the *Barb* shooting maraschino cherries, with whole strawberries for hits. The ship was depicted sinking stern first, with crisp bacon colored with saffron for the flames and the inscription—JAPPY NEW YEAR 1945!

After one rousing chorus of "Auld Lang Syne," the corks started popping as the champagne was opened. With our glasses full, toasts commenced, but after a few I sensed a letdown. Many men refused a refill; they didn't like champagne. Beer, yes.

# Chapter **18** Corks in the Bottleneck

*3 January 1945. Time 0400.* We transited Tokara Strait hell-bent for leather at flank speed to get clear before dawn. Normally, it required about four hours to ease through this well-watched strait, avoiding patrol boats and radar coverage. With a potent minefield below, submerging would be a last-ditch resort.

Because we were in two other wolfpacks' areas Elliott ordered independent submerged patrol for the day en route to our own. We had to stick to our entry route, for another wolfpack operated along both sides of Okinawa and the Ryukyu Islands. Just west of them on the long axis of the East China Sea lay a vast field with over 15,000 mines, held by intelligence. It was over 500 miles long and divided the area.

Being some five hours astern of the Loopers, the *Barb* took this opportunity to catch up and patrolled on the surface.

*Time 1135.* Just as I was winning a game of Acey-Deucey from Max before lunch, Radioman John Delameter stuck his nose in. "Sir, radio from Naval Group China, convoy near Danjo Gunto, course is 240 true, speed 12 knots." Quickly, the officers gathered around the plotting table in the control room as Jim plotted the respective positions. Twelve would be a top speed for a convoy; the course would be toward Shanghai.

"Paul, tell Ragland to bring a sandwich and lemonade to the bridge. We better eat now. The search is big."

A blinding, bright day unfolded. This would help our search. "Dick, kick her up to flank speed, course west." After two hours of a high-speed, wide-swinging search plan, Lego spotted a ship on the horizon. We had contact! The ship kept coming alone, however, so we secured the tracking party.

We picked up U.S. radar interference. The high periscope watch said it was a submarine. "Exchange recognition signals." Reply—the *Sea Poacher.*

What a pleasant surprise! One of my classmates and very best friends, Frank Gambacorta, was her skipper. We had been shipmates on the *S-42* in Panama. When we closed and exchanged messages, he was searching for the same convoy. We parted, going in opposite directions.

An hour and a half later, Torpedoman Arthur, a lookout, reported a ship barely visible astern. Again radar reported U.S. interference. Again it was the *Sea Poacher.*

"Bridge, Radioman Wallace Lindberg with a message from *Sea Poacher:* 'We think you're going to find something. Do you mind if we tag along?' "

"Lindberg, send back, 'Not at all. Your position five miles north of us would help. Good hunting.' "

Shortly thereafter a drizzle moved in. A plane forced us down for half an hour; the *Sea Poacher* was gone when we surfaced.

At 0400 the next day we gave up, having covered every possible course and speed of the convoy. I had wanted so much to have the opportunity to work the convoy over jointly with a stout friend and teammate. The *Barb* had been so lucky so often in the past. I never minded sharing our luck with others, knowing, as a skipper, the empty feeling of searching for days and days with zero results.

For the next several days in our individual areas, bottling up the straits, traffic was nil. Each sub got in some gunnery practice, sinking or exploding floating mines that had broken loose from the minefields. Some 15 were sighted. These were perilous for they were active, and radar rarely picked them up.

*6 January.* With the *Barb* patrolling south along the 20-fathom curve off Wenzhou, the Loopers received a report from Chinese aircraft that a convoy had left the straits heading for Shanghai. The wolfpack searched for 12 hours, yet nothing passed through. Something seemed fishy about these contact reports from Chungking. The flyboys couldn't be that far off in their navigation or time. The Loopers were keeping Zone How time— one hour later than China and one hour earlier than Japan. All Japanese ships always used Tokyo time, Zone Item.

*7 January. Time 0540.* "Captain, radar contact 20,000 yards. We're turning to head for them at full speed."

"Coming up, Dave. Station the tracking party." The weather was rough and nasty, the sky overcast. A quick glance at the radar scope and Lehman told me that the contact had been confused with the Tungying Island group. Gradually, it was separating into individual ships. "Jim, as soon as you can count the ships, send out a contact report. Looks like the blob is heading for Keelung (Formosa or Taiwan)."

*Time 0803.* The *Picuda* had made contact; *Queenfish* couldn't arrive before the convoy reached Keelung. "MAN BATTLE STATIONS TORPEDOES!" We submerged 14,000 yards ahead of the convoy in heavy seas, and for an hour we were handicapped by rain squalls that hid the targets and ruined sonar conditions. The seas made depth control poor. Trying to approach on the pinging of the destroyers and intermittent light and heavy screws was worth little. I needed a visual observation but got none even with 10 feet of periscope out half the time. We could not control the *Barb* at radar depth.

*Time 0902.* We made ready bow tubes. "Sonar reports two sets of heavy screws crossing ahead."

I swung my periscope wildly; nothing in sight. "Visibility appears to be about 4000 yards. The haze is lifting. Jim, setup on a large tanker. Bearing— mark. Down scope, angle on the bow 130 starboard, range 850 yards. There's a large freighter 1500 yards beyond and five destroyer escorts, two stack, with white bands around the stacks. Use speed 11 knots, the same as tracked."

*Time 0905.* "If he zigs toward us, final setup and shoot. Up scope. Darn the luck, he's zigging away. It's too poor a shot and not worth it to alert the convoy to the *Picuda.* Down scope. As soon as they're hull down, we'll surface for another end around."

The *Picuda,* meanwhile, had submerged 17,000 yards ahead of the convoy. Hank Sweitzer, the exec called out, "Sonar is very poor."

Shepard added, "And visibility stinks, 3000 yards."

In half an hour nothing was seen; two destroyers passed overhead, pinging. Later, sonar heard two sets of screws.

"Up scope. Here comes a tanker out of the haze. Shoot four bow tubes. Bearing—mark. Angle on the bow 30 port."

"Range 2800 yards checks; speed 11 knots. Shoot now."

"Up scope. Final bearing—mark. FIRE!" Swinging around as four torpedoes were going out, Shep called out, "Hey! I've got a freighter at less than 500 yards coming in to ram. Take her deep!"

WHAM! WHAM! Breaking-up noises commenced, immediately followed by 14 depth charges.

The tanker, the *Munikata Maru* (10,045 tons), was heavily damaged and sent to Keelung escorted by destroyers.

At 1115, with unlimited visibility, the *Barb* now saw the trailing frigates and a few masts of the convoy change base course to 150, heading directly for Keelung. It was impossible to beat them to Formosa. Disappointed as the trailers vanished, we surfaced and turned tail back to the China coast. We took some consolation in the *Picuda*'s having torpedoed the tanker.

Unfortunately, we had no information on this convoy, and visibility, speed, and the broad separation of ships prevented even a radar estimate of its size. Postwar records show it was the HI-87 convoy bound for Singapore: 10 tankers, the escort aircraft carrier *Ryuuhou,* the Navy ship *Kamoi,* 4 destroyers, and 6 frigates. Without weather information or unlimited visibility, we never even dreamed of what actually followed. All of the Loopers would have been positively devastated had we known.

HI-87 bypassed Keelung and headed for Takao (in southern Formosa). Running into a heavy fog bank in the straits, they anchored in the open roadstead at Shinchiku, Formosa, 10 miles north and well east of the 2200 mines laid at 13-foot depths in May 1944. We could have attacked the convoy at night, submerged or on the surface. This was spilled milk over which we did not cry, however.

In the evening Elliott ordered the corks for the bottleneck to sweep to the northeast, paralleling the China coast shelf. I requested the shallow position along the 20-fathom curve, determined to put the stopper in the next convoy. My request was granted.

Sitting on the bridge after dark, I cogitated on the dearth of convoy traffic sighted considering all the fighting in the Philippines. Only one contact in eight days, though others had been reported. Their refuge must be the China coast. Patrolling on the surface even in the shallowest diving water, we were still over 20 miles from the mainland. Smoke from a convoy close in could not be seen. I eased the *Barb* in closer.

**8 January.** As the scouting line reversed to the southwest, we ran the *Barb* in before dawn along the 10-fathom curve, hoping.

*Time 0430.* According to Japanese records, HI-87 ships, riding at anchor, suddenly heard depth-charge attacks nearby. The frightened captain of the *Kaihou Maru* up anchored to move out of the roadstead and collided with the destroyer *Hamakaze.* Searching escorts found nothing. Toward noon the fog lifted and HI-87 got under way for Takao, leaving the damaged *Hamakaze* at anchor.

Alone after lunch I kidded Ragland, "After your last patrol, your third in *Barb,* did you know you had orders to go to new construction?"

"I did?"

"Yes, but your shipmates came to me and asked me to stop the transfer because you owed them so much money from your gambling debts."

"They did?"

"Yes, they did—and I did. Now Ragland, you need a winning streak, or the option of no more gambling, or else you'll never leave *Barb.* When are you going to pay them?"

"After this patrol, captain."

"Will your pay cover all your losses?"

"Sir, if I tell you, will you promise not to tell them?"

"Okay, shoot."

"Lego told me I would get orders to new construction. I had enough money to pay them, yet I wanted to make this last patrol with you. So I told them I was broke and leaving. They told me they'd fix that."

"They sure did. You hoodwinked us all. Honestly, what am I going to do with you?"

"Captain, I heard that you can't have more than four patrols in command, so this is the last one for me too."

"Ragland, you handcuff me with your loyalty. Suppose the admiral lets me have a fifth. What do you do?"

"Then I don't pay them all I owe. You've promised you won't tell."

"You're an absolute charlatan. Now I'm an accomplice. Why don't you strike for a legal rating?"

Paul Monroe came in just then and sat down. "What's on your mind, my diving officer whiz?"

"Just that, captain. This is my sixth patrol, and orders are in for my going to new construction. Tom King will take over my jobs. In all our combat, I've been in the control room diving or standing by to dive."

"And doing an outstanding job. So?"

"I'd like to see the action topside just once, then I could describe it to my wife."

"Understood. For our next night surface attack, you'll be on the bridge with me as my TBT operator. I've seen you talking with Tuck Weaver when he had the job. You're a perfect watch officer, cool and collected, no problem there. If we're forced down, instead of staying in the conning tower, I'd like you to drop below and keep an eye on Tom. I've depended a lot on your handling *Barb* submerged when we're being bashed. I'll tell the officers that this is a one-shot change. Jack can stand by in the conning tower."

"Thanks a million, sir."

*Time 1300.* "Captain, William Fannin has sighted a convoy's smoke coming out from the coast, broad on our starboard bow. I've gone ahead flank speed. We can't cut in, the water is too shallow."

"Great, Bill. Station the tracking party. Coming up."

A long look through the high periscope revealed at least five separate wisps of smoke. "Jim, I'll leave the high periscope to you for tracking. There are at least five big ships. Your range now is 15 miles, 30,000 yards. Until we have a mast sighted or radar contact, take your bearings on the lead smoke. They're on a southerly course. Send off a report: 'Now estimating speed at 10 knots.' Tell the Loopers we'll send reports every 20 minutes and hold off attack until one makes contact. Have them home in on our radar; we're probably the only one with an accurate navigational position due to radar fixes on the coastal islands."

On the bridge, visibility was excellent, the sea a bit rough with whitecaps, which would help. Looking through his binoculars on the bridge rail, Jack said, "I can make out at least seven separate smoke sources now. What's your plan, sir?"

"Barring aircraft interference, we'll close until we can clearly make out the masts and funnels of some ships. We'll end around at about 20,000 yards. They must be headed for Takao or Manila, evidently having run down the China coast. For Keelung, they'd be heading east, for Hong Kong, southwest. This allows the pack only the early evening for attack before we're in the minefields. Our job is to get ahead, cross over onto their starboard side, and make a submerged attack. This will drive them farther away from the China coast into the arms of the *Queenfish* and *Picuda.*"

The convoy commander from the First Japanese Surface Escort Force got his convoy, Mo-Ta 30 (Moji to Takao), under way in single file on New Year's Day to pass through the slot in the heavily mined Tsushima Strait. This time he had frigates *26, 36* (his flagship), *39,* and *67,* and four patrol craft with sonar and depth charges.

Most of the ships of this important convoy were armed. The most important and largest—the *Anyo Maru* (9256 tons)—was loaded with critical military supplies, kamikaze pilots, and troops for the reinvasion of Luzon. Built in 1913, she was a beautiful ship with a speed of 15.5 knots, length 463 feet, beam of 60 feet, and a 30-foot draft. He planned to protect her especially with his four smaller escorts. Next in the column was the *Hisagawa Maru,* a passenger-freighter (6886 tons) loaded with troops, horses, and vehicles. Following her came the *Shinyo Maru* (6892 tons) on her maiden voyage. Completed 7

The *Anyo Maru* was sunk by the *Barb* in the Formosa Strait, 8 January 1945.

December 1944, she was loaded fully with troops and weapons and ammunition. Behind her was the tanker *Manju Maru* (6515 tons) and the freighter *Rashin Maru* (5454 tons), with 1042 troops on board.

In the lesser-tonnage category followed the tanker *Hikoshima Maru* (2854 tons) and the freighter *Meiho Maru* (2857 tons). En route, the tanker *Sanyo Maru* (2854 tons) would join them, coming from Dairen (near Port Arthur) with aviation gas. The 1200-mile journey would end at Takao, January ninth.

Clear of the straits the convoy crossed to Korea and thence to the inside passage along the China coast, protected by a special heavy air cover for these vital cargoes destined eventually for the Philippines. The convoy anchored at night along the China coast.

***8 January. Time 0800.*** Mo-Ta 30 departed from its Fuchou anchorage, where it had been well protected from the sea by the surrounding islands. This would be its last day and night under way down the Formosa Strait before it arrived in Takao.

Clearing the harbor, the long column of ships steamed south. When the peaks of Hai-Tan Island were 15 miles away, the commodore signaled to his long column of ships: "Execute formation DOG—base course 140 true. Zig plan B—three columns. Starboard marus, *Shinyo, Sanyo, Hikoshima.* Center marus, *Anyo* (with four escorts), *Rashin, Meiho.* Port marus, *Hisagawa, Manju.*"

The afternoon dragged on for the *Barb* at flank speed. A warm, wallowing day with the seas astern and no aircraft was an answer to our prayers.

Coming down the back stretch, the corks were forging ahead. What a deadly race! The flow of information going to the Loopers could not be beat. As an older, thinner-skinned boat than her sturdier sisters, the *Barb* had one advantage. Her bridge structure above the hull, called the sail, had not been lowered. Being higher, it resembled a junk when seen from a distance. As the race went on for hours, it seemed incredible that we watched the masts and funnels of the eight ships amassed without being attacked. Infrequent quick bearings from the high periscope, radar ranges, and my angles on the bow from the bridge provided neat solutions of course and speed.

At 1500 good old *Barb* was out in front, starting a wide sweep around the final turn to position herself from port to starboard for the blistering pace down the home stretch. "Come on Queerfish! Come on Peculiar!"

Laughing, Jack dropped his binoculars from his sky search. "Captain, you sound just like a jockey."

"I am, and I want this filly of ours to show the Japanese how she can strut her stuff. She's unbeatable."

"Sir, you'd better relax. Both submarines were way away."

At 1600 the *Barb* was still ahead after rounding the final turn. Poised on the starboard bow of the convoy, her head was held high. She was determined to prevent the convoy from gaining the shelter of the shallows of the China coast again.

*Time 1612.* "Bridge, Radio. *Picuda* has contact!"

"MAN BATTLE STATIONS TORPEDOES! Jim, send out the message to the Loopers; '*Barb* diving to attack starboard flank eight-ship convoy, speed 10 knots, base course 140, eight escorts, no air cover. Caution, we're in the U.S. blind-bombing zone.'* After the convoy had cleared the shallow China coastal shelf, entering waters where a submarine could submerge, a depth charge was dropped about every half hour. I mused on the efficacy of this. Were they trying to frighten the foreign devils away? Did it enhance their courage or self-esteem?

*Time 1618.* "Bridge, *Queenfish* and *Picuda* acknowledged our message."

"Let's go, Jack. Take her down."

Now submerged at long last, we headed in at full speed.

"Jim, it would be a snap to get into the center of this group and wreak havoc, but we have a more important job. They must be bent toward our packmates while we reload torpedoes. My plan is to smack the large four-

---

* Due to the laying of mines in the Formosa Strait, the area between latitudes 22-30N and 25-00N was a zone in which U.S. aircraft could bomb any target without first identifying it as enemy. Submariners, therefore, entered this zone at their own risk.

goal-post transport with the four escorts astern of the destroyer with three fish. Then we'll let fly three at the engines-aft freighter or tanker leading the starboard column. She's new. It appears she's fresh out of her building yard on her maiden voyage. That done, we'll turn and shoot three from the stern tubes at the second ship in the starboard column. Pass this to all rooms."

"Aye, captain. Dave says plot shows we'll be shooting in 40 minutes if the convoy doesn't change its zig plan. The Loopers have only until midnight to work on this convoy. By then we'll be in the minefield. Maximum depth during combat will be only 130 feet. Twilight ends about 1900."

*Time 1710.* "Up scope. Bearing—mark. Down. Range 3200 yards. Angle on the bow 30 starboard. All ahead one-third. Make ready all tubes. Coming in nicely. Put four cases of beer in the cooler. Jim, be a touch smarter in handling the periscope. I don't want it to be seen. Four-and-a-half-second exposure is what we've trained. Keep your finger on the button. When I say 'down,' I mean down instantaneously, not fumbling for the button."

"Sorry, sir."

"Open the outer tube doors forward. Plot! Any zigs for the next five minutes?"

"Yes, sir, right now."

Squatting on the deck to catch the periscope handles as they cleared the well, I ordered, "Up scope—down. They're zigging away. Plot, you're right on the button. Good work. This'll be the final setup and shoot. Set it in fast and accurately. Set torpedo depth eight feet, divergent spread from aft forward, 50 yards between torpedoes. Everybody ready?"

"Yes, sir!"

*Time 1724.* "Up scope! Angle on the bow 70 starboard. Range 2500 yards. Bearing—mark. Down."

"Set."

"FIRE 1! FIRE 2! FIRE 3! Sonar get on those torpedoes! Jim, we have a full overlap of the big freighter in the port column." The *Barb* shook with joy as the departing torpedoes jolted her and her shipmates.

*Time 1725.* "New setup. Up scope. Range 1500 yards. Angle on the bow 60 starboard. Bearing—mark—down. FIRE 4! FIRE 5! FIRE 6! Left full rudder."

"Sonar reports first torpedo salvo—all hot, straight, and normal. First torpedo, second salvo, is running left of course, not circling. Last two are hot, straight, and normal."

Dick yelled, "Hitting time!"

"Up scope!" As explosions shook the ship I reported: "Smacked the transport just forward of the funnel. . . . Second hit slightly abaft the transport bow. . . ." The third hit, a tremendous explosion, ducked my periscope. Being intent on the coming stern tube setup to which we were swinging, I idly remarked, "Now that's what I call a good solid hit."

Overhearing someone mutter, "Golly, I'd hate to be around when he hears a loud explosion. What does it take to make him afraid?" I let go my grasp on the scope handles. That comment and the tinkle of glass from a shattered light bulb snapped me out of my fixation on the coming shot. The full force of the explosion dawned, as I noted the shocked expressions on the faces of the fire control party. Men had grabbed the nearest support to keep from being thrown off their feet. The depth gauge now read 82 feet versus the 60-foot attack depth ordered. The *Barb* had been forced sideways and down.

"Captain, forward torpedo room reports cases of canned goods burst open and are being cleared preparatory to reload. After engine room reports section of superstructure probably ripped off. Sonar reports high-speed screws all around."

"Rig ship for depth charge and silent running. Paul, use two-thirds speed and bring me up to 62 feet. I need a look. Get me a single ping sounding; we may need every foot of depth we can find."

"One hundred sixty feet and mud, sir."

*Time 1737.* We heard many breaking-up noises as bulkheads bent and collapsed and several thuds as ships or large parts of ships hit bottom. "Sonar reports all screws going away." We exploded with cheers.

"Up scope. Stern of the transport is sticking up at a 30 angle with two escorts alongside taking off survivors. The bow is in the bottom mud. The brand-new engines-aft freighter exploded. There is nothing left but an enormous smoke cloud and flat flotsam; no lifeboats, nothing alive, nothing. The large freighter leading the port column is on fire amidships just above the waterline. Her overlap must have caught the third torpedo that missed ahead of the transport."

"Captain, can we get a setup for the stern tubes?"

"Hold, Max. The whole formation has turned away and appears to have stopped. Right now the range is too great. All escorts appear to have scampered over to the unattacked side except the destroyer who's searching for us in the wrong spot!"

*Time 1745.* I could feel aggressiveness surging through my veins, sensing that the escorts were more scared than we were.

"Okay, gang, let's go after them. Commence the reload forward. We'll set up on that second freighter in the starboard column as they settle down. Same plan; bend them to port exactly as we have so far. They must not get to the China coast!"

*Time 1747.* "Up scope. The fire on the freighter is spreading, but she's under way. Uh-oh! The destroyer is speeding up and heading this way. Down scope. Secure blowing down tubes." My aggressiveness evaporated as she shifted to short scale for the depth-charge run.

"Paul, take her down to 140 feet. Small down angle; you'll be 20 feet above the mud. ALL AHEAD FLANK! RIGHT FULL RUDDER! I'm giving him a high-speed turbulent knuckle to ping on. When we've come right 75°, we'll stop and coast as long as you can hold it. Let me know when you need dead slow speed to hold your depth."

"Sonar says he's coming right at us."

"Captain, this would be a nice spot for a down-the-throat shot if we had any torpedoes loaded forward."

"Right you are, Max. If, dog rabbit. All stop! Rudder amidships. Steady as you go." We held our breath, hoping the ruse would work. The high-speed whine of the destroyer's screws, audible through the hull, passed to port and astern.

Praise the Lord, he did not drop! Paul was soon asking for speed. The situation was well in hand as the destroyer faded astern. Breaking-up noises continued.

With darkness imminent and the seas still rough, the *Barb* secured from battle stations. Our fish reloaded, we ate tuna sandwiches, then surfaced and raced back to the battle.

*Time 1914.* "Message—*Queenfish* attacking, sir."

"Great!" The *Picuda* would attack next, then the *Barb*. The others had ended around and were ahead of the convoy, now reduced to six ships. They would sweep in to attack from the bows with a high relative movement. The *Barb* could not waste time ending around: I had a new concept of attack that I wanted to test.

Normally, Japanese convoys had a tendency to place their escorts ahead and broad on the bows, with a trailer escort astern. For a dark, moonless night, normal wolfpack tactics called for the pack to get ahead of the convoy for an attack on the surface. One packmate would then pounce from port or starboard. Attack completed, the next would attack from the opposite side. That completed, the last of the three would attack from either side, depending on the reaction in the convoy. After a sub attacked, she would make another end around and the procedure would be repeated. All this took a lot of time, which we did not have.

The concept I had been fiddling with on paper fitted the *Barb's* present situation perfectly. We were 20,000 yards—10 miles—astern of the convoy, making 10 knots plus. At flank speed an end around would require two-and-a-half hours. The night was dark, with only the light from the stars, so all ships appeared as indistinguishable blobs, the escorts only smaller ones. To save precious time, which was dominated by the position of the minefields ahead, I intended to bring the *Barb* in on the starboard flank, then we'd fall in astern of the starboard escort line—about 2000 yards from the starboard column of ships—and attack the rear ship at will.

Sinking or damaging ships in the rear of a convoy should not have greatly affected the organization of the convoy. To the contrary, hitting the van ships caused pandemonium. If our strike were successful, without counterattack from the escorts, the *Barb* would then move up the escort line to the next ship, repeating the maneuver until grave problems such as illumination arose: she didn't want to be canonized.

With the loss of the most important ship, *Anyo Maru,* and the ammunition ship *Shinyo Maru,* the convoy turned northeastward toward the north Formosan ports, although the minefields (which would deter enemy submarines) were closer. The cliffs, however, did offer some protection. The convoy wheeled around again and re-formed the remaining six ships into two columns. With escorts aplenty, the Commodore put one ahead, three on each side, and one trailer. Then he closed the mountainous Formosan coast to 10 miles, though in the dark it appeared much closer. All ships increased speed.

*Time 1934.* "Message from *Queenfish.* Attack from port completed. Fired six torpedoes at lead ships each column and four at second ship port column. Incredible misses. Convoy unalerted. Taking trailer station during reload. Hold one . . . *Picuda* reports attacking from starboard bow."

I pressed the intercom. "Barbarians, *Queenfish* missed on her attack, *Picuda* now attacking. We're next, coming up on the starboard flank. MAN BATTLE STATIONS TORPEDOES!"

*Time 1956.* Paul jumped with joy. "Did you see that flash when the torpedo exploded on the lead ship? Look! *Picuda's* socked the second ship!"

"Paul, keep your binoculars on those two ships. I have a pot full of escorts to watch, eight within 5000 yards."

"Bridge, Jim. I've been watching the *Picuda* attack on the radar scope. She's swung around and is very close to an escort which may be after her, heading out. We can move in."

"Understood. Make ready all tubes. Hitting the trail ship to starboard will take the pressure off *Picuda*. Left full rudder, come left 45°."

"Captain, the ships hit aren't sinking or slowing."

"Too bad; shift over to the target bearing transmitter. We're heading in. Bluth, Lego, as we leave the escort line, concentrate on the escorts astern. I don't want any escorts following us. I'll watch the ships and the trail escort. The convoy appears to be turning left, Jim, closing the coast."

Most of the convoy made a 90° left turn to close the coast. The *Rashin* and *Manju* were able to maintain speed. At five miles off the beach the convoy resumed its course of 225 true. Frigate *67* assisted the *Hisagawa*, which was foundering without pumps. The damaged *Rashin Maru* ran ahead of the convoy, leaving the *Meiho* as a trailer to port. The starboard column was still intact, led by the *Manju*, then the *Sanyo Maru*, and last, the *Hikoshima Maru*. The two escorts were now to port of the *Sanyo*.

Unseen, the *Barb* was in attack position at 2150 yards. "Open outer doors, tubes 1, 2, and 3. Paul, keep the hairline of your scope steady on the middle of the last ship. Jim, we have a nice overlap with the trail ship in the port column. The other ships on the port side seem to have disappeared. The escorts hang around us like a bunch of flies, as if we're one of them. The seas have calmed, sheltered by the coast. With electric fish and no phosphorescence, they won't know where they came from. Standby! Final bearing—mark."

*Time 2012.* "FIRE 1! FIRE 2! FIRE 3!—ALL AHEAD FULL! RIGHT FULL RUDDER! Rejoin escorts. Setup on ship ahead."

*Time 2014.* "Bridge, all torpedoes normal, 15 seconds to hitting time." Paul yelled, "Hit amidships! Hit forward! She's nosediving! Gosh, what a sight."

*Time 2015.* "Captain, hit amidships in the port column. Ship overlapping! What a flash, there's smoke all around her." Paul just stared.

I watched our next target as we moved forward up the escort line without opposition. "Captain, Jim. The radar pip of the target has disappeared. She sank. The pip of the overlapping ship has diminished to half size. She must be sinking."

"Right, Jim. Pass the word 'well done' to all hands. We saw the target sink. The other ship's spot is clouded with smoke—we can't see her. Check later to make sure she's sunk. I need your radar to shift to the starboard, middle ship, our next target. The strategy of joining the convoy as an escort is working to perfection. The escorts are dropping depth charges sporadically. Soon as you have a setup on the next target, I'll take her in. Paul! Snap out of it."

The damaged destroyer *Hamakaze,* left by convoy HI-87 at her anchorage 10 miles away at Shinchiku, heard the explosions and got under way.

"Captain, we've got a good setup. The two ships that are left seem to have increased their speed to 12 knots and changed course to 190 true, closing the coast even more. Depth here is about 160 feet."

"Maher, give me a radar range to the three closest escorts."

"Escort dead ahead is 550 yards. Escort to starboard is 980 yards. Escort astern is 1500 yards. The two biggest escorts are near the far side of the target. Others are scattered with apparent confusion, sir."

Pressing the intercom, I announced, "Everything is okay, so on your toes, Barbarians; we're heading in again. There are only two ships left. Open outer doors on tubes 4, 5, and 6, set depth eight feet. Shankles, come left 50°. Lookouts, stay on the fore and aft escorts. Paul, put the hairline of your scope on the target's midship. Range?"

"Fifteen hundred yards. Setup checks!"

"Final bearing—mark."

*Time 2033.* "FIRE 4! FIRE 5! FIRE 6! QB sonar on the torpedoes. Right full rudder. All ahead full. Plot, Dave, have your assistant Higgins break out a chart and check our navigational position. These cliffs look like they're hanging right above us. I don't want to run aground."

"Sonar reports all torpedoes hot, straight, and normal."

*Time 2034:30.* Three perfectly timed hits were followed by a stupendous eruption that far surpassed any Hollywood production. The rarefaction that followed the viselike squeezing, first-pressure wave wrenched the air from my lungs. Somehow I formed the words, "ALL AHEAD FLANK!" The high vacuum in the boat made people in the control room feel like they were being sucked up the hatch. Personnel in the conning tower who didn't have their shirts tucked in at the belt had them pulled up over their heads.

On the bridge the target now resembled a fantastic, gigantic phosphorous bomb. The volcanic spectacle was awe inspiring. Shrapnel flew all around us. "LOOKOUTS, TAKE COVER!" Pieces sparkled over 4000 yards ahead of us. We alternately ducked and gawked. The horizon was as bright as day.

A quick binocular sweep showed only the one ship ahead remaining, and a few scattered escorts turning away. None of the escorts near the exploding ship could be seen. Had they blown up?

"Forward torpedo room reports that some missiles are striking the hull."

"Aye, tell Stretch Laughter it's flying debris."

At this point I was ready to haul ashes and take a respite. Not so Paul, who in five patrols in his diving station had never seen a shot fired nor a ship sunk. He really had his guns out. Frantically, he pleaded that we couldn't let that last ship go. Besides, he loved to hear the wham, wham, wham of torpedoes and to see millions of bucks blowing sky high.

It was a good sales talk. "Jim, with the escort line evaporating, *Barb* has become the escort. Commence approach on the last ship. Put us on a course to pass her at no more than 2000 yards. Then I'll cut in for a stern shot."

"Aye, sir."

The destroyer *Hamakaze* had been underway for only one minute when she saw at 15,000 meters the incredible explosion and immense ball of fire as the shipload of aviation fuel changed night into brilliant day. With 14 knots her maximum speed and an unreliable hull, she stayed on her course for Makou Naval Base in the Pescadores, where she ran aground.

Frigate *36* scraped her bottom in her haste to get away. The *Manju Maru* and a smattering of escorts raced on, everyone for herself, leaderless. The exploded target remains sank.

The *Barb* had a bone in her teeth as she caught up to the last ship. Now on the port quarter, she was minutes away from a sashay in to 1400 yards. Then, with a swirl and swish of her kilts, she'd turn away and let fly with three torpedoes.

"Max, have you a setup? Paul's on her constantly."

"Perfect, sir. Everything checks. No zigs."

*Time 2055.* "Bridge, urgent message from *Queenfish*. 'Hey! Save one for me!' "

"Aye. Check fire! Shankles, resume previous course. Leave the outer tube doors open. Jim, we'll go on ahead to see if any ship got away. Send Elliott, 'Green light. *Barb* attacks completed. Good luck.' "

A message came in from the *Picuda* saying she would follow the *Queenfish*. We could attack as long as there were ships and torpedoes left. We had had our share, however. We passed the last ship abeam to port at 2160 yards. What a temptation!

Investigating a pip 10,000 yards ahead, we confirmed it to be two escorts hightailing it for Takao. Radar found no others.

*Time 2125.* The *Queenfish* requested enemy course and speed. She knew the *Barb* had been hitting. Evidently, she was still some distance off. We replied, recommending she catch up to the last ship. Unbeknown to us, the *Picuda* was well ahead and out on our starboard flank. Waiting for the finish of the *Queenfish* attack, she was also set up to attack this last ship.

"Bridge, radar shows one escort on the ship's starboard beam and another escort coming up fast on the same quarter. This last ship will be well protected."

"All ahead one-third. We can't wait any longer for Elliott or she'll get away. Setup and give me a range. Paul's steady on."

"Twenty-six hundred yards. That other escort is only 3000 yards on her quarter. Better shoot."

"Max, shoot when your generated range is 2200 yards, in about a minute. Paul's giving continuous final bearings."

*Time 2153.* "Two hits in her starboard quarter. CHECK FIRE!" We had had only 20 seconds to go before firing. Shocked, we determined that the escort coming in must have been the *Queenfish*.

The *Picuda* had had a similar shock, for this was the ship she planned to attack next. With no merchant ships left in this area, she went to flank speed and headed for the slot in the minefield, hoping to find something.

In the *Queenfish*, Elliott had noted the escort on the starboard beam of the ship and another broad on her bow. Not realizing this was the *Barb*, he considered firing two fish at her, but gave up the idea to concentrate on the close escort. A third potential tragedy had been narrowly averted!

The target stopped; the *Barb* stopped. Would she sink?

*Time 2157.* Bedlam commenced. The target was settling, but she courageously opened fire in all directions with automatic weapons—25 mm, 37 mm, and at least one gun, 3 inch or larger. The escort did likewise, firing toward the disengaged side.

"Jim, hang in there. I believe the *Queenfish* has been forced to dive with all the wild firing going on. So long as the pip is disappearing, we won't attack, but if she doesn't sink, we will. We'll remain stopped until she does." It was weird to be laying to there, listening to the rattle of the 25-mm guns, the poomp, poomp, poomp of the 37-mm ones, and the blasts from 3- and 4-inch guns. The gun smoke's pungently foul odor hung heavy throughout the ship. We were protected by that smoke from their guns.

"Captain, there's no change in target pip."

"That does it. *Queenfish* can't attack submerged. All back two-thirds. Jim, we'll back in to 1500 yards to make sure a two-torpedo *coup de grâce* will sink her. Shoot tubes 9 and 10."

*Time 2202.* "Target pip disappearing again."

"ALL STOP! CHECK FIRE! We'll stay here until it's definite."

Shore batteries now joined the fray from at least six points along the coast. Fire from their ships—all tracer—was high and erratic. Projectiles flew thick and fast. The shore batteries' fire was novel: their shells burst as they struck the water a few thousand yards west of us. It seemed that the Japanese believed they were being bombed instead of torpedoed.

On the after part of the bridge deck, Paul was manning the after TBT for a stern tube shot. As his first baptism under fire, it was unnerving. "Captain, Captain, they're shooting at us! We've got to get out of here!"

"Paul, which way are they shooting?"

"They're shooting from left to right. No, right to left. Some are passing over us. Let's go!"

I thought I'd have some fun with him. "Paul, it's so dark, I can't see which way you're standing. Are they shooting from port to starboard or starboard to port? I don't want to turn *Barb* smack into the line of fire."

Dead silence. I could see Paul's shadow approaching me. He kept coming until his face was about six inches from mine. "Captain, this is one helluva time to be specific!" We both laughed.

"Okay, Paul, that ship has her lower decks awash. She's not going any place and certainly isn't worth the expenditure of more torpedoes."

"Bridge, pip has almost disappeared. Our lifeguard station for tomorrow's carrier bombing raid on Formosa is close to our submerged attack position this afternoon."

*Time 2219.* "Right. Set us on a course for our lifeguard station. Secure stern tubes. Ahead full. Reload torpedoes and put four more cases of beer in the cooler. We'll splice the main brace tonight at 2245. Tomorrow, we'll main brace again with the cake at 1100. Scratch one convoy. Secure from battle stations."

*Time 2245.* Splicing the main brace, everyone cheered. It was a day to remember forever. In the wardroom, the usually quiet but always affable Paul related the best day of his life. Holding the *Barb* above the mud in the afternoon action, he surfaced into the fever-pitch battle of the evening, scoring 10 hits with 12 torpedoes. We celebrated the great teamwork all around. Joining the general rejoicing throughout the boat, Bluth and Lego had shipmates spellbound with their "Tales of Two Lookouts." I wished everyone could have seen the action.

*Time 2356.* The *Picuda* reported that her attack was completed. She had gone down to the edge of the minefield slot and had found one ship. She attacked and missed with a four-fish salvo. No other ships got away. The *Picuda* confirmed: the pips of both ships disappeared after our first night attack—definitely sunk.

Unbeknown to the *Picuda,* the last ship she attacked foundered and beached in a few hours. This *Rashin Maru*—damaged by the *Picuda's* first attack—had left the convoy earlier. Thus, all the merchant ships were either sunk or beached. The *Hamakaze* and frigate *36* extricated themselves; *Rashin Maru* was salvaged.

Lying in my bunk, I critiqued myself about the attacks and what we needed to sharpen. My main concern was the muttering I had overheard. "What does it take to make him afraid?" Why wasn't I afraid?

A skipper should know his ship and her capabilities better than anyone else. He's at the periscope; he's on the bridge; he sees the enemy and he has studied him. Years of study, training, and being saturated with a sense of duty to country have molded him into readiness for command and eager acceptance of responsibility.

Fear is a natural characteristic of all living creatures, necessary for self-preservation. To win, however, fear must be controlled, enabling expertise to determine when to fight and when to run away—to be able to fight another day. As experience teaches, the subconscious almost automatically weighs the odds.

Could it be that my brain was limited to a bucket of fear? If so, my bucket of fear was filled with concern for my wife and youngster so far away. Out on patrol, submerged or on the surface, I was faced with an enemy that I could grapple with intelligently. Having my bucket of fear fully allotted elsewhere gave me an edge—an urge to get the war finished as quickly as possible.

*9 January.* Surfaced on our lifeguard station, the *Barb* awaited the dawn carrier strike on northern Formosa. Subs were nearly always positioned to assist planes that ditched, but I had heard of no reciprocity on these strike-and-run attacks. They knew our positions and that we stayed on the surface unless forced down briefly. Contact reports from them sent on our assigned radio frequencies could have helped us. Ships they missed or damaged and locations of convoys were important, but all we received was word on downed planes.

*Time 0750-0905.* They came in groups of three to six planes above the heavy overcast. Radar counted 30 in the first wave and 30 in the second. The third wave had too many to count, and then they were gone. Good luck, flyboys. Having no word on their return path, we had to hang around until late afternoon. Two Japanese planes below the clouds forced us to submerge, once for eight minutes. No planes were ditched.

At 1100 the main brace was well spliced, and we ate an indescribable cake. At least 10 men had supplied Phillips with their sweethearts' recipes. The inside ran the gamut from heaven to hell—that is, from angel food to devil's food—with other varieties in between.

The icing almost crushed the angel food, with two ships exploding, three sinking, and a damaged one beached. Paul, Lego, and Bluth, the master narrators, were given the honor of cutting the cake. Doc, who pressed the firing plunger for the torpedoes, monitored the surgery.

For a breather, I climbed up to the bridge to see how Bill Walker was finishing his first officer-of-the-deck watch alone. "How's it going, Bill?"

"No prob, sir. Much better than having a junior officer looking over my shoulder and telling me what I should do."

"Bill, what's the first thing you do as a subschool grad when you join a submarine?"

"Go to work, sir."

"No, you hang your lieutenant's stripes on the gangway and look, learn, and listen to those who are qualified. When you're qualified, you put them back on. That's submarining."

"Understood, sir."

After that sweet cake, I needed a cigarette. I sent for Ragland to bring me one. "Sir, I filled this new silver cigarette lighter. Did you know it's got 'love' engraved on it?"

"Sure, my wife sent it for Christmas. Now I can use it in place of matches." I had just pulled out a cigarette when it came: "CLEAR THE BRIDGE! DIVE! DIVE! Two jap planes coming in low and fast."

One look and I automatically flipped the cigarette and match overboard. "Oh, my God!" I thought as I dropped down the ladder. "That was the lighter!"

Grabbing the periscope, which was already raised, I swung to the oncoming aircraft. Then the scope went under.

"Hudgens, bring her back up to 60 feet. Just a couple of fighters looking for our dive bombers. They passed well clear." A quick look showed all clear. "Surface her, Bill. I'm going to my cabin to have a good cry."

What was I going to tell Marjorie? I had promised her no more lies, but that was about being out on patrol. She'd never understand this faux pas.

A rap on my bulkhead brought me out of my chagrin. It was Bob Phillips. With his success at constructing our magnificent cakes, I had recommended that he consider putting in for a rate change from cook to baker.

"Captain, I've thought about what you said, but in a different way. I know I'm a lousy cook. Even I don't enjoy eating what I cook. Shifting my rate to baker would only be worse. I'd understand if you transferred me after this run to get *Barb* a good cook—the men deserve one."

"Bob, do you want to be transferred?"

"No, sir! I love the *Barb*. We have a wonderful crew, all friends."

"Let me wise you up, Bob. I don't want you to leave."

"Why not, sir?"

"Because you are a terrific morale factor. Nearly all the men are writing home or to their sweethearts for recipes. They buzz around the galley like bees to honey. It's good training for future husbands. You're an asset."

For the next few days the *Queenfish* formed a scouting line with us. She stayed 20 miles to seaward. Riding the 20-fathom curve, we danced and heaved as the high seas moved on to the coastal shelf. Topside, the weather was ugly. Drenching spray flying over the lookout platforms slapped those on the bridge numb. We were forced down daily by marauding aircraft playing a cat-and-mouse game, but so far we were winning.

The immense groups of junks interested me. If only I had a Chinese interpreter and a radio to send with him, the information I wanted would have been forthcoming. I decided that the *Barb* should join the junk fleet for their morning fishing. Would they report us to the Japanese? Did they have Japanese on board watching for submarines? We would soon know, if a plane appeared.

Donning my soggy foul-weather clothing, which hadn't had time to dry, I went up on the bridge until we could trust the junks. At noon they headed back in. No planes.

*12 January.* Junketing with the 20-fathom curve fleet off Fuzhou, I felt sorry for the families perpetually living on board. The tedium of heavy weather—being bashed around until your brain felt thick—discouraged unnecessary activity. The clammy, musty chill mitigated any accomplishment.

Going through the boat, I sensed that my men were getting stale and losing that sharp edge I depended on. They looked at me as if they expected me to pull a rabbit out of a hat. Hell, it wasn't my fault—or was it? We had to do something besides roll our guts out. Sitting on the bridge endlessly scanning the empty horizon, bits of the childhood poem "Columbus" flashed through my mind:

> If we sight naught but seas by dawn
> What shall I say, brave Admiral say?
> Why you shall say at break of day
> Sail on, sail on, sail on and on.

"Captain, what causes these large areas of discolored water in the East China Sea that we occasionally encounter? Look at this one ahead."

"I don't know, Max. Let's investigate this one for the Hydrographic

Office. Pass the word for a trim dive and then take her down. I'm going below." In the control room I called for Paul to take the dive, explaining our experiment. Max followed through, and the *Barb* submerged like a lump of lead.

Paul stopped her at 110 feet by blowing main ballast tanks, then gradually trimmed her up. We were 26,000 pounds heavy. The bathythermograph showed a straight isotherm—no temperature variation—but it was nearly 10° colder than the temperature recorded prior to entering the discolored area. Suddenly, passing out of this spot, we broached and had to flood in 26,000 pounds, reverting to our original trim.

The consensus on our experimental dive was that the discoloration was caused by an underground fresh-water river bursting vertically through the sea bottom. Whether the color was due to sediment or sea bottom disturbance was a question. One thing we learned was that we wouldn't want to run into one of these spots while making a submerged approach. The excitement of the plummeting dive eased the staleness; we surfaced, and the hunt resumed.

*Time 1500.* Swish poked his head into the wardroom. "Convoy contact message coming in from *Queenfish*." Jim put the position on the chart.

"It's 30 miles east of us. Dave, check for radar interference. He's had no stars for days."

"Lehman has *Queenfish* interference on 160 true. We better home on him. His position is way off."

"Jim, Elliott gave convoy course as northwest, speed 14 knots. Let's lead the interference by 45° until we make contact."

*Time 1520.* Made radar contact at 16,400 yards. The *Barb* was well off her track, working ahead. The *Queenfish* must have dived to attack. Visibility was poor in the mist.

*Time 1532.* Price spotted a floatplane with one bomb at two miles; it was heading for us close to the water. Down we went. A minute later the bomb struck astern. We headed for the convoy at flank speed. Circling, the plane stayed with us. Every time the scope was raised that blasted plane was between us and the still-unseen convoy. "How does he fly in this weather?" we wondered. Soon a destroyer was sighted crossing ahead at 5000 yards.

*Time 1600.* "Up scope. There's that plane again . . . WHOA! Here comes the convoy out of the fog. We're abaft their beam and the range is about 6000 yards. One large transport, one large freighter, an escort aircraft carrier, and four more destroyers. They're moving fast. Bearing—mark.

Down scope. Range on a destroyer 6200 yards. Angle on the bow 110°
starboard. We'll have to trail. What a heartbreaker! Max, what's their
course?"

"Two nine zero true, speed 14.5 knots."

*Time 1759.* Having trailed submerged at seven knots, we surfaced after
dark with the convoy 21,000 yards ahead and sent out a contact report to
all subs. The *Queenfish* replied that she too was trailing, having been left
outside due to that plane.

*Time 1803.* "Aircraft radar contact one mile! His APR radar is steady
on!" Tom reached for the diving button. I grabbed his arm.

"Don't dive. He can't possibly see us in the blackness. We learned in
the South China Sea after many bombings that they cannot bomb with
their radar unless they have sight contact. If we dive we lose this convoy.
We will not dive."

The plane came to within three-quarters of a mile and circled for six
breathless minutes, then departed. Our theory was somewhat substantiated.

As seas increased, the convoy faded. The Loopers gave up. For the next
few days we maintained our scouting line, sweeping north of the straits'
blind-bombing zone, allotted to our army aircraft in China and to naval
air from carriers supporting the Philippine invasion. China air reported
several ship movements with ships' courses, yet we found nothing other
than occasional patrol planes. So we returned to hunting our various areas.

Before dawn on 16 January, the *Queenfish* reported attacking a single
tanker with two escorts. She fired eight torpedoes in three undetected
attacks and all missed. This was certainly a mechanical error in the
torpedoes or the fire control system. I felt bad, for with the great risks the
boats take in attacking, misses are crushing to the morale of those on
board. Submariners don't mind attacking with a 50-50 chance of survival,
but Russian roulette—no way. Out of torpedoes, Elliott signaled his
departure for Pearl Harbor. Ty Shepard now took command. The *Picuda*
and *Barb* blockaded the straits and searched for the baffling China air
contacts with zero results.

Just before midnight on 19 January, 10 miles north of the blind-bombing
zone, we were ambling along at eight knots, guarding the Japanese aircraft
radar frequencies. Tom had the watch. Suddenly, the rat-tat-tat-tat of
machine gun fire shattered the quiet. "WE'RE BEING STRAFED! DIVE! DIVE!"

Jolted from my bunk by the horrible sound, I burst into the control
room. "ALL AHEAD FLANK. FLOOD NEGATIVE!" In a flash my eyes noted
the stern planes on full dive and the bow planes rigging out on full dive.
Then I passed on to Machinist Peterson on the diving manifold. All main

ballast tank vents were open, negative, our down express tank flooding. With one hand on the high pressure air bleed valve, he was watching the red and green lights on the hull openings indicator board. In this split second the bridge hatch was opened. More machine guns rattled as the lookouts came tumbling down to man the diving planes. "He's strafing us again! Take her down to 60 feet!"

"Green board: pressure in the boat!" The *Barb* was obediently assuming a down angle when her rear end got a lift. Four bombs exploded close on the port quarter as the plane crossed over—strafing—20 feet above the bridge. Bulbs shattered and the emergency lights went on. Picking myself up off the deck, I ordered, "Report major damage. Tom, tell me!"

"Captain, no indication until strafing—high and to starboard—when we were caught by surprise. While we were clearing the bridge, his rear gunner opened fire."

"Up scope. Strange, no float lights to mark his attack, no Jap radar. Jim, report this attack to our China air force."

On the surface again, I rued the bad luck the *Barb* had had since I threw my lighter—my "Love"—overboard. Like the ancient mariner who prayed to remove the albatross from around his neck, I did too.

# Chapter 19 The Galloping Ghost

## of the China Coast

*20 January 1945.* Frustrated by reports from our Commander Naval Group China in Chungking that all traffic had holed up, I broke out all of the contact reports from China. It was obvious that the vast majority originated in the Lam Yit area. These nearly impenetrable islands offered sanctuary and safe haven for anchored or coastal shipping. Many other reports covered the Amoy area inside the blind-bombing zone, forbidden to subs. All of these coastal contacts, however, had to emerge when they rounded the elbow off Hai-tan Island, a little south of Fuzhou. Or did they? China air's contacts invariably gave courses that would pass near Turnabout Island, a mile east of Hai-tan.

"Jim," I said, "have Higgins break out the biggest chart he has of Hai-tan. Aviators may get fouled up in their posits, but they can't be wrong this often." Scrutinizing the chart, I drew a pencil line from Amoy, tight in along the coast, to the top of the restricted zone. Tracing my way in and out among the island galaxy of Lam Yit, I was brought to a shuddering halt. I realized that the mile-wide channel between Hai-tan Island and the mainland is 10 miles long and only six feet deep. Could they possibly have dredged it? If so . . . I ran my pencil through it and all the way, tight in, on the coast, to Shanghai. By that route no sub would even see their smoke! We sent a message immediately to Commander Naval Group China and China air informing them of our failure to find the reported contacts. We requested all available information on possible dredging of Hai-tan Strait. If such dredging had been done, it would have enabled convoys to escape submarine surveillance. Hence, the riddle was solved.

After dinner the next day, the *Barb* received a reply from ComNavGrp China. "COASTWATCHER REPORTS SUCH A CHANNEL HAS BEEN DREDGED AND MAJOR SHIPS, EVEN BATTLESHIPS, USE IT." What a surprise! No one had ever mentioned that Naval Intelligence had coastwatchers other than in the Australian area. Keeping such information secret from submarine skippers hampered us and wasted much effort. We would use this new information for the remainder of the patrol.

While our air power was hammering Formosa, no shipping moved around naval Keelung. Most traffic stayed along the China coast, probably anchoring at night. No lights had been observed along the coast. This daytime movement stopped sub attacks. The crossing to Takao, southern Formosa, took place only in the restricted zone. Anchorages were probably at Shanghai, Wenzhou, Samsa Inlet, Fuzhou, and Lam Yit—all well mined and a day's run apart.

With seas off the shelf running 10 to 20 feet high, our prospects appeared poor unless we could find an opportunity at night for torpedo boat tactics. Another cheery message noted recent unknown mining north of Wenzhou. If our assumptions were correct, the present convoy for which we were searching was anchored at Fuzhou and would be en route to Wenzhou the next morning. As the junks had not reported us, tomorrow we would junket closer in.

*22 January.* Like water on a drooping hydrangea plant, our new plan lifted our spirits and my striking force revived. Word traveled through the crew as fast as salts through a sailor. "Deadeye" is fed up with sitting on his hands and is going via the junk route after the enemy in their sanctuary.

A live hubbub filled the crew's mess at breakfast. Machinist Charles Johnson had been up on the bridge at daybreak and now reported his findings to the crew. The seas had flattened somewhat. Captain Gene was perched on the lookout platform, closely observing the surrounding junks. For today the *Barb* had joined the massive fleet of over 200 junks just north of the Seven Stars Islands. She was meandering at one-third speed, weaving amidst the fleet. The Old Man swept the western horizon every few minutes with his binoculars. That old smile was back on his face— something was up. Johnson had even waved at passing junks, and their crews waved back.

Roark bummed a chart off Quartermaster Broocks, and the men spread it out on a mess table. Excited, Max put his thumb on the Seven Stars Islands south of Wenzhou, and they all started taking measurements.

"Not much water, only 60 to 90 feet."

"The 20-fathom curve is over 25 miles from the coast."

"How far do you think he will take us in?"

"Far enough, you bet. I'm going to put on my Jesus shoes."

In the control room, Swish monitored the line of crewmen going topside for a look. "Don't stay too long," he said. "Give the rest of us a chance."

My plan was to stay with the junks as they reversed course and moved coastward in the afternoon. I figured that if we closed the coast a couple of miles inside the 10-fathom curve, we could see the smoke of the northbound convoy that had probably anchored at Foochu the night before.

On the morning watch, Max and I were discussing the "what ifs?" of the plan. Ever practical, Max asked, "Suppose we do spot that convoy and track it. What then?"

"That's where the plan gets a bit fuzzy."

"If it doesn't get too hairy."

"Depends on the situation."

"Will you attack?"

"Not in daylight—unless visibility stinks and the convoy has few escorts."

"Nighttime?"

"If they are under way, yes."

"At anchor in Wenzhou?"

"Doubtful, until I have accurate info on their minefields. I want at least even chances to see who's smarter when I bet our lives on the outcome of combat."

Troy Durbin, up for an airing, spoke up. "Captain, that group of junks on the port bow is avoiding something. Looks like a mine."

"It is. I've decided not to sink them with gunfire; some enemy ship might hit one."

I was itching for the junk fleet to return homeward early. If only I could speak one of the Chinese dialects. Though they worked like clockwork, I couldn't detect their leader. At noon they did an about-face and headed in automatically, the *Barb* with them. Some of the seamen and junior petty officers on the lookout list had an easy day because off-watch senior men offered to stand their watches in order to be close to the increasing excitement.

*Time 1421.* Lookouts Charles Swearingen and Powell cried out simultaneously, "Smoke on the horizon dead ahead!"

Sure enough, small wisps of smoke slowly mushroomed in our path. "Jim, we've got our convoy from Fuzhou."

"And you've also crossed the 10-fathom curve. Sounding shows only 48 feet."

"Come on, don't be a kill-joy. We need to go just a couple of miles farther in to get their course and speed."

"What about aircraft?"

"Compromise, Jim. We've a solid overcast. No one's going to dive on us out of the clouds. If they come in low, we'll spot them and be out of here in five minutes at flank speed. Put four engines on the line now, and send up Murphy to man the twin 20-mm gun. He's a good shot and might just bag a plane. Station the tracking party."

Murphy arrived with his loaders and set up the twin 20s. Regretfully, I ordered the bystanders below: too many people topside if we got caught. The puffs of smoke now showed there were six or more ships. Radar picked them up, but with the interference of rocks and islands at extreme ranges, counting was out.

Jim used the high periscope to take bearings on the largest smoke column. In a few minutes he reported, "Plot says this is a different convoy. It's going south. Surprised?"

"You bet!"


Heading south, the convoy commander led his valuable Moji-Takao 32 convoy out of the San-men Bay an hour before civil twilight. His flagship—the destroyer *Shiwokaze*—was modern and capable of defending against enemy attacks on his 10 merchant ships. Other escorts were the frigates *31, 132, 144,* and *Manju,* and the subchasers *19* and *57.* Crossing from Tsushima, they successfully repelled the enemy subs with two probable kills.

Requested to speed up if possible, he had saved a day changing to the San-men anchorage, thus bypassing Shanghai. For one hour of the morning's darkness, he had permitted all ships to use their running lights to avoid collision. Not zigzagging, the *Shiwokaze,* with her excellent radar, could lead the column in single file out of San-men Bay without trouble. With such an early start, Mo-Ta 32 would just make it to the well-protected staging harbor of Namkwan before dark.

The Takao-Moji 38 convoy—heading north, with the convoy commander in the destroyer *Ikuna*—got under way from Fuzhou Harbor at 0800. With 75 miles to go, they should be snuggling into the anchorage at the same Namkwan Harbor behind the Incog Islands at 1600.


Meanwhile, at Namkwan, chief of the Chinese local spy network Chang Tsou had selected six of his best pirates to go aboard the next southbound convoy.* General Tai Li's organization throughout China was answerable

_____
* From conversations with Admiral Milton E. Miles, Commander Naval Group China.

only to General Chiang Kai-shek. He particularly wanted ship, troop, and munition counts en route to the Philippines.

By midday Ta-Mo 38 had passed Samsa Inlet and was traversing the narrows of the Hsiao-an Channel in single file. In three hours they would reach Namkwan Harbor. The anchoring coordinates had been signaled. The destroyer *Ikuna* would be anchored closest to the beach, then the six ships would anchor in a northeast-southwest line parallel and close in to the coast. Frigate *26* would anchor at the north end of the ship line, frigate *39* at the south end, and frigate *112* would patrol the Incog Islands, guarding the entrances to the harbor.

The *Barb* closed to 18,000 yards to track the convoy accurately and to determine ship types. Through the high periscope we could see the tops of the masts of at least six large ships. We stopped four miles inside the 10-fathom curve; the depth was 42 feet. Our junks wandered on. We sent a contact report to the *Picuda,* giving the convoy's course, speed, and position. She replied that she was 30 miles north of the tip of Formosa, proceeding at flank speed to Tung Yung Island. She would search south and east for the convoy outside the 20-fathom curve, which would cover a convoy headed for Keelung.

*Time 1512.* Radar held contact on the disappearing convoy to 28,000 yards, even after the smoke was gone. The *Barb,* traveling at flank speed, headed out for water deep enough in which to dive. I needed to capitalize on our knowledge. "Jim, at 10 knots, when will they arrive at their Samsa Inlet anchorage?"

"Not before 2100, sir."

"Great. We can greet them as they exit the narrow slot of Hsiao-an channel. I don't believe they can reverse course in the channel without going aground. It will be like shooting fish in a pickle barrel. We have their speed, and their course will be exact."

Jim cautioned, "If the overcast remains heavy enough to hide the moon."

"Right. Secure tracking. Lay out courses to end around the Seven Stars, the Tae Islands, and the Piseang Islands outside of island visibility. If we avoid possible minefields, what time will we arrive at the southern exit of the channel at flank speed?"

"Roughly 1930. It'll be dark at 1815."

"Excellent. With 100 miles to go, let's start galloping." Gallop we did, in and out among the various unlighted junk fleets still moving coastward after dark.

*Time 1830.* South of the Piseang Islands—where minefields had been reported—we headed in toward Sansha Bay in Samsa Inlet. Approaching the 10-fathom curve, we cleared the edge of the local junk fleet. Even though they were an obstacle, I was sorry to leave them. We had depended on their routes to keep us clear of minefields.

Navigating by radar was a necessity because of the numerous isolated rocks; a few awash had unlighted buoys.

Quartermaster Bluth was with Dick and me on the bridge. He tugged my sweater and said quietly, "Don't look now, but we just passed a floating mine 10 yards abeam to port, sir."

"Shh—it's probably garbage."

Radar picked up a junk ahead. We closed to identify. Once sure, we patrolled on his seaward quarter, using him as a minesweeper until it was time to turn for the channel.

*Time 1955.* "Captain, we're in position. Shuang-feng Island is 2200 yards northeast; Chih-chu Island is 2000 yards southwest. Water depth is 30 feet."

"All stop. Well done, Jim, snaking us in here. Keep checking our position, and tell me the minute you detect any current. With the big ships in this convoy, they're bound to have some destroyers or frigates with radar." Such ships would be in the lead, and I intended to let them pass, believing our stationary *Barb* was a rock. We would torpedo the sitting ducks as they came by.

Fortunately, the tide was slack. The *Barb* stood still. Time stood still. I fidgeted—not daring to leave the bridge. There was no sound from the islands. Nothing on radar. "Bridge, radar gets a real fine cut on the Chih-chu lighthouse."

"I can see it. It's not lighted. I don't think these islands are inhabited." It was like being at the bottom of the Grand Canyon at midnight with no stars. I felt I could reach out and touch both canyon walls at the same time. The overcast cut out all but the faintest of light from the moon.

*Time 2120.* "Captain, we've covered all possible speeds for that convoy. They've either anchored or slipped out to northern Formosa."

"Okay, Jim, scratch this pickle-barrel shoot. Notify *Picuda,* No joy at this posit. Let's gallop."

The *Barb* went out the same way she came in—the safe way.

Jim asked, "Captain, when we reach the 20-fathom curve, where's the galloping ghost of the China coast going to gallop tonight?"

"Jim, let's ease around the Piseang Islands. I don't think we'll have to go out as far as the 20-fathom curve. After that we'll head back in and search the coast. I feel sure we've got them bottled up in an anchorage between Piseang and Seven Stars. Check the old *Coast Pilot Manual*." We definitely had the convoy bracketed.

Mo-Ta 32 steamed into Namkwan Harbor in the late afternoon. Seeing the long line of anchored ships, the convoy commander signaled his flock to anchor on parallel lines 500 yards apart, with five ships in each line. The destroyer *Shiwokaze* would head the inboard line, frigate *132* the outboard line, and frigate *144* would tail. Noting that frigate *112* of the other convoy was circling the close-by Incog Islands, he assigned his frigates *Manju* and *31* to patrol the center and northeast flank to protect the two convoys of 27 ships total.

"Jack, send an urgent message to ComNavGrp China asking for immediate reply as to knowledge of any coastal minefields between Piseang Isles and Seven Stars. Also ask *Picuda* if she has any info on mines or the convoy."

Getting negative replies from both, I studied the chart below. All of the officers hovered over the wardroom table as Jim spread out the chart to stake out our search. The bulkhead clock struck four bells—2200—reminding us to get going.

"Jim, how much time have we got?"

"Twilight is at 0550."

"Let me have those dividers." Setting them for 10 miles, I stepped off from our position around Piseang, turning back to the coast and along up to Seven Stars. "Eighty miles plus 25 more to escape to good diving water after the attack is too tight. If I were Japanese, I wouldn't mine Piseang out to the 20-fathom curve where the junk fleets fish. Cut that in half and it will save us about 20 miles."

Max piped up, "The whole coast looks good, particularly both sides of Fu Yan Island." Everyone nodded.

"Agreed? We go in. Wipe those rueful smiles off your faces. I know no sub has ever done this before. That's our great advantage—surprise! We must paralyze the enemy until we've finished all of our Sunday punch, our last four torpedoes forward and four aft. We must get away before he counts his strength and realizes he can wipe us out. Believe me, he won't know what hit him. When he finds out, we'll be gone. Any questions?"

"How do we spot anchored ships near the beach?"

"Dave, take that piece of plexiglass you have that fits the PPI scope. On it trace the coastal contours, rocks, and islands shown. Compare that with the largest scale chart of the coast we have. We'll investigate closer any blurb that doesn't match."

Taking Jim to my cabin, I informed him—for his ears only—that I was sending an "eyes only" personal message to Captain Shepard in the *Picuda,* inviting them to join us. Together, we could smash the six-plus ships. Jim agreed.

I wrote to Shepard: "I am positive the convoy has anchored. *Picuda* is now about 20 miles from us. *Barb* will search the coast and can wait one hour if you care to join us." I warned Chief Radioman Hinson not to reveal the message or Shepard's reply to anyone.

In five minutes Hinson gave me Captain Shepard's response: "DROP DEAD!"

*Time 2326.* North of Piseang we moved in at full speed on three engines; the fourth was on charge, topping off the battery. Visibility was lowering. I took this opportunity to talk to the men, for no one was asleep. "Shipmates, we've got this convoy bottled up along the coast. We're going to find them and knock the socks off of them. When we attack, we'll strike and strike hard with eight torpedoes. We'll overwhelm him, topple him, keep him off balance until we've skidded out of the harbor. This surprise will be *Barb*'s greatest night, a night to remember. If you have any questions, I'm coming through the boat now."

Starting in the forward torpedo room, I asked James White where his torpedoes were. "Tubes 3, 4, 5, and 6, sir."

"Let's move 5 and 6 up into 1 and 2. When we shoot, I doubt the water depth will be over 30 feet. I don't want any mud rakers. Set their running depth at six feet."

Passing by the officers' small galley, I asked Ragland, who was busy brewing more coffee for the long night, "How's it going?"

"Captain, if you think you hear the dishes rattling, that's not the dishes, that's my teeth."

"Calm down, Ragland. We're going to be all right. Don't forget, you could have gotten off if you had paid your gambling debts. Are you sure you want to go with me on our next patrol?"

"I'll tell you tomorrow!"

All the way through the boat no questions were asked, no conversations started. Men merely signaled with their fingers in a "V," thumbs to forefingers in an "O," or thumbs up. The control room was like a morgue; the normal joshing had vanished. Back on the bridge, soundings came up monotonously every five minutes: "90 feet . . . 70 . . . 50 . . . 40 . . ."

Jim came up, indicating he wanted to talk privately. We moved to the back of the bridge. "Have you had a reply from *Picuda?*" I waved my index finger back and forth, indicating nothing doing. "Captain, don't you think you should put the men in life jackets?"

"No, Jim, I don't want to frighten them. I need them on their toes and thinking. Look, this is your first patrol in *Barb*. I've never seen the men so quiet before. The control room was so silent, if I had dropped a pin it would have sounded like a depth charge. I know the gung-ho capabilities of each of these men. They have faith in me; I have faith in them. I'm not going to let them down. No, Jim, no life jackets; it's too alarming. The odds are with us, believe me."

*23 January.* We continued our relentless search up the coast among our entourage of several hundred darkened junks. With constant helming, Jack maneuvered the *Barb* frequently to avoid collisions. The silence was broken only by the mechanical giving of necessary orders. The night dragged on. No one cared to sleep in the ambience of imminent combat.

*Time 0112.* "Captain, Dave. This chart-plot matching is showing an uncharted smear northwest of Incog Light. Checked this on the radar A-scope and got saturation pips at 29,800 yards. Both the radar operator and I say they are definite ships. The only doubt is, our radar has never before had saturation pips on ships at such a range."

"Well done, Dave. Continue the search. Watch for patrols. They can't all be asleep."

"Captain, blinking light ahead. No, it's gone out."

"Probably a junk, Lego. Dave, check Incog Island. See if you find anything moving."

"Yes, sir, something there at 14,000 yards moving counterclockwise around Incog."

"Log her posit and the time. I'll circle right here until the situation clears."

*Time 0240.* "Bridge, must be an escort. We're picking up Japanese radars sweeping from her and a couple of other bearings. Took her 21 minutes to disappear on the other side of Incog. The captain apparently is more concerned with using his radar to keep himself off the rocks."

"Probably. The moon has set. Visibility is poor. I'll have to revise our fire control party, assuming the convoy is on the other side of Incog."

Heading up the coast again toward Incog, we cleared the junks. None ahead. I preferred to have them, or to know the reason for their absence. We knew of recent mining in this vicinity; mines could have been laid from Incog to Tae Island. Still, a more effective field would be from Incog to Pingfong, closing the eastern harbor entrance. The anchored ships were too close to that point, so no minefields then existed.

"Bridge, picked up another escort patrolling off Pingfong and a third due east of the ships. Whoops! Here comes our circler again at 7000 yards."

"Jim, I can barely make her out. Looks like a frigate. She's gonna be a damn nuisance, being so close when we turn into the harbor, unless . . ."

"Captain, you're cutting out. Unless what?"

"Unless we give her the 'revolving door treatment.' Let me know just before she disappears. We'll go ahead emergency and skin around Incog when she's on the other side of the island: shoot and shove off."

*Time 0300.* "She's disappearing!"

"All ahead emergency!" The *Barb* galloped forward, intent on outfoxing the foxes. The harbor opened as we rounded Incog. Timothy Maher, Lehman, and Dave, watching the radar, whistled. "My God, captain, the harbor is chuck-full of ships at 10 miles."

"Get me a count and their formation. Jim, fortunately we have a flexible fire control party for night surface attack, and now we're going to flex it. I'll take your place in the conning tower. You take navigational plot below. Secure target plot and send Dave to radar since there's no zig plan to these. Send Tom to the bridge to assist Jack. I'll keep the conn."

"Understood."

*Time 0320.* "Radar counts about 30 ships total, including the three escorts. The ships are anchored in three lines parallel to the coast, with a few smaller ones closer in. On one radar bearing we counted 12 ships. The lines are about 500 yards apart."

"Fantastic! All ahead standard." Jack and Tom took over surveillance from the bridge. I went below and switched on the intercom. "Men, we've successfully entered Namkwan Harbor undetected. We've got the biggest target of the war in front of us. Our approach is starting. Make ready all tubes. I figure the odds are 10 to 1 in our favor. MAN BATTLE STATIONS TORPEDOES!" The gongs resounded to cheers. Action at long last.

Max had an idea. "Captain, the tide is ebbing. I'd like to use a speed for the anchored ships of 1 and ¼ knots to take care of the drift of the torpedoes."

"Logical. Do so."

"And with the seas, wind, and tide, the ships should be heading 050 to use for a course." Also smart. Jim brought up his chart for perusal. I marked the navigational position for firing, a mite less than 3000 yards from the inboard ship line. This was six miles inside the 10-fathom curve, plus 19 more miles to the 20-fathom curve, our home in the deep. "Fire control, listen! We'll attack with bow tubes from the southeast on a 90 track, swing right for stern tubes on a 60 track—praise the Lord!—then gallop."

"Jim, the stern tube shot will be on our escape course. We'll have about an hour and a half run at flank speed to get clear. I'm electing to retire through the area marked 'Unexplored' on this large-scale chart. It contains sufficient 'rocks awash' and 'rocks, position doubtful' to make any over-ambitious escort think twice before risking a chase. Also, it crosses the masses of junks, which should be a definite and final barrier to pursuit."

"That course will stop ships!"

"I hope you're not including *Barb*."

"Me, too! What countermeasures do you expect?"

"Possibly searchlights, hot pursuit, gunfire—no depth charges for sure. Stealth, stupefying surprise, and a sprinkle of serendipity are *Barb's* hallmark."

"What's that last?"

"It's the faculty of making fortunate and unexpected discoveries by accident. Luck is where you find it, but to find it you've got to look for it."

*Time 0352.* Took a quick run to the bridge to check the phosphorescence. Through binoculars it was easy to see the lines of ships quietly at anchor. The center escort was off our starboard quarter about 3000 yards. "Tom, watch him." Through the telescope on the TBT the individual ships in the first line were quite clear. The second- and third-line ships formed a complete overlap from end to end. No one had ever had such a perfect target.

"Jack, get on the TBT scope! Now look at that big ship just left of center."

"Got her!"

"That's our first target. I'm putting the scope on our second target, the largest ship to right of center. Take a look."

"Check!"

"When we shift to stern tubes, you put that ship in your crosshairs on the after TBT. Now get on that first target and stay on it until we shoot. Conning Tower, use TBT bearings and radar ranges. Jack's steady on. How's it look?"

"Checks!"

"I'm going below."

Dave exclaimed, "What a beautiful target! It measures 4200 yards of continuous overlap. We can't miss even with an erratic torpedo. Range is coming on, sir."

*Time 0402.40.* "All ahead two-thirds. Open the outer doors. Dick, insert 150 percent divergent spread. Final bearing—mark. FIRE 1! FIRE 2! FIRE 3! FIRE 4! Dave, Max, shift target to the right for the stern tubes. Dick, our problem is to keep too many torpedoes from hitting the same ship. Make your next insert a 300 percent spread."

*Time 0403.* "Stern tube gyros are approaching zero."

"Final bearing—mark. FIRE 7! FIRE 8! FIRE 9! FIRE 10! ALL AHEAD FLANK! Sonar, on the torpedoes. Sounding?" I dashed to the bridge—all binoculars were trained on the ships. "Tom?"

"Don't worry, captain. That frigate hasn't seen us."

"All torpedoes hot, straight, and normal." Perfect! Breathless seconds dragged, "awaiting the rape of solitude."

*Time 0406.* "TWO HITS MAIN TARGET! She's settling to the bottom. Hit in the second line behind the main target!"

*Time 0407.* "Hit large freighter, third line! SHE'S ON FIRE! She's flaring up in flashes! The fires are being snuffed out! She must be sinking!"

"Sounding 29 feet."

*Time 0408.* "Hitting time after tubes! Hit first line! What a geyser! Another hit in the first line! A large freighter. Good Lord! She's belching out a huge cloud of smoke!"

*Time 0408.36.* "Hit second line. My God, the whole side of the ship blew out toward us! She sank!"

*Time 0409.40.* "TAKE COVER! FAR SHIP IN THIRD LINE HIT AND EXPLODED! MUNITIONS! PROJECTILES 6 TO 12 INCHES ARE FLYING ALL OVER! SEARCHLIGHTS ARE SWEEPING! LET'S GET OUT OF HERE! SHE'S SUNK!"

"Bridge, they've turned on air search radar. Probably think they're being bombed. . . . Hold it! That close escort is heading this way and speeding up, range 6000 yards. The one off Pingfong has her surface radar steady on, range 10,500 yards. Looks like we're in for a race. The one at Incog must be behind the islands."

"Keep on them until we get to the unexplored area, then I'll need the radar to stay off the rocks. Let me know if they're closing." The whole harbor, which had lit up, was now full of smoke. I couldn't see anything astern.

Jack and Tom were jumping for joy. We could hear the cheers below. Smiling, Jack said, "Captain, if you bring *Barb* safely out of this one, the Medal of Honor is a cinch."

That bothered me a little. Napoleon once claimed he could win a war on a trunk full of medals, but that's not submarining. The *Barb*'s goal was maximum damage to the enemy. That meant sharing our contacts with any sub close enough to get in. Submarine crews chasing medals wouldn't do that. It was no good. A wolfpack would fall apart operating that way. That was why I wanted to stay with the Loopers. "Jack, please forget medals."

"Bridge, closest escort is gaining. Range 4200 yards."

"Engine Room, tell Chief Williams to crank up every revolution he can squeeze out. We must have more speed. After torpedo room, reload your last four torpedoes. We may need them. Jim, stand easy on battle stations. Have the 40-mm and twin 20-mm gun crews standing by in the control room. How's the navigation coming along?"

"Right on track."

*Time 0420.* "Dave, escort range?"

"Thirty-six hundred yards, closing."

"Get Williams on the phone. Chief, I need more speed!"

"Sir, the engines are at their top speed now. Any more and the governors will cut the engines out."

"Well, tie down the governors and put 150 percent overload on all engines."

*Time 0430.* "Dave, escort range?"

"Thirty-two hundred yards. Still closing, but not as fast. We're making 23.5 knots, a new world record for submarines."

"Captain, Engine Room. The bearings are getting hot!"

"Let them melt. Jim, how close will we pass the rocks, the rocks awash, and the Strawstack rocks?"

"Three thousand yards, tubes 7 and 8 are reloaded."

"That's a record. Cut in to 1500 yards from the rocks. Jim, I can see she's a frigate. I doubt she'll start shooting until she's closed to 2000 yards. Then she'll have only her bow gun. She must illuminate like they did to us in the South China Sea. The minute she does, we'll shoot two torpedoes aft and turn in toward the rocks, hoping she'll take a shortcut or give us a broadside and pile up."

*Time 0436.* "Junks ahead 900 yards!"

"Oh, my God! We can't stop now! Get me a last range on the escort. Then shift the radar to sweep back and forth 30° on either bow. I'll try to avoid the junks. God help them."

"Range 2700 yards."

The *Barb* highballed toward the 20-fathom curve, maneuvering wildly. Within minutes, the escorts stopped chasing and opened fire on the junks. Their inferior radar evidently couldn't distinguish between a junk and a submarine. No searchlights were turned on to attract a possible torpedo. At 0438, as the frigates commenced firing, the Tae Island lighthouse keeper illuminated. A fleeting thought—I must write a letter thanking him for his courteous navigational assistance in these crucial moments.

*Time 0446.* We untied the governors on the main engines, slowed to flank speed to cool down the bearings, and secured from battle stations. We sent a message to the *Picuda* on the results of our coastal foray. Separately, I sent another "eyes only" to the skipper. I offered to make a joint foray along the coast that coming evening or the next day, though we had only four fish. His response again: "DROP DEAD!"

*Time 0511.* The galloping ghost of the China coast crossed the 20-fathom curve with a sigh. I had never before appreciated how much water this is. But this would be the first place they'd search for us, so I decided to stay on the surface after dawn, half an hour away. The cannonade astern continued; poor junks, absorbing punishment meant for the *Barb*.

A track chart of the 23 January 1945 torpedo attack, drawn aboard the *Barb*.

*Time 0633.* "Radar has a plane coming in fast, seven miles!" I exercised a skipper's privilege to change his mind. We dived. "Sounding?"

As radar went under, the last range was 2.5 miles.

"Depth of water is 40 fathoms, sir."

I sat on the edge of the diving planesmen's bench and opened the ship's log. I wrote: "January 23, 1945—0635. Life begins at 40—fathoms."

I yawned as a great letdown set in. I told Swish to take her down to 150 feet to cool the *Barb* off and let everyone get some sleep.

"Captain," Swish said, "you need some shuteye, too. Do you realize you've been on the bridge for over 24 hours. You didn't even come down for a wee."

"Swish, I can assure you, I didn't do it in my britches. Tell me, how was it down here while we were in the harbor?"

"Captain, the few who were discombobulated calmed down when I said, 'No strain. If anything happens, we can get out and walk.' It was that shallow. Even less than 30 feet at times as I watched the fathometer. We got away from Namkwan like a two-year-old at Santa Anita in spite of one command you failed to give, sir."

"What was that?"

" 'Put four cases of beer in the cooler!' After we fired the stern tubes, and with all the explosions, I knew you were too occupied and had forgotten. So I got a few of the shaky ones to put four cases of beer in the cooler to cool themselves off."

"Swish, you're not a chief, you're a prince. Thanks!"

I went to the intercom. "Now hear this. Well done to each and every one! Eight hits, no errors. Be proud of a night none of us will ever forget. *Barb* did it and will live it forever. Your doughty Chief Saunders came through in the crunch and put four cases of beer in the cooler. We'll celebrate at 1600. So turn in all along and recharge your own batteries."

After diving we heard distant thunder below astern: explosions, gunfire, bombs, depth charges—all indiscernible but angry. The *Barb* slept.

But the rest of the world didn't! As debris from the exploding ships splattered the nearby coast, reports started flying thick and fast! Local residents ran out of their paper-windowed houses to watch the conflagration and battle. Soon overcome by the acrid fumes of the spreading smoke cloud that enveloped the harbor, they ran back in and buttoned up their abodes. Ten miles north of Namkwan Harbor at Pingyang, a pocket of resistance still held by the Chinese Army, the troops manned their bulwarks, anticipating an attack. Even there the sounds of the battle to the south were disruptive.

One of the merchant ship captains of Mo-Ta 32 bolted out of his bed in the room above the Harbor Tavern at the first explosion.* He knew the sound well. As more followed, he knew they were not bombs. How was it possible for a submarine to penetrate this port? What had been overlooked? He threw on clothes, raced downstairs barefooted, slid the front door open, and went outside as the ship exploded. In the fireball light he recognized her: the *Taikyo Maru,* an ammunition ship with the Moji-Takao 32 convoy heading for the Philippines. In the light he also saw his own ship afire, flaring up and settling as the water extinguished the blaze. He heard a voice say, "I pray not many men are killed." Whirling around, the captain saw it came from the tavern owner, Chang Tsou. With typical Japanese courtesy, the captain thanked Tsou, not knowing that the chief of the spy ring's concern was only for his pirates. Tsou wondered how this would affect his spy network.**

Radio Tokyo announced to the world on the evening of 22 January that the U.S. Third Fleet was bottled up in the South China Sea. That evening Admiral "Bull" Halsey slipped through Luzon Strait under an overcast and reentered the Pacific. Coincidently, the *Barb* told the *Picuda* at that same time that she had a convoy bottled up on the Chinese coast (Namkwan). The *Barb* struck first, just before dawn, against some 30 vessels (11 escorts and 16 ships, according to Japanese records) and laid claim to 3 ships sunk, 1 probably sunk, and 3 damaged. She escaped unharmed from what some called her kamikaze attack. In a post-dawn strike on Formosa, Halsey's Third Fleet laid claim to five tankers and five freighters sunk.

This time Japanese aircraft succeeded in striking back. A bomb hit the carrier *Langley,* and suicide planes crashed into the heavy carrier *Ticonderoga* and the destroyer *Maddox.* These kamikaze victims had to retire to Ulithi under escort.

*Time 1130.* In the *Barb* the last distant thunder below was logged. All gun firing, depth charging, and bombing ceased, and the ensuing silence awakened me. I dashed this message off to China:

FROM: BARB. ACTION: ComNavGroup China and China Air
YOUR LATEST INFO RESULTED IN EIGHT HITS IN POT OF GOLD X FOUND

---

* Related by General Tai Li to Admiral Miles.

** Postscript: In May 1991 I interviewed a retired Chinese merchant ship captain as to the feasibility of this interchange. He said the Japanese trained many of their captains to speak Mandarin Chinese fluently and used them as spies when their ships visited Chinese ports. Ashore there, they were indistinguishable from the Chinese. The world never knew how deeply Japanese spies had penetrated China before their invasion.

> YOUR CONVOY PLUS OTHERS AND POSSIBLE LARGE WARSHIPS ANCHORED
> AT NAMKWAN HARBOR LAST NIGHT X THREE SHIPS KNOWN SUNK X
> TERRIFIC EXPLOSION X CAN YOU GIVE US TYPES AND EXTENT OF OTHER
> DAMAGE X MANY THANKS FROM BARB X COAST NOT SUITABLE FOR SUB
> OPERATION X SUGGEST AIRCRAFT MINES X

In Chungking Admiral Miles relayed this message for action to Com-
SubPac Admiral Lockwood and for info to admirals Nimitz, Halsey, and
Ernest King (in Washington).

Miles then sent the *Barb:*

> WONDERFUL JOB GETTING SINKINGS IN THAT SPOT X AGREE WITH YOU
> ON MINING X THERE ARE NO PLANES AVAILABLE NOR HAVE THEY EVER
> FLOWN OVER THAT AREA X ATTEMPTING TO GET DAMAGE FOR YOU BY
> EVERY MEANS AVAILABLE X

Months later, Miles informed me privately that six spies had gone aboard
the largest ships posing as fishermen selling their catch. One had even
climbed up the anchor chain. By bribing the watch with free fish, they
were permitted to sleep on deck overnight. In the morning, they left before
the convoy got under way, having obtained information on troops, cargo,
and equipment. Miles further stated that the *Barb* had damn near ruined
the spy system. Three of the six pirates on different ships had been killed
by our torpedoes. Of the three returning, one wanted a camera to take
pictures, one quit, and one said nix to spying at Namkwan.

The U.S. Naval Group China operated under an agreement made with
Chinese General Tai Li (Chief of Underground Activities). This established
the Sino American Cooperative Organization (SACO) headed by Tai Li
and Admiral Miles. Within a day, reports on the attack tumbled in from
Tai Li's agents and from Chang Yee-chow, supervisor of the pirates who
cooperated with and assisted Admiral Miles.

The *Barb's* dispatch via Miles to Tai Li came in first. A second message
from the pirates was sent to a U.S. coastwatcher who radioed NavGrp
that the *Barb* had sunk three destroyers, damaged one, and sunk four
other ships. Third, U.S. Navy Lieutenant Carl Divelbiss, who had charge
of the coastwatchers of Section Two, Intelligence Net Four, reported in
from Changchow. "We had the extreme pleasure of being advised that out
of the northbound convoy reported by Pinghai on the 20th, a submarine
on the 23rd definitely sank 3 of the destroyers and damaged 4 other ships
of the convoy." Fourth, a report came via Chang Yee-chow that all 11
ships of the Japan-bound convoy had been sunk by the sub. I was also

credited (falsely) with sending a message to the coastwatcher: "Next time I'll put wheels on my keel."

Still another report came from U.S. Naval Unit Eight covering the Wenzhou environs. "On 23 January Allied planes and 30 or 40 warships were reported in engagement with a Japanese fleet south of the Yuhwan Peninsula at Namkwan." Radioman Robert Sinks of Fredericksburg, Texas— part of the SACO team of coastwatchers in China—was stationed at Changchow in charge of relaying messages he received from them and their pirates to headquarters at Chungking. He reported a message from Marine Sergeant William T. Stewart—their best and most daring coast- watcher—who was based at Ping Hai Island. Sergeant Stewart, who worked alone and frequently cruised along the coast using the friendly junks of the pirates working for him, reported that the submarine attack sank four ships and damaged three. The following day Sinks received a message from the pirates stating that a large number of bodies of dead soldiers were washed ashore. Then General Tai Li sent a message to the pirates to check the pockets of the clothing of all dead soldiers for possible intelligence information. Afterwards, headquarters considered that Stewart was taking too many risks, so they had him relieved, commended him, and sent him back to the States, where he received the Legion of Merit for that tour of duty.

**Time 1600.** Having surfaced after lunch, the *Barb* continued her retire- ment to the east with the seas rising. Promptly at eight bells, Swish passed the word, "Splice the main brace! All hands on deck, below deck; you can't show a leg, so shake a leg."

Chief Williams was awarded the honor of cutting through the convoy of ships embellishing the cake. His engines had established the new world's speed record for a surfaced sub of 23.5 knots and an average of 21.6 knots for the escape.

With a flourish, he presented me with the first piece. The interior was marbleized. I glanced at Phillips. "How come, baker man?"

"Captain, there were so many recipes and so many helpers, I would have had to make cupcakes! My solution was to use them all!"

**24 January.** While the *Barb* wallowed in a full gale, the ambience was no less stormy at Admiral Nimitz's headquarters in Guam. At 0800 the intelligence and operations briefing was held for a select few. Admiral Halsey's Task Force (TF) 38 strike had been on the table, revealing that TF 38 was no longer bottled up in the South China Sea as depicted by Radio Tokyo. That was no problem. The *Barb* strike, however, was a different kettle of fish. It would not be made public for months so as not to compromise the subs currently in the area and their new modus operandi.

Admiral Lockwood was adamant that submarine operations not be publicized for at least 60 days afterward. For instance, known losses of our subs were not listed as "overdue and presumed to be lost" until two months later. Admiral Nimitz understood this. As one of our earliest submariners, he would never forget the secret briefing at Pearl that went awry and cost us 10 subs. A politician had informed the press that the Japanese were not setting their depth charges deep enough to sink more of our submarines. A war crime!

"Admiral, how are we going to explain Radio Tokyo's rebroadcast of the China report of the coastal sea battle?" one of the briefers asked.

"Tell them the truth," said Nimitz. "We know of no sea battles between Fuzhou and Wenzhou, and therefore we have no comment."

"Admiral, that still leaves the void as to what caused the flap. Could we mention that we have only subs in the East China Sea?"

At this Lockwood took umbrage. "Absolutely not. The press would just keep digging. Chester, you remember that redhead you met when President Roosevelt wanted movies of a sub returning from a successful patrol? Well, he's still running the *Barb*, though I tried to cool him off on my staff. If I know him, he might just try the same thing again."

"Concur, Charlie. If we let the press blow this up as a rampaging sub, the Japanese may be smart enough to bait him in. He'd run into a hornet's nest. With Loughlin's report of the convoy destruction earlier, confirming four or five ships sunk by *Barb*, plus Miles's relay, she can't have very many pickles left for self-defense. Charlie, when you return to Pearl in February, be sure you have Admiral Towers debrief Fluckey on this patrol. Let's adjourn."

Admiral Nimitz's staff was correct. The reporters were perturbed that he seemed to be lying to them, even though they were abiding with complete censorship. Something was odd.

*25 January.* Back at his Pinghai post, Sergeant Stewart sent his pirates posing as fishermen aboard the remnants of Mo-Ta 32, which had anchored in a nearby bay. On their return, he radioed:

> CONVOY OF 4 DESTROYERS, 5 TROOP SHIPS AND 2 TANKERS ANCHORED
> HSING-HUA BAY X SHIPS HAD BEEN DAMAGED AND THEIR SPEED REDUCED
> TO 8 KNOTS X DEPARTING 0600 TOMORROW FOR KEELUNG, FORMOSA
> X INFO OBTAINED ON BOARD BY PIRATES X

Before the *Barb*'s attack on the 23rd at Namkwan, Mo-Ta 32 had 5 destroyer types, 2 subchasers, and 10 ships. Now reduced by one destroyer, two subchasers, and three ships, the remainder showed damage and would have to be diverted from Takao to the closer shipyards at Keelung. The missing were either sunk, heavily damaged, or engaged in rescue work. The voyage from Namkwan to this anchorage would require a day and a half.

**26 January. Time 0800.** Elsewhere, the flap created by the *Barb's* attack continued to spread. In Washington, D.C., Secretary of the Navy James Forrestal and Fleet Admiral Ernest King seated themselves to hear the daily Japanese Naval Activities Summary. Most interesting was the report from the East China Sea:

> Unconfirmed Chinese reports said that 50 American and Japanese ships battled for 9 hours in the East China Sea within 300 miles of Shanghai, 23 January, in the biggest naval engagement since last October.
>
> Japanese forces broke off the battle at noon and fled toward their homeland, some 650 miles to the northeast, the Chinese Army newspaper *Sao Tang Pao* at Chungking said.
>
> The newspaper said the engagement began at 0300 China time (0400 *Barb* zone time, 0500 Tokyo time) south of Wenzhou and Pingyang (10 miles north of Namkwan Harbor). Pingyang is 250 miles south of Shanghai and 200 miles north of Formosa. Gunfire was audible at Pingyang, it added [the *Barb* fired at 0402].
>
> According to press reports, Pacific Fleet Headquarters made no comment on the report.

The assemblage laughed when Admiral King reminded them of the *Barb's* attack message relayed to him by ComNavGrp China, then added that he looked forward to the day when the full story of her inspiring attack could be released.*

**29 January.** Blockading Formosa Straits, the *Picuda* was searching off Keelung while the *Barb* worked between the 10- and 20-fathom curves around Hai-tan channel. At 0337 the *Picuda* contacted a convoy. The *Barb* contacted a different, smaller convoy 85 miles away at 0450. Coincidence? We exchanged contact reports, content with our finds.

At exactly 0540 both the *Picuda* and *Barb* fired torpedoes in their different night surface attacks, miles apart. The *Picuda* fired six, all misses; six minutes later she fired her last four and sank the *Clyde Maru.* The *Barb* fired her last four torpedoes at the small convoy she had contacted

---

* According to Captain Walter Karig at the briefing for Public Information.

running along the 10-fathom curve. The small transport with a larger freighter trailing and one escort to seaward were passing off the northeast part of Hai-tan Island on a steady course and at a steady speed. One minute after the *Barb* fired at the freighter, the ships became invisible in a sudden rain squall. Two timed hits slowed the freighter; 10 minutes later she disappeared off radar in less than 55 feet of water. Sunk? Beached? Could this be the *Katsuura Maru* (1735 tons), believed to have hit a mine in this same locale, beached herself, and sunk? She was declared unsalvageable on 2 February 1945.

Out of fish, the remaining Loopers went home.*

---

* To settle the confusion surrounding the *Barb*'s attack of 23 January 1945—especially in light of missing Japanese records—the author revisited Namkwan Harbor in June 1991. His account of that trip appears as the epilogue, part 1.

# Chapter 20  *Barb* to Pearl—A Pearl to *Barb*

There's nothing quite like the frustration of being out of torpedoes and snared in the grip of a gale. Due to the extensive minefields stretching over 300 miles along the axis of the East China Sea, we had to take the long way home. Our operation order required our exit via the closest strait to Kyushu, which meant heading directly into the ferocious weather. We tried to race up sea, pounding along at two knots, gulping green water down the hatch. The boat was a mess; so were our temperaments. No one had an interest in splicing the main brace, so we postponed it until the sun shone again. Knowing that the *Barb* would have a one-night fueling stop at Midway and a few days at Pearl Harbor before going on to California didn't help matters. Nor did the three or four drifting mines per day we passed. Many wondered to themselves how many we missed at night.

Once we finished the statistics for the patrol report, the officers assisted me in working up an extensive mining plan, which envisioned mining the China coast at 29 locations from the Yellow Sea to Amoy. Ten wolfpacks would mine simultaneously on a given night. The crushing blow would drive the ships out into deep water where wolfpacks could chop them up. Captain Wally Ebert, a top-notch, extremely intelligent submariner on Admiral Miles's staff, was having similar thoughts. On 31 January ComNavGrp China sent the following message to Admiral Lockwood (info Admiral Nimitz):

> DUE COAST HUGGING POLICY JAPANESE SHIPPING SUBMARINE MINING
> COASTAL AREAS BETWEEN SHANGHAI AND HONG KONG RECOMMENDED
> X THIS COMMAND CAN SUBMIT SUITABLE PLACES X NIGHT SURFACE

MINING SUITABLE AT ALL FEASIBLE LOCALES X OTHERWISE THIS SHIPPING
WILL CONTINUE UNMOLESTED EXCEPT FOR EXTREMELY HAZARDOUS
ATTACKS SUCH AS MADE BY BARB X

*2 February.* In the wee hours before dawn, the *Barb* cleared the strait
in calm, sheltered, mined waters at breakneck speed.

The blue Pacific seemed like coming home after the miserable weather
and molested shallows of the East China Sea shelf. The main brace dutifully
spliced, we ate our half-size celebration cake. Senior cook Lawrence
Newland explained that flour was in short supply because a batch was full
of weevils and had to be dumped overboard. No matter: only one ship
sank in the heaving sea of icing.

*10 February.* The *Barb* moored and received a grand welcome at Midway.
The commodore apologized for a false alarm that had kept us in a holding
pattern 20 miles south because of the possibility of an attack, which didn't
materialize. Fortunately, our patrol report was typed and ready for mim-
eographing, because Admiral Lockwood wanted advance copies run off
and forwarded to Guam and Pearl.

At dawn, the *Barb* quietly crept off to Pearl. Her crew sewed feverishly
on her battle flag and ship pennants. En route we received the nightly
message from ComSubPac, providing news of subs reporting in and those
presumed lost. Elliott Loughlin had made his report as wolfpack commander.

NIGHTLY NEWS FROM THE LOOPERS X ON ONE JANUARY IN COMBINED
GUN ATTACK ON 300 TON PATROL BOAT PICUDA AND QUEENFISH DAMAGED
HER X BARB LATER BOARDED AND REMOVED EQUIPMENT AND THEN SANK
HER X SEVEN HOUR COORDINATED ATTACK BY LOOPERS ON FORMOSA
BOUND EIGHT SHIP CONVOY X BARB IN ONE SUBMERGED ATTACK AND
TWO NIGHT ATTACKS FIRED TWELVE TORPEDOES FOR TEN HITS WHICH
SANK ONE LARGE TRANSPORT ONE LARGE FREIGHTER TWO LARGE AM-
MUNITION SHIPS PROBABLY SANK ONE UNIDENTIFIED AND DAMAGED ONE
LARGE FREIGHTER-TRANSPORT X PICUDA HIT TWO UNIDENTIFIED
FREIGHTERS IN TWO NIGHT ATTACKS X QUEENFISH AFTER MISSING ON
THREE NIGHT ATTACKS FINALLY SANK ONE TANKER ON FOURTH ATTACK
X ONE SHIP ESCAPED ESCORTED BY ONE DESTROYER AND AT LEAST EIGHT
SMALLER ESCORTS X FURTHER REPORTS FORTHCOMING ON BARB IN
NAMKWAN HARBOR AND PICUDA AND BARB SINKINGS LATER X

Cheers roared throughout the boat when Jim read this message over the
intercom. I quickly took over the speaker. "Men, please listen carefully.
You and I know what *Barb* has accomplished on this patrol. There is no

boat better than ours, but I don't want anyone to brag about it or toot *Barb*'s horn. Let others toot it for her. Shipping is getting very thin, and this coast-hugging tactic that keeps ships out of sight of submerged subs will make things worse. Almost two-thirds of our subs are now returning with empty bags. There is bound to be some resentment from those less fortunate. Human nature is like that. We are not in competition with anyone but the enemy. We will do our utmost, and I trust that our efforts will be recognized factually and truthfully. Be careful in what you say and do. Don't exaggerate. Your conduct ashore will be more noticed as a *Barb* crewman. She expects more from you after this eleventh patrol. Be proud."

*15 February.* The red carpet was out as never before for the *Barb* as she entered Pearl Harbor. Her large battle flag flapped from the raised periscope. Pennants for ships sunk and damaged and for the naval weather picket flew from the antenna. Ships blew their whistles as we headed for the submarine base pier. The top brass and a band, plus a large crowd, were on hand. The news was out.

The normal patrol discussion with the brass was omitted. They had studied the advance copy of my report from Midway. We hoped to depart for San Francisco and Mare Island Navy Yard on the 18th. Admiral Lockwood, back from Guam, set departure for the 19th. He explained the delay to me privately: "Gene, I want you to debrief your last patrol to Admiral Towers, Fleet Admiral Nimitz's deputy here. His only open time is at 1000 on the 18th. My deputy Admiral Babe Brown will be with you because I'll be gone."

"Admiral, couldn't your Johnny Corbus in operations do it better? He'll know what Admiral Towers' interests are."

"Nope, as skipper you'll be more authentic and can better field the questions. My staff is preparing slides of your attacks and track charts. They'll be ready the day after tomorrow. Till then enjoy the Royal Hawaiian and take it easy—you've lost a bit of weight."

We were a bit surprised that a relief crew was on hand to take over the *Barb* for our four-day stint in Pearl. Each man threw clothes and his bundle of unread mail into a duffel bag, then boarded the waiting bus to the Pink Palace. The priority of pleasures was first, to read their mail; second, to shower, uninhibited by water restrictions; third, to consume fresh milk and lettuce from the open mess; and finally, to nap in a huge twin bed twice the width of their bunks, with an innerspring mattress instead of kapok.

Not that they didn't think of girls; there just weren't any around. I telephoned Jan, Helen, and Leilani Hull. They invited all of the officers to a barbecue and swim that evening, provided we could find some meat (which our commissary officer dug up). Otherwise, we would have a pineapple salad. Helen sparkled; Leilani, a lovely doll in pigtails, gloried in submarine stories. She had a memory befitting an elephant and reminded me that I still owed her my diving story. I was happy to oblige.

"Okay, Leilani, here goes. In July 1942, all the *S*-boats on the scouting line off Panama that defended the canal against a Japanese task force had been sent to Australia. This left the *Bass* and *Barracuda*—both broken down for lack of parts—and the still-active *Bonita*. The strategists then sent the *Bonita* to moor to a buoy at Cocos Island, 700 miles west of the canal. Our scouting line was now filled with tuna boats who would make contact with the enemy and then radio the *Bonita* to come out and sink the Japanese battle force. Ha!

"I was the engineer and diving officer. Arriving at Cocos, an uninhabited island, we found the mooring buoy. It had been anchored by a submarine rescue vessel whose divers had run a four-inch firehose from a creek to the buoy to provide fresh water. The depth at the buoy was 110 feet. What a bonanza!

"We hooked our hose to the buoy valve—no water. A hose coupling must have parted on the bottom. Leilani, do you remember that after my dive in Bermuda in November 1941 I sent for instructions for the antique diving gear we had?"

"Yes."

"None came. So, at quarters Shorty called for volunteers. None. 'Gene you're the diving officer. Dive!' Now, this cove is always full of fish, though we hadn't spotted any barracuda or sharks. The hand air pump working, deep-sea leads looped on my ankles, in swimming trunks and brass helmet with faceplates, over the side I went. Under orders.

"Small fish nibbled my toes. As I went down the tether line, with one hand on the firehose, a layer of 10-inch trigger fish actually pinched me at a depth of 10 feet. I saw a kaleidoscope of other beautiful fish all the way down, but the water level inside my helmet was rising. Frantic, I kept signaling for more air. As my feet touched the bottom, the water level was at my chin, then my lips, which were tightly compressed. More air, more air! Kneeling alongside the firehose, I could not look down. The white sand bottom was bright and clear, yet my head was tilted back. After the first snuffling of salt water up my nostrils, I almost gagged. Give up? Abort my mission at 110 foot depth?

"Never. I could still feel along the hose. A few yards along, there was a coupling. I could feel warm water coming out into the cooler sea water. The coupling had a bayonet-type lock that I was able to hook in tight.

What a relief. My 10 minutes below seemed like an hour. Hauled aboard on signal, I was greeted by cheering. When the hose was held up, water came gushing out."

"Did you go ashore?"

"In our wooden dinghy. We tried showering under a cliff waterfall about 20 yards inland. Lovely, until we put our clothes on again. Yelping, we then dived into the sea. We discovered why Cocos is uninhabited: it's covered with hordes of hungry red ants. Pirate loot from the cathedral in Lima may be buried there, for on the beach and in a creek one finds the names of pirate ships carved in the rocks. All searches have been driven away by the red ants.

"When we returned to Coco Solo, the instructions for our diving gear had arrived, together with a missing nameplate for the box. 'UNDER NO CIRCUMSTANCES USE THIS GEAR BELOW SIXTY-FOOT DEPTH.' What a shock!"

*18 February. Time 0900.* Admiral Brown's office was set up with a slide projector and screen. The admiral and I went over the slides in detail, which covered the patrol adequately. Both of us had a copy of the patrol report. He had a typed speech to introduce me. I didn't need a script or notes, the patrol was so fresh in my memory.

After a second cup of tea, I tightened my belt a notch. Admiral Towers was reputed to be brusque and domineering. I could not fully understand why a virgin debriefer like myself should compete with senior pros on the staff. Admiral Brown was there to carry me over the threshold with his speech. Then I was tossed into the lion's den, protected only by what was in my head.

At the Makalapa headquarters the admirals and senior officers were already seated when we were ushered onto the dais at the stroke of 1000. Admiral Brown launched forth with a brief sketch of the *Barb's* eighth, ninth, and tenth patrols, adding that these had already been approved by the Secretary of the Navy for a Presidential Unit Citation. Now her skipper, Gene Fluckey, would debrief her eleventh patrol in wolfpack with Loughlin's Loopers in the Formosa Straits. He left the platform, taking the vacant seat in the front row next to Towers.

The silence was not deafening, the reason being that I grabbed my pointer like a Samurai sword, strode forward, acknowledged Admiral Towers, and pounded the pointer on the deck with a bang, calling up the chart slide. I thought I heard Admiral Brown mutter, "First depth charge," but perhaps it was my imagination. There seemed to be some feedback from the microphone around my neck: as I described the Loopers' sortie from Guam, my voice changed to a falsetto. Nerves? Where was the imperturbable Fluckey? Some admirals' smiles brought my voice back to normal.

For the next 45 minutes my listeners joined me in the Formosa Straits with the Loopers, attacking and following the *Barb's* escapades, which seemed to grow as vivid in their minds as they were in mine as I relived them verbally. They shared in the frustration of missed convoys, the discovery of the dredging of Hai-tan channel, the excitement of attack, the quivering danger of mines and pursuing patrols. They heard the explosions of sinking ships and the messages that flew over the wires following the *Barb's* attack on the anchorage. They sensed the confusion that reigned as the Japanese—and others—tried to figure out what had happened. Finally, I summarized the damage we had done to the enemy war effort and brought the *Barb,* out of torpedoes, home to Pearl.

When I finished, no one moved. My eyes darted back and forth across the audience. There was a dead, unpunctuated silence. Admiral Brown broke it, standing up. "Gentlemen, any questions?"

Admiral Towers queried, "How many days were you submerged on this patrol, that is, submerged over a total of eight hours?"

"Zero, sir. During the Formosa Straits submerged attack on the eighth and the resulting escort holddown, we were down about five hours. After the Namkwan attack we dived to avoid a plane and slept for less than five hours. A submerged day in the *Barb* is rare. Surfaced, we contact more ships. The price is more bombs and depth charges, but it's worth it."

Other admirals also had questions. "Captain, did you have your men wear life jackets in Namkwan?"

"Sir, the exec suggested such, which I vetoed. I believed the best life jacket was to be calm and alert, not apprehensive."

"Have you any less-risky suggestions regarding eliminating the China coast traffic?"

Admiral Brown grimaced. Holding out my hand to pass the question to him, he shook his finger negatively for me to proceed.

"Sir, I have recommended the formation of a submarine battle, or task, force of 10 wolfpacks. These would be taken from low-production areas. Their task would be the simultaneous laying of minefields on a specific night at some 30 locations along the China coast, from the Yellow Sea to the Gulf of Tonkin. It could be the most devastating blow to their lifeline. Certainly it would drive them out into deep water where the wolfpacks could chop them up."

Silence.

"Well, Admiral Towers, gentlemen, I appreciate the time you have given up in your busy schedules to listen to a firsthand report of the *Barb*'s eleventh patrol. That's all there is to it. There is no more. Thank you."

Admiral Towers asked the flag officers to stay for a few minutes. Admiral Brown asked me to wait outside. Ten minutes later he joined me, smiling. "Gene, do you know what you've just done?"

"Admiral, I told you Johnny Corbus could debrief it better. They were so quiet and had so few questions."

"Gene, they weren't quiet. They were stunned! You are going to receive the Medal of Honor. This was the purpose of the debrief Nimitz wanted. Now it is approved."

Now I was stunned—and somewhat troubled. The criteria for awarding the Congressional Medal of Honor include the general provision that the recipient has acted without regard for his own personal safety. It's one thing for a Marine to throw himself on a hand grenade, sacrificing his life to save his buddies. It's quite another thing to suggest that a ship's captain has recklessly endangered not only his life but the lives of everyone in his crew, not to mention his ship. That was what bothered me about the decision.

It made what I thought was a very intelligent operation sound foolhardy. We suffered no injuries or damage. We received no close shells, bombs, or depth charges. I wouldn't have taken the *Barb* into Namkwan if I had thought we didn't have at least a 50-50 chance of coming out alive. I had weighed the risks and knew we had the advantage of a deep penetration on the surface that no one had tried before, including an hour's run to diving water, a chase in which I fully realized we needed a head start. My error was that the Japanese ignored the rocks awash. Aside from possible success with two down-the-throat shots, the junk fleet had saved us. Regretfully, they took a 10-hour murderous onslaught that was meant for the *Barb*. My last resort, if all else failed and we got into a gunfight with the frigates, was to shoot at their depth-charge racks. I explained all that to Admiral Brown and added, "I think that the Medal of Honor should be reserved for dead men who have done a valorous deed. Then the risk is certain."

"Gene, I think you're wrong. I have another pearl which will be more to your liking, however. Admiral Nimitz has approved a second Presidential Unit Citation for the *Barb*'s eleventh patrol alone. Your debrief closed it out."

"Sir, that I do appreciate. This she deserves. I assume all of this is for my ears only until President Roosevelt has signed the citation?"

"Correct."

"Now, admiral, in view of what you have said, my important objective is a fifth patrol as *Barb*'s skipper, which Lockwood promised. I want a graduation patrol."

"Charlie told me that you shall have it when you return from *Barb*'s overhaul in May, in spite of the psychiatrists' insistence on no more than four patrols in command. You have some unfinished business in the Okhotsk Sea alone or on patrol with the Loopers, right?"

"Right, sir, with perhaps a modification, depending on the degree of strangulation of Japan."

After lunch we took a last dip on the beach at Waikiki, savored drinks at the Outrigger Canoe Club, and sent letters home. I broke the good news to Marjorie that I would fly home in early March for a month's leave. Then I wrote to my daughter, telling her that we would drive across country in April and shift her school to Mare Island, California, until mid-May.

*19 February.* The *Barb* got under way for California, leaving practically no broken hearts behind. From the high periscope streamed her traditional Homeward Bound pennant of red, white, and blue—a foot in length for every month she had been away from home.

It was with regret that she detached herself from the Loopers under Elliott's outstanding leadership. To have had the privilege of fighting alongside such splendid packmates as the *Queenfish* and *Picuda* was a source of pride and inspiration to us in the *Barb*. We hoped that after our sojourn the Loopers would again form for a deeper invasion of enemy waters.

# Part V Intermezzo

# Chapter 21  Be It Ever So Humble

*20 February 1945.* En route to the navy yard at Mare Island, messages started coming in that elated the crew.

> COMSUBPACFLT AND COMMANDER IN CHIEF PACIFIC FLEET MOST HEARTILY GIVE THEIR CONGRATULATIONS TO GENE FLUCKEY COMMA THE OFFICERS AND MEN IN THE BARB UPON HAVING COMPLETED WHAT IS PROBABLY *THE BEST CONDUCTED PATROL OF THE WAR* X TWENTY HITS WERE MADE WITH TORPEDOES X EIGHT SHIPS SUNK ONE PROBABLY SUNK AND AT LEAST FOUR DAMAGED X DURING A DARING AND EXCELLENTLY PLANNED NIGHT SURFACE ATTACK ON THIRTY OR MORE ANCHORED SHIPS WITH RADAR EQUIPPED ESCORTS SCREENING IN A SITUATION WHERE THE PRIMARY DIFFICULTY WAS TO KEEP FROM GETTING TOO MANY HITS IN ONE SHIP COMMA THE BARB FIRED EIGHT FOR EIGHT HITS SINKING THREE AND DAMAGING AT LEAST TWO MORE X THESE SHIPS WERE ANCHORED NINETEEN MILES INSIDE THE TWENTY FATHOM CURVE AND SIX MILES MORE INSIDE THE TEN FATHOM CURVE X ALL OTHER SHIPS WERE SUNK AND DAMAGED IN WELL PLANNED AND BEAUTIFULLY EXECUTED SUBMERGED AND SURFACE ATTACKS X ALL HANDS ARE ENCOURAGED TO READ THE BARBS PATROL REPORT FOR A FULL APPRECIATION OF AN OUTSTANDING PATROL WHICH IS A MODEL FOR WOLFPACK COOPERATION AND UNSELFISHNESS ON THE PART OF THE BARB X

This message was sent to ships in the Pacific and to Washington, where it went to the President. It was relayed on to the Atlantic.

The following night another one came:

> COMMANDER TASK FORCE SEVENTY ONE [COMMANDING SUBMARINES

SOUTH PACIFIC IN AUSTRALIA] SENDS TO ALL SUBMARINES PACIFIC FLEET
VIA COMSUBPACFLT INFORMATION TO BARB X OUTSTANDING PERFOR-
MANCE BARBS LAST PATROL IS AN INSPIRATION FOR ALL HANDS X SEVENTH
FLEET SUBMARINES COMMANDER SENDS ADMIRATION AND CONGRATU-
LATIONS FROM THE BOATS DOWN UNDER FOR DARING AND SKILLFUL
OPERATIONS X

Nor did this end the messages.

THE ADMIRAL COMMANDING BRITISH SUBMARINES HAS READ WITH GREAT
ADMIRATION THE DESPATCHES ON BARBS RECENT PATROLS AND ON BEHALF
OF ALL BRITISH SUBMARINES WOULD LIKE TO ADD HIS CONGRATULATIONS
ON THE OUTSTANDING SUCCESS ACHIEVED BY BARB X

Admiral Lockwood's private message to me rounded out the collection.

The U.S. Naval Representative in London sent the surprising word that
Prime Minister Churchill intended to award a medal for the hazardous
rescue of the prisoners of war during our South China Sea patrol.

Regarding awards, nearly everyone in the *Barb* knew that when an
enemy naval vessel or a merchant ship of over 1000 tons was sunk during
a war patrol, the patrol would be considered successful, and an award of
the combat insignia would be authorized to all hands. Those already so
awarded would add a gold star to their insignia. If during a patrol more
ships are sunk, the medal awarded the captain moves up the ladder, from
Bronze Star to Silver Star to Navy Cross. For the Navy Cross, that captain
then has the privilege of recommending three people for the Silver Star
award, four for the Bronze Star, and six for Letters of Commendation with
Ribbon.

Having personally been awarded the Navy Cross on each of the *Barb*'s
first three patrols under my command, I had already recommended a total
of 43 shipmates for awards. Now I was in a quandary, for I did not intend
to accept the Medal of Honor. I still believed it should be reserved for the
deceased. My plan was to trade off the Medal of Honor recommendation
for a Navy Cross, with additional awards allowed for the crew.

Again I warned the crew not to toot the *Barb*'s horn. Bragging would
only cause resentment. Champions not only fall, they get careless or cocky.
Then something or someone knocks them off their self-made or media-
made pedestal.

*24 February.* Following the Yalta Conference, the chiefs of staffs of the
Armed Forces were back in Washington running their normal business.
The personal aide to Fleet Admiral King was Commander R. E. "Dusty"
Dornin, a distinguished submariner who had sunk 10 ships as captain of

the *Trigger.* He was also a close friend and classmate of mine from the Naval Academy. Dusty sent me a private message:

> FROM COMINCH X TO USS BARB X EYES ONLY COMMANDER FLUCKEY FROM DORNIN X CONGRATULATIONS ON THE GREATEST PATROL EVER X FORTUNATELY I WAS ABLE TO RETRIEVE THE APPROVED PRESIDENTIAL UNIT CITATION FOR BARBS EIGHTH COMMA NINTH COMMA AND TENTH PATROLS BEFORE IT WAS SIGNED BY PRESIDENT ROOSEVELT WHO IS IN POOR HEALTH X NO PROBLEM IN CONVINCING THE BOARD TO INCLUDE BARBS ELEVENTH PATROL WITH THE OTHERS SO ALL YOUR PATROLS IN COMMAND WILL HAVE SUCH CITATION X GENE YOU OWE ME ONE FOR THIS X DUSTY X

I just couldn't believe what I had read. Dusty's apparent intent to be helpful had robbed the *Barb* of a second Presidential Unit Citation (PUC) and deprived her crew of being entitled to wear an additional Gold Star on that highest ribbon awarded to the entire crew of a warship. It put me in a difficult position as the protector of my crew, a dilemma I had not anticipated and one practically impossible to correct. If I cried "foul" it would only provoke animosity among those few who tended to be green-eyed over the *Barb's* record. Dusty was completely familiar with the medal awards system, having received three Navy Crosses himself. The *Trigger* had received a PUC while under the command of the greatly admired Roy Benson. Under Dusty, a second PUC was awarded to *Trigger* for his first two patrols, but not his third. There was no doubt in my mind that Dusty was a winner, extremely cognizant of the PUC award system, and wasn't about to have *Trigger* surpassed.

Successfully preempted, knowing most submariners would have given their eyeteeth for a PUC, my decision lay in letting sleeping dogs lie. Dusty said that I owed him one. I'd return the favor with a good kick in his backside when I saw him in Washington—and we'd still be good friends.

*27 February.* The Golden Gate held out its arms to welcome the *Barb,* dwarfed as she passed under the bridge. Traffic there came to a halt as patriots parked, debarked, and waved at the *Barb* returning home with her large battle flag flying. Other ships and a convoy on the move blew their whistles in appreciation of successful returning warriors. The United States had never lost a war its people believed in, and this one would be no exception, judging from this heart-warming reception.

After a day of conferences, half the crew shoved off on a month's leave; the second half would go later. Jim Webster's wife, Doris, arrived, and he relieved me. Along with me the married officers—Max Duncan, Dave Teeters, Dick Gibson, and Jack Sheffield—departed on the first leave period. We had sympathy for the poor bachelors in the *Barb* who had only the glories of San Francisco to contend with.

Admiral Lockwood's private comments on the eleventh patrol were sent by Captain Buddy Yeomans:

> The all time all timer!
> Aggressive, tenacious, daring, unique, extremely injurious to the enemy.
>   In other words: Plus quam perfecto, as the Romans would say.
>   Award of combat insignia authorized.
>
> E. E. Yeomans
>
> P.S. Congratulations from all of us and the whole U.S.A.

Arriving home to a joyous reunion with Marjorie and Barbara, I thought I could forget the war for a month. It was as wonderful to be home as I had imagined, though I had forgotten how to relax. It was just as well, for Marjorie dragged me off to the ration board to get my stamps, then to the Safeway to use them. It was positively incredible! We stood at the tail end of a queue a block long. Marjorie bemoaned the everyday waiting. I had not realized the inconveniences the war put on civilians.

"Gene, stop fretting—there's nothing you can do about it." I watched a little old lady coming up the line with her package of meat. Smiling, she stopped and held out her hand to shake mine. Politely, I introduced myself and my spouse.

She held my hand. I could feel a piece of paper between our palms as she whispered, "It's the next number coming up. I always take two, to give one to you boys returning home. Now hurry down to get your meat."

We did. The number was being called a second time when I roared up to the counter, spent my allotment of ration stamps on a standing rib roast of beef, and stood by for a ram from Marjorie. "Honestly, some days I wait an hour. On others the meat is gone when I get to the head of the line. If this wasn't your first day at home, I'd probably wring your neck. Now I know why some call you Lucky Fluckey. What am I going to do with you? Just love you I guess, but I did want you to see what my problems are."

Life was sweet and every moment precious and treasured, even when we were with others. There were 13 women in the Submarine Wives' Club, all of whose husbands were at sea or "overdue and presumed lost." Marjorie invited them to our small apartment in Annapolis for drinks, then on to dinner and dancing at the North Severn Officers' Club.

Aside from a bureaucratic skirmish over the number of guests I was permitted to bring to the club—which we got around by having each of the wives pay her own way, after which I reimbursed them—the evening went beautifully, if sadly in obvious ways. The women questioned me endlessly. Since everything was rationed, they had brought their own drinks; all of them were determined to have a good time. I danced with each of them and they danced with each other. Their happiness was not forced— it was genuine pleasure in having even one man around.

Five of the women knew they were widows. As each snuggled close, dancing with me, my heart did flip-flops. I knew four others were widows, but they had not yet been notified. Damn the war! Already over half my submarine school classmates were buried in steel coffins at the bottom of the ocean. The horror those women had yet to face brought tears to my eyes as they danced with their eyes closed, dreaming of dancing with their husbands. I was struck with the thought that I was dancing on skeletons. I bit my lip and listened to their loving babble.

A fortnight disappeared, much too quickly. Then, in mid-March, a phone call came from Captain Johnny Davidson, Submarine Detail Officer in the Bureau of Personnel. A distinguished submariner, he had been the much-admired captain of the *S-44* in Panama when I was in the *S-42*. On 19 February 1943, as skipper of the *Blackfish* in European waters, he torpedoed and sank a German Navy ship. Only one other U.S. sub, the *Shad,* had so distinguished herself. Was I being transferred from the *Barb?* Beware the ides of March!

"Congratulations, Gene. You're to receive the Medal of Honor." The old reservations arose in my mind again. "There is one problem, however. President Roosevelt wants to present it personally, but is too ill to do so. Consequently, the ceremony is tentatively being set up for mid-April, so you'll have to return to Washington then."

"Johnny, please tell them that I am refusing the Medal of Honor because it should be reserved for dead men. They're trying to make a very intelligent operation sound foolhardy. Besides, to return in mid-April, I would have to interrupt my exec's leave in Tennessee. No, it's not on."

"Okay, Gene. I think you are wrong, but I'll pass the word." Two days later I received naval orders in the mail to report from 21 to 23 March to Admiral Richard Edwards, Vice Chief of Naval Operations. I had known him as the commander of the submarine base, New London, when I was at school there. I carefully rehearsed what I would say to the admiral in arranging a swap for a Navy Cross, a simple trade-off. I just wanted to recommend double the number of awards to the crew. I jotted down some notes.

*21 March.* Admiral Edwards was blunt and to the point: "I want you to stop giving the Navy trouble. Charlie Lockwood told me you gave him a hard time over the Medal of Honor. The *Barb* has just finished the greatest of submarine patrols. Our highest honor honors her as well as you. You don't make the rules. You're going to get the Medal of Honor whether you like it or not. So relax and accept it gracefully."

I pulled my sheet of notes out of my side pocket to present my case. He shoved his wastebasket my way. "Fluckey, throw those notes in there. I'm told you think that you should be dead. You think we're making your Namkwan harbor attack sound foolhardy instead of the smartly executed operation that it was. You're wrong on both counts. First, we need live, smart heroes in command positions to inspire our talented sailors. Second, no one who has read your report and understands your planning and judgment considers the attack too risky. Your battle plan was sound—absolute perfection in strategy and tactics. I say a hearty congratulations on a job magnificently well done. Now are you satisfied?"

"Well, for the *Barb,* yes, sir." I basketed my notes.

"Good. I knew you'd cooperate. You're a great team player, once you understand the objective. Do you know what one of our rather large problems is going to be after we finish this war?" Before I could answer he went on. "What are we going to do with some of our wooden-headed heroes?"

Then he passed me my schedule. "Today and tomorrow you will be with public relations for interviews and photos for future release to the press when the *Barb* story can be declassified. For your ears only, the President may not get well. Secretary Forrestal and Admiral King will present the Medal of Honor to you on the 23rd at a small ceremony in the secretary's office. Be sure to bring your wife. After that you're free."

We shook hands, and he added, "Charlie says you cadged him into promising a fifth patrol in *Barb.* Good hunting."

As I left, it occurred to me that Admiral King's office was next door. Dusty would be there. When I poked my head in, he tried to hide beneath his desk. "Dusty, come on out. I see you and I'm going to torpedo you."

"Gene, I didn't know. I swear, I didn't know," he said as I neared him.

"Dusty, you are a bum, a lout, a lowdown dirty dog. You radioed that I owe you one. Well, here it is." Spinning him around, I gave him a solid boot in the rear. "But I still love you."

"Gene, I'm truly sorry. I thought old Helpful Henry was doing you a favor. Still, there's really no difference. All the patrols are included."

"No difference! You know better than that. *Trigger* has two stars in her PUC ribbon. The *Barb* crew will now have only one, because Dornin stole the second star from them. Fie and a malediction on you that will ruin your sleep for the rest of your miserable life."

We chatted briefly about more mundane matters, then shook hands and traded brotherly slaps on the back, and I departed.

*23 March.* The Great Day was very quiet. Barbara was in school, and Marjorie and I stopped by my Dad's apartment to see if he was up to attending the presentation. On crutches, he thought all the walking would be too much, but he wanted us to stop by afterward. He had never seen the Medal of Honor.

The ceremony went off well with Navy photographers and no reporters. Secretary of the Navy Forrestal made the presentation for President Roosevelt. The local newspapers played it on the front page because I had been born in the District of Columbia.

The following day we packed up, took Barbara out of school, and headed for California, traveling like immigrants in our 1941 Plymouth. Driving across the country was a real pleasure with gas rationing: empty roads, less city traffic. Yellowstone Park had more bears than tourists and a choice

President Franklin Roosevelt was terminally ill, so the Medal of Honor was presented on 23 March 1945 by Secretary of the Navy James Forrestal, flanked by Chief of Naval Operations Fleet Admiral Ernest King and Commander and Mrs. Eugene Fluckey.

A farewell gathering at the Top of the Mark Hopkins, in San Francisco, 14 May 1945: (*clockwise*) Dick Gibson, Tom Gould, Max Duncan, Gene Fluckey, Tom King, Bill Walker, Dave Teeters, Phyllis Teeters, Trilby Duncan, Marjorie Fluckey, Margaret Gibson.

of empty cabins. We lunched at a nearby diner, and while I was paying for our hamburgers, the owner-chef spotted the Medal of Honor ribbon on my khaki uniform. "Soldier, they tell me I'm too old to fight, but I've studied and read all about the military. When that there ribbon comes into this railroad car, your money is no good. Please have something else."

Before I could respond, Barbara piped up, "How about some ice cream?" To which the chef replied, "You shore can, Honey, and here's some for your folks." For the first time in my life, my money was no good. I was losing control of my life.

As March blew itself out, we rolled into Mare Island. Our new home was half a Quonset hut, 15 feet wide and 30 feet long. Jammed inside were two bedrooms, a bath, and combination kitchen-dining-living space. For $30 a month we at least had happiness in being together, humble lodgings or not.

That same day Max and Trilby Duncan drove in along with Dick and Margaret Gibson, whom they had encountered by chance in New Mexico. Closely following were Dave and Phyl Teeters and Jack and Marjorie Sheffield. What a grand gang! On April Fools' Day the bachelors and Jim started their leaves.

The overhaul was progressing nicely. The *Barb* now had a 5-inch gun aft of the conning tower in place of the 4-inch gun forward. I wanted something to shoot when she was being chased on the surface, such as at Namkwan. More important was the ST radar for ranging, now installed in the periscope. This type of radar was a godsend. It took the guesswork out of estimating the masthead height of a ship when measuring a range by periscope during a submerged attack.

For fun on weekends, two or three couples would pile into a car, then take off for the Top of the Mark Hopkins, Fisherman's Wharf, or any of the zillion spots around the Golden Gate. Another great attraction was Mom Chung's Golden Dolphin Society. Mom Chung, a lovable and renowned Chinese surgeon, laid on a feast at her home every Sunday for her far-flung members passing near the area. The backbone of her membership consisted of three groups: Golden Dolphins (submariners), Bastards (aviators), and Kiwis (notables of the performing arts and business figures). The Fluckeys and Duncans were inducted into this extremely interesting group. Wonderful people like Helen Hayes or Sophie Tucker popped up for gatherings of the clan. Mom always arranged some occasion to celebrate, even if the date was wrong.

One day in early May, with the rejuvenated *Barb* soon due to head back into combat, Mom surprised us by announcing a birthday celebration on the following Sunday. Present would be two people who were born on the same day in October. As they would be far apart at that time, they would be Mom's guests of honor.

On that special Sunday, with the party rolling along, Mom introduced the honorees to the crowd. "Here they are—Gene Fluckey and Lili Pons." Stunned, I excused myself from André Kostelanetz, who had been teaching me how to drink a Nickolayev cocktail, and stepped forward. Giving Mom a peck on the cheek, I told her how honored I was to be included with such a lovely Libra.

Mom seated us together at the head of the table set for 50 guests. Long an admirer of this famous virtuosa, I was delighted to have the opportunity to talk to her without interruption. I had a personal question that had always puzzled me. Dared I ask it?

"Lili, as a shower singer myself, I don't have to worry about my audience, because my two cocker spaniels always howl. Yet, may I ask you a personal question about your singing?"

"Of course."

"Then tell me your feelings at the beginning of a performance when the audience is quiet and you are about to sing. You must know they expect absolute perfection. Does this give you a problem?"

"Give me a problem, Gene? I vomit several times during that day and can't eat. When I'm on the stage, I think that if a croak comes out I'll die. That first note is the most important note of all. Then it comes out clear as a bell and I'm in for the night. Afterwards, I go out and wrap myself around a 14-ounce steak and sleep soundly."

"Bully for you! My admiration increases. So many believe that voices are God-given and performances a snap. They forget the years of training and study that go with it and the burden of being an idol."

At the end of a fabulous feast, a Bastard and a Kiwi brought forth a candle-lit cake. The group broke into "Happy Birthday." Fearing my breath might be a blow torch, I asked Lili to do the honors. She insisted we blow together. I blew so hard it knocked over a couple of candles.

Thus ended a perfect day.


Now the *Barb* was girding her loins for combat and readying herself for sea and weapon trials. Plans for her twelfth patrol—my graduation patrol and final one in command—kept grinding around in my mind. I wandered over to the new shipbuilding section of the yard. For each class of ships the flow of materials was smooth: the workers knew where every piece fit, so it was a fast and efficient operation.

In the repair section for battle-damaged ships, on the other hand, every problem was unique. Reconstruction and supply problems created a logjam of work for many classes of ships. I wondered which was more injurious to the Japanese industrial effort: sinking or damaging.

Nightly news dispatches from ComSubPac showed more subs coming off patrol with little or no bag. Hunting was poor. My thoughts went back to that cable station in Kunashiri Strait in the Kuriles. Twice our intended attack had been frustrated by the perpetual fog pocket covering the area. The station held a treasure of two million dollars' worth of equipment, but more important, its destruction would force ship routing to be radioed, thus giving our ULTRA intercept decoders a chance to uncover the routes.

Looking at the vacant space on the *Barb* forward of the conning tower where the old 4-inch gun had been, I recalled a wardroom conversation a year earlier.

During the *Barb*'s seventh patrol, off the east coast of Formosa, I was bored stiff with the inability to discover action, as were the junior officers. When the captain suggested a rescue of General Jonathan Wainwright, who had surrendered at Corregidor and been removed to Formosa, I volunteered to take the rubber boat and go look for him. The skipper said he was joking.

Annoyed, I requested permission to take one of the *Barb*'s three high explosive, self-scuttling charges ashore in a rubber boat to blow up a railroad bridge between north and south Formosa. The request was denied.

At that I turned the conversation to the possible use of other weapons such as rockets in submarines. Tuck Weaver was interested, but the skipper discounted that idea because rockets were too inaccurate. The 4-inch gun was far superior, rendering rockets useless.

But were they? Our gun had been useless in attacking the cable station, even with accurate navigational positioning. We had a ready-made space for a rocket launcher forward. If the exploding pattern of the rocket salvo was small enough, rockets just might be our ticket to scratching that target.

Realizing that I needed to be educated in rocketry, I wrote a letter to the most astute submariner I knew, Commander Harry Hull, a more senior friend from academy days who was the "can-do" gunnery officer on ComSubPac's staff.

I described my target. If Harry concurred that rockets were a practical solution, I would promise one million dollars' worth of damage for a launcher and 100 rockets. The *Barb* would arrive in Pearl on 24 May and depart on patrol 8 June, so I asked him to expedite my request, knowing only Harry could do it. I decided on a letter rather than a message, because the broad staff distribution of messages might give some doubter a chance to squash the idea for reasons I was beginning to distrust. No submarine in the world had ever carried rockets. The war was awakening those whose philosophy had been, "Be not the first by which the new are tried, nor yet the last to lay the old aside." Some officers still resisted the irresistible, everlasting tide of change. I didn't want to see our request strangled to death with endless staff studies. Harry could override those problems.

The *Barb* spent a day outside the Golden Gate for test-depth dives and 5-inch gun training. Another day at a degaussing pier in San Francisco made her less sensitive to magnetic mines or torpedoes—a real joy to contemplate. Having shaken off her magnetism overnight, she would return to Mare Island the next morning.

Over half the officers and a quarter of the enlisted men were married. With time running out, almost all the wives drove down from Mare Island to grasp every moment possible with their husbands. A large group dined at Emilio's and encountered Sonja Henie, Ray Milland, and other movie stars. Marjorie's father came up from Palm Springs for the jovial night.

The next night, our last, a splendid ship's party was laid on at Mare Island, with music and dancing. By 2130 all the married couples became "sleepy" and crept off to their nests. The bachelors held forth until midnight.

*16 May.* Cars were packed early as families began the trek homeward. The men would be too busy after 0900 to see them.

# Honolulu Star-Bulletin

## 14 PAGES—HONOLULU, T. H., U. S. A., MONDAY, JUNE 4, 1945—14 PAGES

Hawaiian Star, Volume LI. No. 16449
Evening Bulletin Est 1882, No. 17369

★★★★    AIRPLANE DELIVERY ON
ISLANDS OTHER THAN OAHU  7¢    PRICE
ON OAHU  5¢

# U. S. Sub Barb Sneaks Into Enemy Harbor, Has Field Day

WASHINGTON, June 4. (U.P.)—The daring exploit of an American submarine which sneaked at night into a harbor jammed with Japanese ships, surfaced under the muzzles of enemy guns, torpedoed vessels all around it and then got away in a miraculous exhibition of broken field running, was revealed today by the navy.

It was a thriller such as usually is encountered in boys' war books but it sounds too incredible to really happen. But this episode was real enough to have earned Cmdr. Eugene B. Fluckey of the submarine Barb the Medal of Honor and his entire crew the Presidential Unit Citation.

Cmdr. Fluckey long had suspected that a certain harbor concealed a haven hiding a large number of Japanese tankers, munitions ships and warships. On a dark night when visibility was poor, Cmdr. Fluckey found the target. The navy said Cmdr. Fluckey took his ship on what appeared to be a hopeless mission because the anchored convoy he was after was hidden behind a protecting screen of escorts concentrated on every logical approach.

The water was so shallow that the 1,500 ton Barb was forced to remain on the surface during her approach and for at least an hour afterward.

Faced with one of the best targets he had ever had, Cmdr. Fluckey decided to attack despite the odds. The escape involved a flight through uncharted water filled with mines and rocks and packed with fishing junks.

But Cmdr. Fluckey figured that the Japanese escort vessels would hesitate to make a run through the rocks and that they would find the junks a handicap.

The Barb moved inside the screen of Japanese escorts and let go with all the torpedoes she could fire in the time available. Then Cmdr. Fluckey ordered full right rudder and moved toward the rocks.

From the bridge Cmdr. Fluckey saw the Japanese ships erupting like a nest of volcanoes. Columns of fire leaped from several vessels. The first target settled in the water and the others burned.

Only the Japanese know how many went down. Cmdr. Fluckey had no time to count.

The Japanese escorts moved up in hot pursuit, tossing a hail of shells at the Barb. Many were close, but all missed. The junks confused the enemy and several became targets for Japanese guns instead of the Barb.

The sub skinned by the rocks and reached open water at dawn. A Japanese plane spotted her and she submerged for the first time since she sighted the target.

One of many newspaper accounts of the *Barb*'s exploits.

I had dreaded the thought of Marjorie driving all the way across the country alone with Barbara and our two cocker spaniels in her condition. Fortunately, the Duncans had sold their car, so Trilby Duncan and Phyllis Teeters joined Marjorie and Barbara. We kissed the girls good-bye and they headed for New Orleans; we marched off to war.

Westward through the Golden Gate, the *Barb* proudly flew her beautiful new battle flag, a gift from the ladies at the Sail Loft shop who worked there after hours. Cars stopped and people got out and cheered. Waving back, I wished that my Dad could have experienced this moment.

# Chapter 22 Overkill

As California slipped over the horizon, so did our dreamland. The metamorphosis was abrupt. "Start your zigzag plan! That Japanese sub that shelled the coast may have a sister lurking below."

"Aye, captain. Lookouts, smarten up. Concentrate on surface search for periscopes."

We were at war once more.

Unsatisfied with our readiness posture, we all took it upon ourselves to bring our skills up to the level where they meshed smoothly into the well-oiled team that gave us the combat edge. Day and night we drilled on every phase of combat and casualties that might be suffered, striving for personal excellence at every station.

Word had seeped out through the yeoman grapevine from ComSubPac staff that I would keep the *Barb* for one more patrol. All of us were happy that we would be together again, for we knew what we could do, and we were ready for any challenge. To be sure, top-notch shipmates had been transferred. Paul Monroe had left, as had Arthur, Bowden, Donnelly, Elliman, Hinson, Houston, Hudgens, Lego, Novak, Peterson, and others. All of them were outstanding submariners. But we still had good men. Turnage had returned. The replacements would be well tutored. The staff personnel office worked well with Jim, because many had requested to come aboard. He looked over the records, picking the best.

Regrettably, the Loopers had dispersed. The *Picuda* had left for overhaul, and Elliott Loughlin had departed *Queenfish* for the training staff. So I searched for an independent area.

With the sad death of President Roosevelt on 12 April, only 20 days after I had accepted the Medal of Honor as the representative of the *Barb,* came no change. Fortunately, the fighting war was run by the military, not the politicians. The President settled any disagreements and conferred with the Allied leaders using his military staff. The civilian secretaries took care of needed production to support the war effort. This was a winning combination, and now President Harry Truman followed along.

On April Fools' Day Okinawa was invaded. The struggle would be a long one, combining the Navy, Marines, and Army. On 6 April the giant battleship *Yamato,* the cruiser *Yahagi,* and eight destroyers were reported bound for Okinawa by two subs patrolling off southern Kyushu. On 7 April this force was sunk by Task Force 58 under the strategic command of Admiral Raymond Spruance. Since then there had been no naval engagements.

These developments affected the *Barb,* since I was examining all the patrol areas to test out my evolving theory of maximum harassment with minimum force. Such a tactic would occupy a disproportionate enemy force to counter the threat. It would also enable incursions by our own forces into areas where the normal protectors had been temporarily moved out to beef up the forces assigned to counter the threat elsewhere.

With the battle raging on Okinawa and our Fifth Fleet controlling the East China Sea, we had plenty of submarines on station. Many were engaged in picket duty, reporting aircraft movements of kamikaze planes being ferried from Kyushu to Okinawa to attack the offshore fleet. Others were engaged on lifeguard stations to rescue downed aviators. These patrols, though necessary, would not develop any new concepts for submarine strategy or tactics.

Carrier strikes on the Inland Sea found little to attack. This intrigued me. Where were the remnants of the Japanese fleet holed up? Should they be in some locale accessible to submarines, I would forgo the development of harassing. But first I needed to check with intelligence and weigh the probability of finding anything. Otherwise, the Okhotsk Sea and my unfinished business with the cable station held the greatest promise for the further development of submarine warfare.

*24 May.* The *Barb* arrived in Pearl Harbor with little fanfare. Our rough schedule on arrival showed four days for voyage repairs, eight days of underway training, four days for loading and minor repairs, then off on patrol.

Leaving Jim to take care of things, I hurried off to discuss rockets with Harry Hull. He concurred with me completely because I had a worthwhile and important target. Harry had selected the 5-inch, spin-stabilized, high-capacity rocket as the best for our purpose. Now we looked at the various launchers. I wanted something simple and reliable that did not require especially trained operators on board. We chose one quickly, for our minds were in unison. As I left for operations, Harry picked up the phone to complete the project. He assured me that the 100 rockets could be obtained on Oahu. It was good to have his brilliant and capable support to plow through the bureaucracy.

When I dropped in to talk areas for the *Barb*'s twelfth patrol with Johnny Corbus in operations, I was dealt a body blow. Johnny told me we had been assigned lifeguard duty off Wake Island.

"What?"

"That's what Dick Voge sent from Guam."

"Johnny, that's not what Uncle Charlie promised me. He said I could pick my area to develop new concepts. This is my graduation patrol. I'm not going to be shuttled off to the sidelines without direct contact with the admiral."

"Gene, you do realize that there is a general military policy that those awarded the Medal of Honor are not to be exposed to danger unnecessarily? Dick must have been thinking of this."

"Well, they can damn well have the medal back. Harry Hull has an okay to get *Barb* 100 rockets, and I have guaranteed one million dollars' worth of damage for them. I want an area in the southern part of the Okhotsk Sea alone."

"Let's do this, Gene. I know nothing of what's up between you and the admiral, but I'm betting on you. He'll be here in two days on a routine visit. I'll hold up on the operation order for Wake Island without sending word to Guam. As you know, Dick's pretty strong-minded—he might win this one."

"Thanks a lot, Johnny; I love you." I borrowed a chart of the Okhotsk Sea to take with me.

Next, I went down to the basement, where Jasper Holmes, chief of the codebreakers, had his office for submarines. "Gene, I hear you're taking rockets this patrol. What's cooking?"

"Problems, searching for solutions, as usual, Jasper. I need your help and I need straight answers."

"Shoot."

"Does intelligence know where the remnants of the Japanese fleet are holed up?"

"No, we do not."

"Please take a gander at this chart. The soundings are based on a

Swedish survey in 1894. It's a volcanic area from Hokkaido up the Kurile Island chain to lower Kamchatka, over to Sakhalin Island—or Karafuto, the lower, Japanese half. The soundings are lousy. On *Barb*'s eighth patrol we corrected many. The shore-side information is equally obsolete. Now, let's focus on Karafuto. The *Coast Pilot* is antiquated, but the Japanese have published that they have a large air base in the northwestern corner of Patience Bay at Shikuka. If you were Japanese, trying to hide the remnants of your fleet from carrier strikes on the Inland Sea, Honshu, and Kyushu, where would you place them? Why not Shikuka?"

Jasper mulled that over. "To the best of my knowledge, we've never had an agent on Karafuto nor has a plane flown over any part of it. There's a shallow shelf, like the China coast, leading up to Shikuka, so none of our subs have ever observed the harbor or the state of construction of support facilities ashore. A big buildup there is logical in case the Soviet Union enters the war. Gene, it's a long shot well worth exploring. It would be a surprise. Harry has told me about the cable station in Kunashiri Strait as your principal target for the rockets. I'll buy that. We get so little intercept concerning the Okhotsk Sea. That alone is worth the patrol."

"Jasper, Dick has scheduled *Barb* for Wake Island patrol as lifeguard."

"What? Wake is just a practice target for new carriers heading west."

"I know. But Admiral Lockwood gave me his word that I could choose my area to develop new concepts. All I want is Kunashiri Strait, the north coast of Hokkaido to La Perouse Strait, nearby Aniwa Bay leading to the port of Otomari, and the whole of Patience Bay. When he gets here in two days, I'll tell him that you back my patrol in the Okhotsk, both from the destruction of the cable station and the possibility of finding the remnants of the Imperial Japanese Fleet. Agreed?"

"Sure, I'll buy it."

"Thanks, Jasper."

A few days later the flag lieutenant phoned to say the admiral would like to see me. I was ready for him with all my arguments, but none was necessary. The admiral hadn't forgotten his promise, nor had he seen Dick Voge's Wake Island assignment for the *Barb*. He told me to forget it.

"Gene, Johnny says you've picked the southern Okhotsk. Harry has given me the dope on the rockets, which I have approved. I've checked with Jasper and he strongly supports you. You've done your homework well. Every turn I make, I find your backers. So be it. We do have a wolfpack going into the Okhotsk Sea soon, but Johnny says that they won't interfere with your small area. You have your wish."

"Thank you, admiral. I was expecting a struggle."

"Then struggle with this proposition, Gene. It's called Project Three.

Submarine still photos and moving pictures of their deeds are, on the whole, very poor, understandably. I want to have a good film made showing true submarining. One great lack is a film from the time a target is sighted until it sinks. I've chosen *Barb* to do it with a professional photographer on board because I know you'll find something. What's your honest opinion? Be frank."

"Sir, it's late to do this now. This project should have been one for the coastal shipping, with little or no escort. Second, the probability of being sighted increases enormously in comparison with my four-and-a-half-second periscope exposures. I should think a camera shot should be at least 15 seconds. Suppose we lose the ship because she evades?"

"Gene, it would be worth it."

"Understood, sir. Rest assured we'll give it a try. Shipping is getting so thin that I spent all last night dreaming of chasing a seven-ton tanker. It's frustrating to have so many subs returning with empty bags. Admiral, I do thank you for this opportunity to show what a harassing submarine can do. The rockets are most important. *Barb* shoves off in two weeks, so if Harry runs into problems, please put your horsepower behind him. I've never seen anything stop you."

"You'll have it. Incidentally, Admiral Nimitz was very annoyed that Admiral King combined *Barb*'s second Presidential Unit Citation with the first one. However, he doesn't wish to contest it because he needs the full support of both Admiral King and Admiral Leahy in current differences with General MacArthur. I am sorry."

"Sir, I understand his concern."

From the admiral's office I dropped in on Harry. "Gene, we've got a problem. There is only one rocket launcher in all the Hawaiian Islands. It's down at La Haina Roads being tested to see if it can possibly be used to defend against kamikazes. Tough break."

"When will they be through with the testing?"

"Their estimate is three weeks."

"Uncle Charlie just told me that he will help you. What model is it?"

"It's the simplest of all, the pipe-rack type they use on the landing ships for saturation bombardments. I'll see if the admiral can expedite the tests to less than two weeks."

Back on the *Barb,* two new officers were waiting for me. Jim introduced me to Lieutenant Chuck Hill and Ensign William Masek, both fresh out of sub school. After a few pleasantries we got down to fitting them into their submarine niche.

"Chuck, since you're relieving Dick Gibson, we'll assign you as assistant torpedo, gunnery, and rockets officer. Bill, we must change your name because we already have a Bill Walker. To avoid confusion, particularly in emergencies, I now christen thee 'Willy.' You are our new commissary officer." His face fell. "Willy, I know you want to head a department that you feel is more important. I've been commissary officer in both destroyers and submarines. Let me tell you something, that job is most important for the morale of the crew. Excellent food, well prepared, and we'll have a happy crew. Furthermore, learn fast, for your life may depend on it. If *Barb* gets stranded, captured, or sunk, and we are survivors, I'm going to tell the Japanese that I'm the commissary officer and you're the captain. So you better know more about your job than I do. You'll also be the assistant first lieutenant under Walker, taking care of the hull, damage control, et cetera. Glad to have you both aboard; you'll not be bored."

Our underway period with a much-decorated training officer, Creed Burlingame, was perfection: eight fish fired—eight hits. Max had become a genius on the TDC. Creed even thought we were a thick-skinned submarine tested to 435 feet. I assured him that although the *Barb* was thin skinned, we had been that deep before, but my criterion for going no deeper was the forward battery deck. Our maximum depth was when it started to rumple or flex, not 312 feet.

About the time Creed had left his successful *Silversides* after his last patrol in June 1943, Admiral Lockwood made the submarine force the "Silent Service" because of the infamous leak. What Creed had used as a relatively safe evasive depth became a bowling alley of depth charges when used by the *Barb*. I asked him not to report the deep submergence test of ours, for someone who wasn't fighting might try to prohibit our using that possible means of salvation, unless, of course, he wanted to come with us on patrol.

Returning a day early to port with everything completed gave me the opportunity to see Dick Gibson off to a new sub. I would miss his happy, ready wit and astute observations. In his five patrols in the *Barb* he had developed into a top-notch officer whose judgment I depended on.

*4 June. Time 0700.* The crew was at quarters, having spent the night on board after returning from our sea trials. Jim told them to take the day off as soon as they had submitted their work requests to the sub base, since loading was scheduled for Tuesday. This included everyone except the skeleton watch.

I called Harry about the rocket launcher. He was not unhappy, but it would be touch and go. The kamikaze project was a top priority because of the damage being inflicted on the landing fleet off Okinawa. Their test program had been speeded up to 16 hours a day to accommodate the

*Barb.* Both launcher and rockets should be on board by late Wednesday or early Thursday. While I admired Harry's ability to eliminate the red tape, I reminded him that our departure was set for noon Friday. I knew that I was becoming oversensitive to the pressure of time and was frustrated by getting unreliable information. But these were not toys I wanted; the game we were playing was for keeps.

Tom King, the duty officer, came in and offered to give me a hand as I laid out charts of our Okhotsk area. I still had a lot of information to sift through, however, so I told him to take the day off until late afternoon. I would take his watch.

As I was working, the gangway watch informed me that a shore patrol detachment had arrived in a jeep with Jesse Penna and wanted to turn him over to the duty officer. I went topside, where one of the Marines saluted and said, "Sir, we caught this man in Waikiki. He claims his ship is the *Barb.*"

"She is."

"Our orders are to return him to his ship. Here's the report slip."

I couldn't believe it. The report slip said that Motor Machinist's Mate Jesse Penna had been arrested for having his hat on the back of his head. "And you drove him 15 miles back from Waikiki for that?"

"Orders, sir."

"That's all. I'll take care of Penna."

As they drove off, I turned to the culprit. "Okay, Jesse, spill it. What did you really do? Everything!"

"Honest, captain, I took my camera, and all I was doing was taking snapshots when those two grabbed me. Maybe I pushed my hat back while snapping."

"Jesse, you neither smoke nor drink. Are you pulling my leg?"

"No, sir, I swear."

"I believe you. So go back ashore and keep that blasted hat on the front of your head," I said as I tore the report in two. "And throw this report in the trash bin at the head of the pier."

I returned to my studies. Harry had given me a single sheet of data on the rocket system. I found it much more limited than our 5-inch gun. One advantage, however, was that the rocket packed much more wallop and weighed about 20 pounds—not bad. On the minus side, the Mark 51 Launcher was fixed in azimuth, which meant that we would have to aim the ship at the target. Ranges were limited to four—4500, 4900, 5150, and 5250 yards—which would have to be set before the launcher was loaded. The desired range was set by adjusting pipe pins to the corresponding launch angle: 30°, 35°, 40°, 45°. Added to that was the minor complication of drift, or deflection, which varied from half a degree left at 4500 yards to 3° right at 5250 yards.

For the cable station, no problem. We had waited around outside the fog pocket before, with a good navigational fix position. Entering the harbor of a city with factories would be entirely different. I would have to be like a jockey coming down the homestretch who suddenly brought his mount to a shuddering halt. Then I would give our filly a flick of my crop to line her up and let fly. Twelve rockets would belch out in four-and-a-half seconds for a 30-second flight. The *Barb* would then reverse course, and we would ride her hell-for-leather through the defenders' gauntlet. I found this extremely interesting.

Tom returned at 1600 to take over the watch. I tossed a chicken and my swimming trunks in a tote bag and took off for a relaxing swim with Jan, Helen, and sweet Leilani Hull. When I was with them, my problems were shucked. I simply lazed in the pool overlooking Diamond Head.

Tuesday, loading day, Photographer's Mate First Class Bob Singer reported aboard for Lockwood's Project Three. We examined the bridge and conning tower to best position him. He was eager to get started on his first war patrol assignment. Tall, slender, and affable, he would be an asset to our team. I explained we would have many drills en route to our patrol area.

Max joined us and reported that all of our gun ammunition was on board: 105 5-inch shells, 400 40-mm, 7300 20-mm, 2400 .50 caliber, and 20,000 smaller-caliber rounds, and a case of rifle grenades. There was still space in the magazine for the rockets. Knowing we'd need all we could carry for our harassment mission, I told him to pick up more 5-inch and 40-mm shells for the space left after we loaded the rockets. Our torpedo allotment included 19 Mark 18 electrics, 3 Mark 28 homing fish, and 4 Mark 27 cuties (homing torpedoes). The homers were experimental. I hoped they were well tested, because they would be extremely useful and save a lot of torpedoes—if we could find any targets.

**6 June.** My wedding anniversary. The *Barb* was loaded except for last-minute fresh foods. I came aboard early to await the rockets and to write Marjorie an amorous letter. Bill had the duty, but I relieved him so he could run off until late afternoon. Out came the charts.

Certain towns in Patience Bay intrigued me. The bay was shaped like a three-quarter moon, the northwest corner of which hid the big airbase at the factory town of Shikuka. Tracing the chart clockwise, I noted the large canneries at Chiri and that the island of Kaihyo To contained a lighthouse and a large seal rookery protected by a company of soldiers each summer.

The phone rang, interrupting my search for targets. "Sir, Bochenko here.

There's a Marine major in a jeep alongside who wants to speak to the captain."

"Send him to the wardroom."

As I arrived in the meeting place, I offered, "Coffee, major?"

"Thanks, not now."

"How can I help you?"

"Captain, on Monday our shore patrol brought one of your men back to the *Barb* on report. The action taken for the offense was to be returned to us within 24 hours. This has not been received. I'm the follow-up after 48 hours to ascertain the action. Perhaps your executive officer forgot to put it in the Guard Mail."

"Major, I take full responsibility for any dereliction. Look at this stack of paperwork. We've come back from overhaul ready to go on patrol. It must be lost somewhere in these papers. What does it look like?"

"Just a regular report slip. The bottom quarter is what is returned showing the disciplinary action taken."

"Well, it's not in this pile. To save me searching, do you have another slip to make me a copy, so I'll know what this is all about?"

"Certainly, commander."

From the copy he held in his hand, he made out a new report. I read it and took his pen to award punishment.

"Major, now I remember. It's the Jesse Penna case. He will never do that again," I said as I wrote SUMMARY COURT-MARTIAL. "His case hasn't been tried yet."

The major took the paper, uttering "Whew," and departed.

Back to the charts in Patience Bay. On the western side lay the industrial town of Shiritori (Makarov), a worthy target. Twenty miles south, the town of Kashiho looked good. A high mountain range on the south rim divided Patience Bay and Aniwa Bay, the latter a half-moon. The city of Otomari (Korsakov), at its northernmost part, was the main ferry and railway terminal for Karafuto.

The railway line kept my rapt attention for an hour. From Otomari it headed north to a junction of two main spurs that headed south and to the Sea-of-Japan side of Karafuto. North of the junction lay a single track along the coast for about 50 miles to a three-spur junction near Kashiho. If the *Barb* could only destroy a train or the tracks it would halt all traffic and movement for some time.

Bill returned. I called Harry. "Sorry, Gene, no rockets today. For sure early tomorrow." Frustration.

*7 June.* On board for quarters, I heard Jim caution the crew to be on board by 0800 the next day because the *Barb* would shove off at 1200. Anxiously, I paced the deck waiting for the rockets.

One rocket came, but not the type I expected. Admiral Lockwood wanted to see me immediately. His aide gave me a fishy eye, causing me to wonder what was up. Uncle Charlie was pacing around like a lion in a cage.

"Sit down, Gene. I know this will disappoint you, but I want you to go back to Guam with me. I need you there."

I shook my head as I recovered from the blow. "Admiral, you promised me this patrol. After this run, I'll go anywhere you want me to, but now it's not right. I've put too much of my heart and soul into developing this harassment concept, and we're ready. *Barb* leaves tomorrow."

"Gene, you don't understand my position. I'm truly sorry this has to happen. I promised you this patrol against the advice of our psychiatrists. Now I know I was wrong. You can't comprehend that you're too tense."

Stretching out my arm, I retorted, "Tense! Look at this hand. Can it be any steadier?"

"It's steady, Gene, but you don't realize that you're too hard on your men."

"Too hard on my men? Sir, go ask them. You find one man who wants to leave *Barb* today and I'll leave. We have so many men pestering us to get aboard it isn't even funny. Now what's really behind this? What's got you so upset?"

"Okay, Gene, you asked for it. I'll be frank. This morning I had a call from the Marine brigadier in charge of security, who was told by the Marine colonel in charge of the shore patrol that you gave one of your crew a summary court-martial when reported for 'hat on back of head.' He says you're the worst martinet he's ever heard of and should be mentally checked. Now don't you think that's a bit much?"

Absolutely stunned, I looked at the admiral, and the tension of the last few minutes erupted. I burst out in a fit of laughing and couldn't stop. Doubled over, I could see the admiral looking sternly at me.

"Stop it, Gene, stop it! This is serious. I'm taking your command away from you." He must have thought I really was out of my mind, for this made me laugh even harder. I held up my hand to quiet him until I could contain myself. My chest was heaving and I could hardly speak.

"Admiral—do you honestly—believe that I would give—a man a court-martial for having his hat on back of his head? A man who has fought alongside me and who would live or die carrying out my command. It's incredible! We don't even have a mast book on board for punishments. I threw it overboard months ago. The shore patrol jeeped that man all the way back from Waikiki to the *Barb*. I tore up the report and sent him back ashore. Yesterday, a Marine major came to see me because I had not returned the disciplinary action in 24 hours. Since I lost the report, he wrote another. The Marines wanted action. I gave it to them."

"Then you never really gave him a summary court-martial?"

"Are you kidding? I wouldn't even think of it." Now the admiral burst out belly laughing, in which I joined. As we simmered down, he chided, "Gene, you've already started your harassing patrol. You've overkilled the tough Marines. You better get out of town before this gets around. Come here, I've got something for you."

Standing up, he came around in front of me, spun me around, and booted me with his knee in my backside. "Get out there and give them hell!"

The operations office was just down the corridor, so I dropped in. "Gene, did you get relieved?"

"No, but the admiral is relieved. Someone filled him with a bum estimate of the situation. He'll tell you about it, I'm sure. What's new, Johnny?"

"Something that will concern you. "Hydeman's Hellcats"—consisting of three wolfpacks with the new FM mine-detecting sonars—are transiting the Tsushima Straits' minefields at this moment. The nine subs will go to their areas. At sunset on 9 June they will start shooting. The Japanese are going to be surprised because this will be the first penetration into the Sea of Japan since October 1943. The wolfpacks will exit the sea via La Perouse Strait in your area on the 24th of June. We hope that they'll come out on the surface at night because of the stronger currents there through the minefields."

"Golly, how *Barb* would like to go into the Sea of Japan! From what a prisoner told us, we could go around the north end of the minefields in Aniwa Bay, in and out of La Perouse. We were scheduled to get the FM sonar in Mare Island, but a shortage precluded. On our eighth patrol we had orders to stay 100 miles east of La Perouse, but on this patrol . . ."

"You are not permitted to go through the strait, Gene. Don't try it."

Next, I dropped in on Harry Hull. He promised the rockets the following morning "for sure!"

Back on board I asked Jim about our personnel situation. "We still have 31 men in *Barb* that have served with you on your last four patrols in command," he said. "We also have a couple of returnees, who had been transferred. Turnage is back. He missed the last patrol, but has been on all previous patrols. This leaves Swish and Syd Shoard as the only ones doing all 12 patrols. Owen Williams is also back." I had tried to get him back previously, but he had just missed us at Midway.

"I'm sorry we lost Houston, captain. From what you said about the fishnets, he was our best diver. Now, you're about the only experienced diver we have."

"Forget it, Jim. In the Sea of Okhotsk, not me!"

*8 June.* Departure day. Everyone was on board, and last letters were being censored. Only the rockets were missing. At 1130 the mooring lines were singled up. Creed and others came down to say farewell and look at our rocket setup—which hadn't arrived. Word came that the motor-sailer carrying the launcher and rockets had left the ammunition depot and was due at 1230. Our visitors left for lunch. Harbor control requested us to get under way. Our escort destroyer awaiting us outside the channel had searched the area. All was clear, except that I refused to leave without our rockets. At 1315 the heavily loaded boat drew up alongside. I asked the officer in charge where the instructions for the launcher were.

"There are none."

"And the instructions for the 5-inch rockets?"

"There are none."

"How are the rockets ignited?"

"The fuse is an electric impulse, not percussion."

"Our magazine for their storage is directly under the radio room. Is there any danger of a possible inductive current setting the rockets off when we transmit a message?"

"I don't know."

"Who does know?"

"Beats me."

"Is it possible to separate the rocket motor from the explosive charge?"

"Just unscrew them."

"Max, Swish, have your men start unscrewing. I want the explosive charges and the motors separated as far as possible, with charges stored in the forward torpedo room and motors in the after room. Stow the launcher in the doghouse on deck. Go!"

At 1400 the *Barb* was under way for her twelfth war patrol, unscrewed. Max climbed up and over the bridge rail. "Captain, they shortchanged us."

"Shortchanged?"

"We only received 72 rockets, not the promised 100."

"How come?"

"The officer in charge claims that's all there were in the whole Hawaiian archipelago."

"Beggars can't be choosers."

Tom King asked, "Now that we're actually under way on your 'graduation patrol,' how many ships do you plan to sink this run?"

"No estimate, Tom, with so many submarines returning with empty bags. Our area is small—not much commerce passing through. But we can accomplish a lot with harassment, which is worth just as much."

"Come on, captain. Let's make a bet. A quart of Midway's best. You must have some idea for a wager."

The more I thought, the more I realized that here was an opportunity to raise the morale of the crew. "Tell you what I'll do, Tom. I'll bet you that quart of brew that *Barb* will sink 15 vessels of some kind—not necessarily ships. Trawlers, luggers, schooners, sampans included."

"You're on, captain." We shook hands on the deal, and the bridge watch eavesdroppers disseminated the word to the crew before the last mooring line was stowed.

# Part VI The Twelfth War Patrol of the USS *Barb* in the Southern Okhotsk Sea, North of Hokkaido, 8 June–2 August 1945

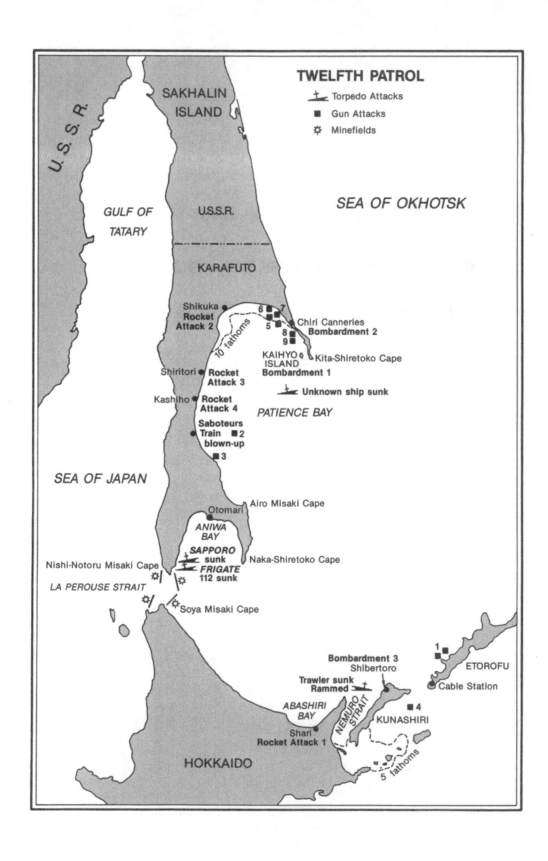

# Chapter 23 Raise a Rumpus

As the *Barb* backed out from the sub base pier, the off-watch sections were all topside, happily throwing coins back onto the pier. This age-old tradition of submariners was meant to ensure their safe return to collect them. Yet that day the hordes of sailors from the base and other moored subs who filled the pier, waving and shouting "So long, good hunting," were not after the coins—they didn't bother with them. Their fantastic farewell was genuine.

Out into the channel, with Ford Island on our right, we passed down the long row of moored battleships, carriers, and larger ships. The farewell became contagious. Jim had to call all our quartermasters to the bridge to answer the many blinker-signal and flag messages of "Good luck." Ships blew their whistles and crews lined their rails, some taking photographs. The climax came when Admiral Halsey signaled, "GOOD LUCK BARB AND FLUCKEY X GOOD HUNTING AND GIVE THEM HELL X HALSEY."

This was too much to believe. Gazing around, I noted the crew on deck and the lookouts snickering. Then I saw it. From the high periscope flew a four-by-five-foot blue flag embossed with large white letters: "Fluckey's 8th Fleet."

Embarrassment overwhelmed me. "Take that flag down before I blow a fuse." What to do with a crew whose enthusiasm I couldn't hold a lid on?

Without the flag, we were once again just another submarine departing to fight the war alone.

Our destroyer escort stayed with us for two hours. After a genial parting message we set a zigzag course for Midway, our oasis in the Pacific and last stop for fuel, repairs, and a decent shower. We arrived on 12 June.

Max and Swish now had the help they needed to be able to mount our precious rocket launcher on 10 minutes' notice. Shortly, we would be ready for the inaugural "BATTLE STATIONS ROCKETS!" It was the dawn of a new era in submarining, a forerunner of the ballistic missile submarines that would follow later.

While the *Barb* was being groomed for combat and her spurs were being attached, I headed for the cable station to relay the good news. The manager was delighted to hear we could now attack the Kunashiri Strait cable station with rockets. Since our February visit, however, he had been studying its position and had concluded that it was not a relay station like Midway, but a connecting station with large copper busbars where the cable landed. Therefore, if damaged, it could be jury-rigged to operate again in a matter of hours. Heartbreaking news, but we had our rockets, and I had promised million-dollar damage. This adversity must be turned into an advantage. The field was now wide open for experimentation.

On departure, I climbed the metal rungs to the bridge. Noting two extra lookouts on the platform attached to the periscope shears, I kept on climbing up past the bridge. The two—Don Miller and Ray Mulry—were innocently looking off at the blue horizon with their hands behind them. I stood in front of them. "You're not going to do a repeat performance of that eighth fleet biz!"

"Aw, captain, it's only a flag."

"I believe you, but I'll bet that it's got my name on it. Hand it over before I have both of you rascals keelhauled. That flag is mine." Sheepishly, they did so, while down on deck the whole crew laughed.

Once we were over the horizon, we commenced our rocket drills. The ocean was kind, only occasionally wetting down the team on deck. The launcher was securely mounted in less than a minute, and the reload crew kneeling alongside the conning tower had it loaded. We were now ready for a firing run. It would not be like a surface battle, wherein the sub is submerged, surfaces suddenly, and the gun crews scramble up on deck to man their guns and open fire.

The *Barb* would already be on the surface at night and would enter the Japanese harbor. Our first test-firing was in daylight to ascertain that all circuits functioned properly and the reload crew was in a safe position when the rockets were fired. It went off well. The night test using red goggles and red flashlights did too. The red goggles were not quite satisfactory, however, in preserving our night vision from the glare of the rockets, so we shifted to Polaroid goggles set at their darkest setting during

Flag made by the crew and flown from the high periscope, without the captain's knowledge, on leaving Pearl Harbor, 8 June 1945.

launch, and these worked perfectly. Certainly, in close contact with Japanese convoys, we seemed to "out-see" the Japanese.

*20 June.* At evening twilight, with Etorofu Island in sight and its snow-clad volcano steaming away, we entered our area, waiting for dark. At 2200, using flank speed, the *Barb* sped through Kunashiri Strait, ignoring the cable station lying quietly in its fog pocket. Halfway through I received a personal "eyes only" message from Admiral Lockwood.

> RAISE A RUMPUS AND DRAW ALL ANTISUBMARINE ATTENTION OFF LA PEROUSE STRAIT X HYDEMANS NINE HELLCATS WILL DEPART SEA OF JAPAN AND TRANSIT STRAIT ON THE SURFACE COMING HOME NIGHT OF TWENTY FOUR JUNE X LOCKWOOD X

Great. He understood the worth of harassment. Yet it changed our plans to head directly for the train-ferry route from the Sea of Japan through La Perouse to Otomari at the top of Aniwa Bay. That route passed through the four-mile coastal strip free of mines where, if not restricted, I had considered entering the Sea of Japan sanctuary.

We were restricted during our patrol in May and June 1944. Since 1943 no sub was permitted within 100 miles of La Perouse. Before that subs could go in and out, latterly, to be done on the surface. The *Wahoo,* our last sub due to exit, was required to transit on the surface over the broad minefields. She disappeared, her fate uncertain. In the past these exits had normally been made at night.

On 11 October 1943 a submarine was seen by a Japanese patrol plane crossing the La Perouse minefield on the surface in broad daylight. Three depth charges were dropped and the sub disappeared. A sinking was claimed.

There is a strong probability that this was the *Wahoo* because the time fits in. No other submarines were within 1000 miles, and her skipper, Dudley "Mush" Morton, was known to take such chances. If so, the *Wahoo* may have sunk due to the depth charges or may have dived to avoid the plane and been sunk by mines.*

"Jim, our strategy can only be to raise a rumpus along Kunashiri—*Barb's* small bit of the Kurile Islands—the eastern section of the north coast of Hokkaido, and inside Patience Bay. Enemy forces we must attract are the patrol boats, frigates, and aircraft responsible for patrolling the minefield and along Hokkaido. Let's get going."

At 2330 we cleared Kunashiri Strait. The *Barb* swung south, close along the island coast headed for Hokkaido, Japan.

*21 June. Time 0015.* "Radar contact, two small craft, 4800 yards!"

Jim, Dave, Willy, and I studied them. We moved in for a closer look at 1000 yards. Luggers or motor torpedo boats?

"Fellows, looks like we're in for a long night. They're coasting along at eight knots in column. Here's the dope. We'll maintain this position, keeping them to eastward. They'll be silhouetted against the first streaks of dawn. When they can be seen through the gun sights, we'll attack. That will be a little after 0300. Everyone topside must be quiet from now on until we start shooting. Reveille for all hands at 0230. Jim, let's go below and catnap. Dave can handle things until reveille, in case they try something. This will be good experience for Willy in his first combat action."

In the control room, Williams was on watch. Swish, being the gun captain of the 5-inch, was waiting. "Captain, when do we take them on?"

* From my personal research with Japanese Navy friends in Tokyo, I learned that the La Perouse minefield was laid from 27 July to 6 August 1943. Four hundred and seventy mines were laid, 60 meters apart, at a depth of 13 meters, ideal to trap a submarine. Even for a brilliant skipper the price of one mistake can be too high.

"Swish, soon as we can see them through the new gun sights on the 40-mm gun they put on in Mare Island. We'll do a split attack at first light. The difference from all other gun shoots that you've done is that this one will be a tiptoe affair until we start firing. I want to wake them up, groggy, so they won't know what hit them. Make sure that after reveille your gun teams man stations slowly and quietly. No clanking, only whispers. I'll give you plenty of time. This includes your ammunition train and everyone topside. We'll have the advantage of silhouetting them against a lighter dawning sky while we stay in the dark side. Complete surprise is the order of the day."

*Time 0300.* "MAN BATTLE STATIONS GUNS." No gongs. The men moved around so quietly it would have delighted some of our training officers. A few of these coaches had insisted on quiet approaches and attacks instead of our normal ones in which we screamed lots of information. On this warpath we were silent redskins.

We eased in to 850 yards. By 0315 we could recognize their silhouettes as 100-ton luggers loaded with supplies and a 37-mm cannon mounted aft. Our guns now had them in their sights as navigator's twilight ended. "Forty-millimeter gun, take the forward target, 5-inch gun the after one."

*Time 0318.* "COMMENCE FIRING!" The 5-inch gun hit with the first shot and checked fire to save ammunition. The 40-mm one fired seven rounds for four hits. Both enemy crews ran out on deck, manned their cannons, and fired to the east—the wrong direction—probably seeing the 40-mm tracers that had missed. The trailing lugger stopped when hit.

Max sang out, "SHIFT THE 5-INCH TO THE FORWARD TARGET!" Three hits sank the trailing lugger's twin. The trailing ship, still afloat, now realized where we were. She shifted her gun in our direction as we reversed course to finish her off with both guns. Her wild, sporadic firing was splashing well astern, which indicated she hadn't seen the *Barb* reverse course. At 800 yards we resumed firing. Three shells from the 5-inch gun and six from the 40-mm one sank her. Close-by, on Kunashiri Island, lights flashed on and off.

*Time 0336.* We secured from battle stations, not desiring any prisoners from this area. We headed down the coast at flank speed. Our objective was to be seen, or to cause a commotion, in Nemuro Strait, the Okhotsk Sea's southeast corner that separated Kunashiri from Cape Shiretoko on Hokkaido.

*Time 1321.* We dived when a periscope was sighted in Nemuro Strait, but couldn't pick the sub up by sonar search. We could do more elsewhere, so we surfaced and headed into Abashiri Bay off the air base near Shari, Hokkaido. Not worried about being seen since no ships were in sight, we watched the activity on the airstrip some six miles away. Bill had the watch.

*Time 1454.* "Plane starboard beam coming in fast, DIVE! DIVE!"

We waited. No bombs. In five minutes we were back at periscope depth. "Up scope—all clear. There's a plane circling the field to land, and it looks like the same plane. It might have been searching the area where the luggers were attacked. Certainly the good citizens of Kunashiri must have reported the attack." We surfaced, and lookouts scrambled up into the conning tower. Singer, our photographer, was all smiles. As a lookout he had spotted the plane, his first one.

Putting my binoculars on the air base, I noted activity was increasing there. Several planes were being readied. Two minutes later Singer sang his song again. "Plane on the horizon, a bit on the left side, heading this way."

Smiling, I said, "Well done, Singer, but let's call it 20° off the port bow to be more specific. All ahead two-thirds. I don't want him to have our position too accurately, if he has spotted us."

"SD radar range 10 miles, closing fast."

"Take her down, Bill!"

The *Barb* slid under to periscope depth. The water was so calm that, though submerged, we could hear the plane whine over us. I raised the scope for a cautious look. The plane continued its antisubmarine patrol of the coast, apparently not spotting our wake. I had no doubt that we had been seen, yet this was our purpose—to attract all the opposition we could. We had been submerged 10 minutes, long enough for the cat-and-mouse game.

"SURFACE! Send an extra lookout to the bridge; business may pick up." Flaherty, a good-looking, ever-smiling Irishman, appeared. "Tom, I give you the whole of the Okhotsk Sea to observe for periscopes and ships; the rest of us are on sky search."

"Aye, aye, sir."

The bridge hadn't finished shedding water when Singer sang out again, "Plane circling way off, two points abaft broad on the port quarter. Maybe it's the same plane."

I reflected on Singer's super-salty language, which left me feeling a bit wet behind the ears. "Two points abaft broad on the port quarter." I had to go back to my midshipman days to dredge that one up from my memory—32 points to the compass, or 8 points in 90°. One point equals

11° and 15 minutes. God knows, my brain couldn't hack that in an emergency. "Singer, you win. Your nautical vocabulary is perfect, but let's stick to degrees so we landlubbers can comprehend."

*Time 1604.* We dived to avoid a plane coming straight at us and rigged for depth charge. Minutes later one bomb fell fairly close-by. It reverberated like a bass drum on the hull. A second bomb exploded astern, port, below. "Swish, make your depth 225 feet. We'll stay here for about a half hour in case some buddies join him. All ahead one-third. Broocks, come right to 340. We'll head out before we surface. I don't want to be spotted again today in this same area. We must show up elsewhere as a different sub. Jim, have Tom King relieve Max now; I'd like to confer with all other officers in the wardroom."

When we were assembled, I explained Uncle Charlie's personal message about raising a rumpus. "We're bait. Our job is of vital importance to Hydeman's nine Hellcats. The prisoner we had on the *Barb*'s eighth patrol told us that four frigates or patrol boats patrol the four minefields in the La Perouse Strait area. These must be attracted well east until after the 24 June night exit. Now we have established the presence of a wolfpack hanging around Kunashiri, Nemuro, and Hokkaido. Tomorrow before dawn we will repay the bombing compliment today by giving the town of Shari our first 12-rocket massage. Upon successful completion, we'll then head for Patience Bay inside Karafuto to be sighted."

All of us studied the largest chart we had of the area. Dave found Shari Mountain—peaking sharply at 5089 feet—only 18,000 yards behind the coastline at Shari. With fair visibility, a line of sight bearing plus radar range would give us an accurate position. Max, Chuck, and Masek had been seeking probable factories. Railroad tracks came north over the mountains to Shari, where they turned west along the coast. Spurs went into some of the mountains, which indicated mining. We knew there were fish canneries and probably sawmills. Max figured logically that the tracks would run through the big factory area. Therefore, the target complex would surround the tracks coming north into the town of 20,000 people, probably starting about 300 yards from the beach front. We plotted the *Barb*'s maximum firing position—4700 yards from the beach, due north of the railroad tracks. We were pleased to note that we would have 20 fathoms of water.

At periscope depth again, we saw a patrol plane still searching the area. By 2010 we were well clear, and it was safe to surface and charge batteries. Believing that most people sleep most soundly about 0230 in the morning, I scheduled that time for the assault.

*Time 2309.* We sighted Notoro Misaki Light burning brightly at the west end of Abashiri Bay, which enabled us to accurately position the *Barb*.

*22 June. Time 0150.* "MAN BATTLE STATIONS ROCKETS!"

The thrill of this inaugural submarine combat cry surged through the boat with cheers. Rocket Captain Saunders came up on deck with his rocketeers. They set up the launcher, then quickly loaded it with 12 5-inch, spin-stabilized, high-capacity rockets, each carrying a 9.6-pound explosive wallop. The rocket range was set at 5250 yards, and the deck was cleared. Most of the men stayed on the bridge behind Singer, who would attempt to photograph the event.

At 10,000 yards from the beach, the command was issued, "Secure the main engines and answer all bells on the batteries. Quiet! Whispers only, topside. Control room, put four cases of beer in the cooler."

*Time 0230.* The *Barb* lay still. We twisted her tail a little to set her heading exactly on 177° true, to allow for the 3° right deflection at this range. Now 4700 yards to the shoreline depth, 20 fathoms, no ships anchored in the harbor, sampans moored to the docks, visibility fair, we were ready. Max checked his final bearing on Shari Mountain. Set.

*Time 0234.* "ROCKETS AWAY!"

An inspiring sight! My polaroid goggles were at their darkest setting to preserve my night vision. All 12 rockets whooshed up in four-and-a-half seconds. Their trail disappeared 20 feet above the deck.

"Right rudder. All ahead two-thirds." Off goggles. In 30 seconds the thunder of the rocket explosions on target was heard as chunks of buildings simultaneously flew up above the shoreline. Then silence. The *Barb* pussyfooted out of the harbor.

Fifteen minutes after the assault, shore-based air search radar was turned on. Evidently, they thought the north coast of Hokkaido was being bombed for the first time.

We headed for Karafuto at flank speed to simulate yet another wolfpack. We would celebrate this first, but far from last, ballistic missile firing from submarines at 1100.

*Time 0811.* "Captain, we've sighted a plane at 12 miles, patrolling." On the bridge we watched the plane until he had a good look and headed in, forcing us to dive to periscope depth. He circled astern for about 15 minutes, gave up, and returned to his patrol. The *Barb* then resurfaced.

*Time 1100.* Max's stentorian call—"All hands, splice the main brace!"—

brought the crew to the control room. Bentley, Phillips, and Ragland proudly produced their creation. The usual model of the *Barb* rested on the blue icing. Pieces of spaghetti, glued together, formed the rocket launcher. The arced shore showed up in yellow icing dotted with several blue smashed buildings and some solid red circles for rocket explosions. Carved from a tin of beets, the red circles were Ragland's contribution.

While we were sipping our beer, Dave called out, "Captain, Karafuto is in sight. Also Naka Shiretoko Misaki Lighthouse on its southeastern tip."

"Dave, close it to six miles so they'll see us. Then parallel the coast north to turn into Patience Bay. Watch for aircraft."

The men looked at each other quizzically as if I was losing my marbles. He wants us to be seen? I explained to the crew that our mission was to "raise a rumpus" and that three wolfpacks were to come out of the Japan Sea on top of the minefields. Our job was to attract the surface and air patrols away for the next few days. Consequently, we had to stay away from La Perouse and pretend to be a wolfpack to the west, off the north coast of Hokkaido. We hoped to divide any forces searching for us by shifting areas and cruising at high speed between sightings, or by finding something to attack in Patience Bay, a simple "divide and destroy" strategy.

By nightfall we had cleared the high cliffs and mountain range and turned into three-quarter-moon-shaped Patience Bay (Taraika Wan), 160 miles long and 80 miles wide. Now we were sweeping along the coast clockwise.

"Captain, two urgent ULTRA messages from ComSubPac.

> ULTRA WARNING X SPECIAL AIR GRP EQUIPPED WITH MAGNETIC AIRCRAFT DETECTION EQUIPMENT HAS BEEN ORDERED TO SEARCH FOR SUBMARINES SOUTHEAST OKHOTSK SEA X WHEN EVADING ANY AIRCRAFT COMMA DE-GAUSSING EXPERTS RECOMMEND EAST DASH WEST COURSES X AVOID NORTH DASH SOUTH COURSES X

In the wardroom I read it to the assembled officers. Dave and Max raised their eyebrows. I looked at them, "Does this remind you of our South China Sea patrol?"

All agreed that it did. On that one our packmate, the *Tunny,* had been knocked out, and we yo-yoed up and down all night long. We had to withdraw and stay on the surface during the day to charge batteries, and we wondered if those night flyers had a secret weapon. This might be it.

Jack came in and tossed the second message to me:

> ULTRA X HUNTER DASH KILLER GRP THREE FRIGATES DEPARTED LA PEROUSE AREA X ORDERED SWEEP NORTH COAST HOKKAIDO X CARE-FUL X

"Fellows, if they split up, we may get a chance to use some of our new homing torpedoes. Meanwhile, let's get some sleep. Business is picking up. Who knows what tomorrow may bring."

**23 June. Time 0346.** "Captain, contact 9800 yards on a large two-masted, double-decked diesel trawler. I'm closing. It's daylight, overcast, and cold. Icy—almost freezing. You'll need a sweater and parka."

While we were closing, yeoman lookout Everard Lunde, smart and a great one for detail, called down from his high perch: "She's got a 37-mm gun aft, a 25-mm forward, large antenna. She's speeding up."

**Time 0413.** "MAN BATTLE STATIONS GUNS!" The gun crews spouted up the hatch like a human geyser. They knew their action was helping other submarines. This innate feeling in the unique submarine brotherhood generated an undaunted enthusiasm in spite of the inherent personal danger in combat. I'd never had to push them.

"Max, without doubt they've reported a submarine chasing them. We're about 90 miles from both their major air bases at Shikuka to the north and Otomari to the south. Give or take an hour, we may have an aerial response. Use your 40-mm gun at about 2000 yards first."

"Aye, we'll go after his guns."

**Time 0423.** "COMMENCE FIRING!"

Max spotted the shots in: first shot over, second short, third a hit. Seven men ran forward. Three more hits put that gun out of commission. With the trawler still making over eight knots, we continued closing.

"Stop her, Max. I want to board and pick up some updated charts."

Max obliged. The 5-inch gun fired three shots for 3 hits while the 40-mm scored 11 hits to take care of the gun aft and possible machine guns.

"AWAY THE BOARDING PARTY!" Tom King's boarding group materialized on deck, weapons, grapnels, and crowbars in hand, ready to strip the ship.

Unfortunately, a fire broke out below her decks, quickly spreading through the superstructure. Black smoke surged skyward, pinpointing our position to any air or sea patrol. Our inability to board was most disappointing.

The fire had to be put out. There was one 5-inch shell hole above the waterline, so to save ammunition, I decided to quench the blaze by making a high-speed sweep alongside. The bow wave could wash sufficient water through the hole to do it. I turned over the conn to Willy for shiphandling practice. He opened out and speeded up. The *Barb* swept by at 16 knots, 10 yards off the port beam. The target filled and sank at 0504.

One Japanese swam over from the quenched flotsam. We took him aboard as prisoner to get some area information. Our new Pharmacist

Mate First Class Cecil Laymon became a surgeon instantaneously. The prisoner had two gaping wounds—one in his left shoulder and one in his left calf—each as large as half a fist. Doc removed shell fragments and stitched up the wounds. With a "well done" to the gun crews, Max, and Willy, I ordered that the *Barb* head for Shikuka. Visions of sugarplums danced through my head. I had high hopes: Shikuka could be the unknown haven for the remnants of the Imperial Japanese Fleet.

I could not imagine a more fitting climax to the *Barb*'s rumpus raising than to ride into Shikuka Roads with a full load of torpedoes and surprise the last major warships. Such an attack would be similar to the one at Namkwan, for Shikuka—sitting on a broad (80 miles wide), shallow, flat shelf—lay 22 miles inside the 20-fathom curve and 10 miles inside the 10-fathom curve—right up the *Barb*'s alley. The warships would be anchored in rows close to the beach and sheltered from the northern gales by the ring of mountains beyond the plain. I rubbed my palms together in happy anticipation of the coming attack.

We were surprised that no planes came searching for us. Surely we had been sighted during the dawn gun attack. Our sole conjecture was that the air bases may have been emptied to supply planes to the Okinawa battles. The previous night's news had announced that we had secured Okinawa on 21 June. Unattacked Karafuto needed little defense.

Water depth was perfect for sweeping this area of the coast. Three miles off we had 20-plus fathoms, until we met the shoals 40 miles south of Shikuka.

Passing Kashiho at four miles during lunch, we saw their local sampans scamper ashore. New factories were smoking away, but no ships. Twenty miles north we passed Shiritori. Here, huge factories lined the coastline for miles, an ideal target for the future, after we had combed our whole area. With Shikuka, 60 miles north, our target for the night, the *Barb* reversed course as a feint. As at Namkwan, our attack on Shikuka must be a surprise, not a trap. We planned our "rocket-shocker" for 0100. The crew was elated.

*24 June. Time 0010.* Having crossed the 20-fathom curve half an hour earlier, we were now feeling our way into Shikuka Roads at 12 knots, taking soundings every minute. Dark night, no moon, heavy overcast—I felt blindfolded in the freezing air.

"Radar contact 33,000 yards!"

"Where away, Dave?"

"Dead ahead. The pips are just beginning to separate into a large group of big ships!"

"Hooray! We've got them! Station the tracking party!" Cheers below rang from stem to stern.

*Time 0100.* "MAN BATTLE STATIONS TORPEDOES!" There was not much of a scramble for no one had slept while awaiting impending combat. Each man was already close to his station. "Jim, range and soundings?"

"Range 16,000 yards, seven fathoms shelving slowly. We're coming in nicely. There are no escorts patrolling."

"Good! They're asleep at the switch. Max?"

"Cold turkey, no current. I'm using speed zero, radar ranges, and bearings."

"One assist. I still can't see anything; phosphorescence is low, rolling swells. Shoot three fish at each of the biggest pips with a 200 percent spread, then two fish from the stern tubes, each at 100 percent spread, at two other ships. I'll stay on the bridge in case they have a hidden sleeper and to watch the circus as we reload. With no patrols to chase us, I plan— barring any surprise interference—to shoot all of our torpedoes except the homing ones so long as there are ships left. We don't need to race off this time. I just wish that I could estimate the size of these ships."

On the intercom I announced, "Men, we have a lot of sleeping beauties ahead which we haven't yet been able to see at this range. After we shoot the bow tubes at 2000 yards, we'll swing around and give them the stern tubes. Then I want a quick reload, fore and aft, as we circle. There are no moving patrols. There may be PT boats."

*Time 0115.* "Range 9300 yards, sounding five fathoms. Will Jack give us bearings? Radar bearings are sharp."

"No. Still nothing in sight. Secure the engines. Answer bells on the battery. Make ready all tubes. At 4500 yards open the outer doors on the forward tubes."

*Time 0126.* Jack whispered, "I can make out some blobs."

"Range 4600 yards. Coming in nicely." The calm was caressed only by the *Barb*'s murmuring bow wave. Quiet anticipation pervaded the boat.

Bluth, a battle lookout, screamed, "Breakers ahead!"

It was like a jerking jolt from an electric chair. I yelled, "All back emergency!" The *Barb* shuddered as the shafts groaned in their contortions. "Sounding? Range?"

"Four fathoms; 4200 yards. What's happening?"

Jack yipped, "They're smokestacks ashore. There are no ships."

"What?"

"Jack repeated, "There are *no ships here!*"

I said nothing. My dreams crumbled. Idly, I watched the phosphorescence churning up both sides of a forlorn *Barb*.

"Sounding five fathoms. Captain, shall I secure from battle stations?" Jim seemed so far away.

"Yes. All stop. Right full rudder. Set course across the bay." My mechanical orders echoed in my ears, devoid of enthusiasm or inspiration. Inside of me something had died. We would return to Shikuka for unfinished business.

Dawn brought with it seals and field ice, slowing us to a crawl. The northern bay was barren, forcing us to behave. Time for additional effective rumpus ran out.

That night, 24 June 1945, eight of Hydeman's Hellcats formed up in columns waiting for the *Bonefish* to exit by La Perouse Strait. Earl Hydeman, a brilliant submariner, had the overall command of the three wolfpacks in the Sea of Japan, as well as being captain of the *Sea Dog* (and he was top dog). Of the record 28 ships sunk by the Hellcats, the *Sea Dog* sank six. My classmate, roommate on the battleship *Nevada,* and best man at my wedding, Bill Germershausen, was skipper of the *Spadefish,* which sank five. Another classmate, Ozzie Lynch, was skipper of the *Skate,* which sank four.

Unbeknown to the Hellcats, the no-show *Bonefish* had been sunk with all hands during depth-charge counterattacks. After dark, Earl could wait no longer. On the surface in a fog he took his columns at 18 knots through La Perouse Strait without encountering any of the normally patrolling frigates. Nor did they know that the *Barb* had raised the rumpus requested by Admiral Lockwood, which had worked.

# Chapter 24 Terutsuki

Since entering our tiny area, we hadn't been bothered by Soviet shipping because "raise a rumpus" had kept the *Barb* away from normal Soviet commercial routes. To date they had provided no help in the war with Japan, and the Japanese respected their neutrality, thus avoiding a second front. During our eighth patrol in the Okhotsk in early June 1944, I'd asked Pearl for permission to land two men on lower Kamchatka to observe the Pacific and Okhotsk exits from Paramishuru Harbor.

These exits were 25 miles from an accessible peak on Lopatka Cape. The men would not be armed. They would carry supplies, a telescope, and a transceiver radio in order to communicate with the *Barb*. Otherwise, the distance required to go around Paramushiru Island from one exit to the other was 120 miles.

With all the armaments, supplies, and convoys the United States was providing the Soviets, I anticipated a little lend-lease in reverse. My request was forwarded through channels to the Soviets and came back negative.

Now, even after victory in Europe, 12 May, we still were not receiving one iota of assistance from the Soviet Union. With such an attitude, perhaps they were even helping the Japanese—but how? Oil? Studying our meager intelligence, I noticed that the Soviets had an oil depot at Urkt, in their northern half of Sakhalin Island. They could name their price when selling to oil-desperate Japan. I decided to reconnoiter.

"Jim, set course for Urkt. We'll take a look, just in case any Japanese tankers are loading oil there."

"Captain, you know we're forbidden to enter Soviet territorial waters."

That was not quite so. International waters recognized nothing beyond the three-mile limit, although the Soviets claimed more—like Kurile Strait between Kamchatka and Paramushiru, which we were forbidden to use. That strait would have saved us some 60 miles of running to cover the two exits from the base there. "Jim, we have no other exclusion, so we'll honor the three-mile limit, nothing farther out. Break out a chart and pray those 1894 Swedish soundings are still accurate in this ice bucket of volcanoes and glaciers called the Sea of Okhotsk. How in the devil could the Swedes have taken all the soundings covering the 582,000 square miles of Okhotsk?"

**25 June.** Now that we didn't want to be seen, the weather was crystalline. We kept well off of the island of Kaihyo To, a convoy turning point and seal rookery just south of Cape Kita-Shiretoko. Then, exiting Patience Bay, the *Barb* nosed north toward Soviet waters.

A bit before noon we dived and headed in. Beautiful coniferous-covered mountains cascading down to a flat river delta filled the periscope. To the left of Urkt stood huge cylindrical oil tanks with pipelines coming down to the beach and disappearing under the water. Four large mooring buoys were anchored offshore for ships to take on oil, but there were no ships. We sneaked out until the *Barb* could surface without being easily identifiable from the beach. On the surface once more, we headed north to search the coast to the tip of Sakhalin.

The tanks were a tempting gun target. The coast was empty. Heading south, we moved out to the 100-fathom curve. My plots from previous patrols in the Okhotsk indicated that the Japanese used this chart line for their convoys. Being some 35 miles from land, their ships could not be seen. Perhaps they assumed that the Soviets reported Japanese movements. The Soviets did not, neither from their many ships afloat nor from land.

**26 June.** Another brilliant, windless, waveless day unraveled, chasing a full moon down off its perch in the sapphire blue vault above. The *Barb* moved slowly through the conglomeration of broken-up field ice shimmering in the sun. On the bridge we removed our olive-drab parkas, revealing our olive-drab sweaters and flattened hair. I mused on our peripatetic meanderings since raising a rumpus. Absolute zeros had resulted from my orders; the crew was keyed up for action. Strike one: Shikuka had been worse than no ships. We almost beached at 12 knots and bent over our swordarm. Strike two: Urkt was another lead balloon. We couldn't afford a third strike for this inning. The 100-fathom curve had to produce results.

*Time 1655.* "Captain, I think I'm hallucinating."

"What?"

"Look southeast, 5° below the horizon. What do you see?"

"It's a convoy in miniature! Well, I'll be damned. We've got a remarkable atmospheric lens. A destroyer is in the lead followed by two medium freighters and one small freighter. On opposite bows are two frigates and on the quarters two patrol craft. A plane is circling the convoy. You can even see the bow waves of all the ships. STATION THE TRACKING PARTY!" At long last we had something we could really put our teeth into.

"Captain, what range shall we use? We've no radar contact."

"Good Lord, how should I know? It's like being in a war-game room. All the ships are in miniature and appear to be on this side of the horizon. No parts of the ships are sticking up above our normal horizon. With the lens effect, the convoy appears to be viewed from about a 5° elevation. Thermal waves are apparent in this sector. Max, set your range at 45,000 yards."

"How about bearing?"

"Hold a minute, Max. Jack, put your binoculars on top of the target bearing transmitter and depress them until you see the convoy."

"Got them, captain."

"Okay. Jim, take Jack's bearings. They won't be exact, but we'll end around on these hypotheses until they're in sight. Pray we don't have any horizontal aberration as well."

*Time 1746.* Lens persisting. Aircraft left the convoy.

*Time 1910.* The *Barb* completed her end around and stopped dead ahead of our minuscule convoy, waiting for it to discard its ethereal existence. Willy, Tom, and Bill had joined Jack and me on the bridge. Shipmates bobbed up and down from below to have a peek at our ghost convoy. The high periscope gave other sightseers a look. When would it really appear?

"Captain, they've ceased zigzagging 30° each side of their base course. With our ability to see them on this side of the lens, isn't it possible they will see us from their side of the lens, lying dead ahead?"

"Good point, Willy. The enemy may have spotted us during our high-speed end around. In view of no darkness tonight, the late sunset, long twilight, full moon, and this sparkling clear day, we may as well attack submerged now. All visitors below. Stand by to dive."

*Time 1920.* Submerged. "Max, just let your torpedo data computer continue to generate the range and bearing of the convoy until we see them. What speed are you giving them?"

"The calculation is about nine knots from this crazy tracking through the lens. If we head for them at three knots and they don't change course, we'll be closing the distance at 400 yards per minute, or 24,000 yards per hour."

*Time 2000.* "Hey, gang, tops of masts are appearing on the horizon. Angle on the bow near zero. Come on, targets."

*Time 2028.* "Setup! Same formation. Angle on the bow zero. ST periscope radar range—mark. Bearing—mark."

"Range 11,000 yards, bearing 320°."

"How does that compare with your generated range, Max?"

"I had only 4000 yards less after three-and-a-half hours of tracking. That means that when we picked up the convoy through the atmospheric lens, his actual range was 49,000 yards instead of your guestimate of 45,000 yards. Good Lord, it was almost 25 miles away."

"We're lucky. It could have been 50 miles."

"BATTLE STATIONS TORPEDOES!"

"Max, what's your torpedo loading, fore and aft?"

"All Mark 21 electric torpedoes, except one Mark 28 electric homing torpedo, in each nest."

"Good. We'll save the homers for the escorts since we only have three of these experimental acoustic fish on board. Up periscope."

*Time 2037.* "Range 6900 yards, bearing 320. Hold every thing! All ships are starting a simultaneous turn toward the beach. Damn it! They're leaving Soviet waters and heading in to cruise close in down the coast. They've left us out in left field. We haven't a prayer in hell of getting a shot at the ships. The closest is the destroyer, which was well out in front and is now on the convoy's port beam. Jim, take a quick look and check her identity in the *Recognition Manual.*

"Shankles, all ahead full, left full rudder, course 270. Max, we'll use the homer forward. Make ready tube four!"

"Captain, positive identification on the destroyer. She's of the *Terutsuki* class, the best antisubmarine destroyer they have."

"She's a smart-looking warrior with those loaded depth-charge racks and those side throwers running up her sides. She'll look even better when this homer hits her props. Her depth charges will blow her stern off and stand her on end. How are we doing, Max?"

"Perfect solution: speed nine knots, torpedo run 30° right is 2000 yards. This is about as close as we can get without her sonar picking *Barb* up.

Water temperature at this depth is the same as at the surface, which exposes us to her and her to our homer. Recommend we shoot."

*Time 2105.* "All stop, open the outer door tube four! Up scope. Range—mark. Final bearing—mark! FIRE 4! Sonar, put the JP sonar on that torpedo."

"Sonar reports torpedo is normal and turning right to the target. She's straightening out and her constant noise is on the exact bearing of target's propellers."

"Chuck, what's hitting time for a 31.8-knot fish?" I stopped, just in case she became erratic. "We can handle a circular run, but not an erratic homer chasing after us."

"One minute and 53 seconds, sir. A minute, 20 seconds to go."

"Sonar reports torpedo constant; noise ceased at 33 seconds, as if a switch cut her off!" A complete failure.

"Up scope! All ahead one-third. Target still has same course and speed. No depth charges. Sonar, use a sound head to sweep for our torpedo, keeping the JP on the target for any change of speed. Damn, she's too far away for a salvo."

"What now, captain?"

"Jim, these ships aren't going to get away. When they're hull down, we surface and end around. Darn, I forgot the beer again."

Within an hour the *Barb* was on the surface at flank speed, raring to go under a cloudless sky and full moon. Twenty minutes later—visually and by radar—we picked up the convoy at slow speed near Anaiwa Misaki and way inside the 10-fathom curve.

"Jim, they're beyond our torpedo range submerged. Let's move in closer on the surface. They know we're around."

*Time 2234.* Easing in on the motors because the engine exhausts were large white tail feathers in the frosty air, we reached 12,000 yards as unwelcome intruders.

BOOM-BOOM! BOOM-BOOM! BOOM-BOOM! "Lookouts below! Put four engines on the line! All ahead flank! Left rudder!"

Now alone on the bridge, Max and I were hypnotized by the flashes and thunder as the *Terutsuki*-class and one or two of the frigates fired broadsides in our direction. We could hear the whine of the shells, but saw no splashes. Overs?

We zigzagged away, waving our tail feathers as a target for their gunnery. The firing ceased and Max flashed his ever-present smile. "Damned inhospitable characters, I would say. Where to now, sir?"

"Jim, pick out a spot down the coast where it's deep enough that the convoy can't get away from our torpedo range of 4000 yards to the beach."

**27 June. Time 0200.** As we circled off Netomu Misaki, it was evident that the convoy had anchored for the night in the shallow bight of Anaiwa Misaki, since it should have arrived by now. Our tight-in attack position had only 80 feet of water.

**Time 0749.** "Captain, we have sight contact on a convoy coming down the coast deep inside the 10-fathom curve."

"Take her down, Dave!" Submerged, we headed in. They wouldn't get away this time. "Station the tracking party." Twenty minutes later, a quick look across the oily calm revealed their plane escort as they came over the horizon. "BATTLE STATIONS TORPEDOES."

**Time 0822.** Two bombs—close! Light bulb shatterers. "RIG SHIP FOR DEPTH CHARGE! Our hull must be visible against the bottom in this shallow water. Up scope. The Terutsuki is leaving the formation, heading directly at us with a bone in her teeth. Range 7500 yards. The plane is circling overhead. Let's get out of here to deeper water. Right full rudder. All ahead full. Down periscope."

**Time 0825.** Another two bombs, close, followed by single depth charges at intervals moving our way. Within half an hour, as the *Barb* reached for deep water where the plane couldn't mark our position exactly, the thunder below was becoming uncomfortable. "Make ready all tubes."

**Time 0855.** Two depth charges, cork-dust poppers. The *Barb* eased over the 20-fathom curve at a 100-foot depth. Two more depth charges—teeth rattlers—exploded broad on the port bow and slightly above it. Our new depth-charge indicator was much more accurate. It was a great assist in evading the thunder below. "Tom, what's the bathythermograph show?"

"From surface to 80 feet no change in water temperature. Here we have a 15° negative temperature thermocline." This thermal protection effectively deflected the sonar beams pinging on our hull, making the depth-charge attack erratic.

"Max, their two-depth-charge attack evidently comes from the Terutsuki's side-throwing Y-guns. She has several of them along each side, so she can shoot a wider pattern when she has a sub snared in between. Real hull collapsers."

Another two depth charges. Dave called up from his plot. "Those couples show astern, to port, one well above, and one at our depth. Must be set at 50 and 100 feet. Plot indicates she has lost *Barb*, probably attacking some poor seal or a whale heading north. We're in the clear."

A bit frustrated, Max added, "Recommend we head east until we can surface unseen and end around to catch her in deep water. Here it's so calm and shallow, that plane can spot us as easily as if were in a fish bowl. This cookie has a real smart in-depth defense. With the coast protecting one side, the plane and Terutsuki team up on the outer semicircle. When we pierce that, we face the inner-defense line. This inner line, with two frigates and two antisub patrol vessels protecting the same flank of one small and two medium freighters, isn't going to be any picnic."

*Time 1039.* Surfaced, with the smoke of the convoy in sight, the *Barb* commenced ending around on a radius of 30,000 yards. "Jack, that plane circles ahead of them with our thus far impervious adversary, the Terutsuki. If the plane sidles over to this quarter, dive. I don't want *Barb* to be seen."

An hour later Jack pulled the plug. The plane had not seen us. My race to the conning tower came to a shuddering halt in the control room. I noted Randall Wilson, a lookout, manning the bow diving planes with his left hand on his forehead and blood streaming down his face. "Roark! Take over the bow planes. Here, Flaherty, help him back to Doc Laymon. Wilson, don't talk! Try to relax and hang in there. Outside of a nose bleed, you have a gash on your forehead that may need stitches."

In the conning tower Jack was glued to the scope. "The plane has returned to his position ahead. Permission to surface, sir?"

"Sure. We can't waste time submerged."

Back in the crew's mess Wilson was stretched out on a mess table. Laymon was lacing a few sutures in his forehead. When he had finished, Wilson sat up groggily with a weak smile and explained that, not being a regular lookout, he had forgotten to tuck his binoculars inside his jacket. "I grabbed the ladder side rails to the conning tower and jumped down. They hit one step and smacked me."

"Well, look out when you're a lookout."

Jim interrupted. "Captain, recommend we end around at 40,000 yards to avoid that plane in this clear weather."

"Okay, Jim, but be careful we don't lose contact. The convoy will be beyond radar and visual surveillance."

Another hour with Tom on watch and we dived again for a plane. We opened out to 50,000 yards. Now crossing the 90-mile mouth of Patience Bay toward Airo Misaki, the southern entrance cape, we knew the convoy could not escape into the Okhotsk Sea. Another 15 minutes was lost before we surfaced.

By supper time the *Barb* was ahead of the convoy's estimated position and the back-track search commenced without letting the enemy out of our Patience Bay trap. I was on the bridge at about 2000, feeling absolutely frustrated. Custer, another temporary lookout, spotted a plane at 12 miles. Bill, the watch officer, and I studied it. It wasn't the floatplane. They had fooled us by bringing another plane into the problem. When it eased in to 10 miles, we dived. Less than a half hour later, we surfaced and resumed our determined search.

The convoy was anchored somewhere in Patience Bay. If it went to Shikuka, the shallow shelf and full moon would defeat us. Other anchorages—Shiritori, Kashiho, Sakayehama—had possibilities for a night submerged attack. During the night, we searched 60 miles of southern coast, to the southwest point of the Bay at Sakayehama, without luck. Our only sure bet was to return and catch them off Cape Airo Misaki.

*28 June. Time 0338.* We dived at dawn. A frigate interrupted our breakfast by starting a patrol 8000 yards to seaward of our position. She could have been trying to draw us out, so we remained close in along the coast in 20 fathoms of water. Two hours later she gave up and disappeared south.

Mid-morning brought a plane patrolling our position. After lunch he gave up. We waited and waited, certain we had caged the convoy inside our rattrap bay. Evening came and went.

*Time 2134.* The last flickers of twilight. We had been submerged 18 hours. I had limited smoking to 10 minutes each hour when submerged. Regardless, oxygen was so low, one could not even light a cigarette. Matches fizzled, lighters sparked. Talking was an effort. Silently, we surfaced. Gulping air as the battery charge started, the *Barb*'s masters sounded like a covey of magpies—all talking at once.

Having wasted a day, we resolved to enter our cage and confront the enemy on his terms, like Daniel in the lion's den.

*29 June.* During the night the *Barb* charged her batteries and "whiskbroomed" the southern and western coasts of the bay. At dawn we submerged north of Motomari, following the 20-fathom line north toward Shikuka. By lunchtime we passed Kashiho.

*Time 1217.* "Up scope—down scope! Ninety feet! Rig for depth charge! My God! I saw the face of a Japanese pilot in a single pontoon plane only 200 yards away, 50 feet off the water. Angle on the bow 90 port; so close I could see his teeth. He was looking straight ahead. The surface is glassy calm. Pray no bombs! I don't think he's sighted us."

We reached periscope depth in 20 seconds. The plane flew south.

*Time 1245.* "Up scope. MAN BATTLE STATIONS TORPEDOES! Here they come!" Cheers mingled with the gongs made a pleasant, complex cacophony. The tense waiting was over.

"Listen, you tigers! Here's the scoop from the scope. Same convoy minus one frigate. Ships are 1500 yards inside the 10-fathom curve, following down the coast. The frigate plus two escorts are about on the curve, with the Terutsuki 1500 yards outside and ahead. We're moving in." I asked Max where he had loaded our last two Mark 28 homers.

"Tube number 4 forward and number 10 aft."

"Good split. Make ready all tubes. Time for a setup."

*Time 1255.* "Up scope. Range—mark. Bearing—mark. Darn, the Terutsuki is swinging out from the formation, giving a zero angle on the bow. The plane's about halfway between us. Hope we haven't been spotted."

"Range 9000 yards."

Photographer Singer was standing by, ready to take movies. As usual in subs, the unanticipated seemed to be happening. "Bob, you'll just have to stand by to stand by. With that low-flying plane and shallow, calm water, I cannot endanger *Barb* more than necessary for a film. I'll call you if we have the advantage, so be ready. It may be all of a sudden."

*Time 1305.* "Up scope. Bearing—mark. Range—mark. Left full rudder. All ahead standard. Angle on the bow is still zero; the plane is weaving back and forth across the Terutsuki's bow. We're coming left for a down-the-throat stern shot using the homing fish in tube 10. Ready, Max?"

"Cold turkey, set!"

*Time 1307.* Five depth bombs! Dust and cork flew, light bulbs shattered, and anything loose took off. The *Barb* shuddered. "UP PERISCOPE!" It acted like a whip antenna. "He must have dumped his whole load. Forget the plane. The Terutsuki knows we're here. Bearing—mark. Range—mark."

"Sir, I'm having trouble on radar separating the destroyer from the plane doing figure eights across her bow."

Max piped, "No matter, I've got her speed and bearing. Down the throat the fish will go."

*Time 1317.* "Up scope. Final bearing—mark."

"FIRE 10! JP sonar, follow that torpedo!"

"Range 1600 yards."

"Sonar reports no sound from torpedo after firing."

"Damn it! Open the outer doors on tubes 7, 8, and 9! Max! New setup!"

Click—Wham! "It's not enough to take five bombs from his zoomie; Terutsuki's starting to lay his eggs. Let me know when his generated range is 1000 yards and I'll get a final setup. Sounding?"

"Aye, 18 fathoms."

Whoom! "What was that shaker, depth charge or torpedo?"

"Gierhart says he doesn't know whether that was the homer that failed to run or an ashcan. Setup?"

"Chuck, spread your fish one-quarter degree left, zero, one-quarter right."

*Time 1320.* "Up scope. Angle on the bow 3° starboard. Bearing—mark. Range—mark."

"900 yards. Checks."

"FIRE 9! FIRE 8! FIRE 7! Down scope. Sonar!"

"Sonar reports all torpedoes hot, straight, and normal, heading directly for target. Target has shifted pinging to short scale for depth-charge attack."

"All ahead full for 20 seconds to give him a knuckle to ping on. Time— ALL STOP!"

*Time 1322.* "Sonar reports all torpedoes passed directly under target."

"IT CAN'T BE! Her draft is 15 feet and the fish are set at 6 feet. What's causing them to run deep? Lock all watertight doors except conning tower hatch to control room. HANG ON EVERYBODY!"

*Time 1324.* "She's passing overhead. The rumble of her screws sounds like an express train rolling along our hull."

*Time 1325.* A cloudburst of close depth charges rained down on us. The depth-charge indicator was acting like a pinball machine: ahead-above-below! "Up scope. Thrilling! Geysers are flying up from a stream of charges shot from her side throwers. Singer! Quick, bring your camera to take the best shot in the whole war of being depth charged. With *Barb* doing a Saint Vitus's dance and ST radar picking up the geysers, as well as the depth charge racks, this is it. Range?"

"One hundred fifty to 200 yards."

"Singer! Where are you? Not much time. I must shoot."

From the back of the crowded conning tower came, "CAPTAIN, FOR GOD'S SAKE, GET THAT PERISCOPE DOWN!"

"Too late now. Open the outer door tube 4. Max! Ready for our last big homer up-the-kilts shot for the simplest homing shot ever? This fish must hit."

*Time 1326.* "Angle on the bow 175 starboard. Final bearing—mark. Range—mark."

"800 yards, speed 10 knots. Straight shot up the kilts. Range checks!"

"FIRE 4! SONAR!"

"Sonar reports no sound from torpedo. Torpedo sank."

"Down scope. Blast it, open the outer doors tubes 1, 2, and 3. Terutsuki will be back for another round. Max, we'll give her another down-the-throat salvo, same one-quarter degree spread. What's our injection water temperature?"

"Thirty-five degrees."

"Pretty chilly. It might possibly be affecting the torpedo batteries, slowing the speed a smidgin if nothing else. There'd be absolutely no effect on down-the-throat shots."

*Time 1327.* Another pattern of depth charges rattled our teeth once more. "Up scope. She's turning around to starboard with the plane doing figure eights forward of her bridge. Down. Those depth charges are so close together it's hard to count them. This time we'll let her come closer so we can't miss."

Max advised, "A shooting range of about 600 yards will ensure that the torpedoes will have the 20-second run that they need to arm themselves to blow her up before she gets on top of us."

"Smart point. Six hundred yards it will be."

As Terutsuki straightened out and bore down on us, a couple of periscope peeks showed she had speeded up to 12 knots.

*Time 1331.* "Up scope. Angle on the bow five starboard. Range—mark. Final bearing—mark. Down scope."

"Six hundred twenty yards checks!"

"FIRE 1! FIRE 2! FIRE 3! JP sonar on torpedoes."

"Sonar reports all torpedoes hot, straight, and normal on same bearing as target."

*Time 1332.* "Sonar reports all torpedoes passed directly under target. Now fading out on same bearing."

"All ahead full for 20 seconds. Damn the torpedoes! All back emergency! Tom, ease her down to 80 feet. I hope she's clocked us at full speed. I'll stop in a minute."

*Time 1333.* "Gierhart says he heard side throwers firi . . ." WHAM! WHAM! WHAM! WHAM! "Indicator shows a bracketing pattern ahead of us, above and below."

"Gierhart says they've ceased firing. She'll pass close overhead."

"Good. All stop. Rig ship for silent running." We could hear her screws passing alongside as they broke off their depth-charge pattern.

*Time 1334.* "There go the sidethrowers again. Stand by." WHAM! WHAM! WHAM! WHAM! "All ahead one-third. Max, I can't hit him! How many evasion devices do we have on board? Bubblers, swim-out beacons, everything."

"Seventy-six."

"FIRE THEM ALL!"

"All?"

"I said all. I need them now, before he attacks again. We have no other weapons for these depths. *Barb*'s going to leave here alive and kicking. Get cracking!"

"Aye!" Max and Chuck sped off.

"Jim, to delay the next attack until we get some evasion devices in the water, let's throw her a couple of full-speed knuckles. These will give her something to ping on until the beacons simulating a sub swim away from the tubes and make their submarine noises. We're only 5500 yards off the beach, so don't let us run aground."

"All ahead flank for 40 seconds, right full rudder."

*Time 1338.* "Sonar reports several other escorts joining the Terutsuki."

"All stop. Left full rudder. One minute."

*Time 1339.* "All ahead flank. Right full rudder."

*Time 1340.* Max returned breathlessly. "Captain, all five beacons are on their way and will be sounding off momentarily, as are the Pillenwurfer bubblers and gassers."

"Sonar reports there is so much noise in the water it has blanked out all targets. Sonar is useless."

"Great! All ahead one-third. Shankles, come left to 65°. We will now withdraw tactically and tacitly from the coast. When we get clear of our terrible Terutsuki, I want a fine-tooth inspection of our fish and fire control. I consider us extremely unfortunate to only get a draw in this beautiful fight after all of the punches we threw."

# Chapter 25 Little Iwo Jima

*29 June. Time 1450.* Happy to be clear of the bubbling caldron we left behind, sad that nine torpedoes had gone awry, the *Barb* nosed up to periscope depth. Terutsuki and a frigate still worked the area over trying to sink the unsinkable *Barb*. Yet even the distant thunder below fell on deaf ears. We simply wanted revenge. The men angrily cleaned up the mess left by the close bombs and depth charges. Surprisingly, they didn't blame me for firing nine torpedoes at the same ship for zero hits. The sonar stations all agreed the six electrics passed under the Terutsuki, running deep, and the homers, failing to run, sank.

A plane appeared sporadically, having joined the search. His distant bomb indicated his finding some submerged or diving target, another whale, or perhaps only effective hold-down tactics.

Jim and I worked over the charts, searching for a place to intercept the convoy in deep water with a thermocline. There, we could penetrate their screen to torpedo the ships. Much to our disgust, we lacked two hours of being able to enter the ring for a fifth round with the Terutsuki. In the four hours of darkness that night, the convoy would anchor deep in the shallows of Airo Bay, then scoot through La Perouse Strait the following day. We were too far north.

In this mood we commenced plotting all sorts of foul deeds. At 1950, all clear, the *Barb* surfaced while the sun was still up.

"Jim, set course to Anaiwa Misaki where we picked up the convoy. It seems to be the principal convoy turning point to and from the big bases at Paramushiru. The Northern Fleet bombardment there two days ago

342

may have driven some ships this way. If we find nothing, it's about 60 miles from Kaihyo To, the convoy turning point to enter Patience Bay. I've been cogitating on invading that island."

"INVADING?"

"Right! It's a seal rookery, worked in the summer time. A caretaker winters there. When the Okhotsk freezes over, he can dogsled to Hokkaido. So we destroy their radio, take a prisoner, and get all the dope on the frequency of current shipping. Also, for good old Project Three, we will have a flag raising, well filmed by Singer."

"Do you mind if I say it sounds a bit crazy. But it is harassment. Where did you get your information?"

"At the intelligence center in Pearl. They probably copied it from *National Geographic*."

"Well, we do need a knowledgeable prisoner from this area. Our present one knows nothing of the northeastern part of Karafuto. The crew will love your idea."

"Enough talk for now, Jim. Set that course to Anaiwa Misaki. We'll bring the officers, chiefs, and leading men into a team talk-fest tomorrow while we search. Let's rest now. What a day!"

*30 June.* We swept back and forth on calm seas, covering the traffic from Paramushiro and Matsuwa in a restless search. After breakfast the officers assembled and studied Kaihyo To as an assault possibility. Their approval was unanimous. When I requested a volunteer to lead the assault force, each one raised his hand, myself included. Jim grabbed my arm and pulled it down. "Captain, you're indispensable."

"Wrong, Jim. Anytime someone in *Barb* is indispensable, our organization is faulty."

"But a captain cannot leave his ship during combat."

"Okay, I give up. Tom is leader of our boarding party for ships, so Willy is hereby designated as the leader of the assault force."

"HOT DOG!" he replied.

"That's an acceptance speech unfit for Californian politicians. Now, each of you, independently, make out your private list of seven men to accompany Willy and give it to me before lunch. Please don't talk to the men you select or anyone else. After lunch we'll confer with the chiefs."

On the bridge all was quiet, the lapping of the bow waves and the roar of the engine exhaust the only disturbance as the *Barb* wallowed around in this saucer-like world of her own, a too-peaceful world. Calm, low swells hissed up through the superstructure. Cloud-free skies stretched to the end of nowhere. Searching hopefully with binoculars, I found myself tilting them 5° below the horizon. Another atmospheric lens? Were others doing likewise? Yes!

My chuckling brought a voice from above. "Captain, you found something?"

"Flaherty, all I find is that every pair of binoculars on this bridge is aimed just below the horizon. A plane could spot us, so look all around."

"But, captain, don't forget that's where the ships are."

How could I when I wished they were still there? Jack came up to relieve the watch and reported, "Soup and sandwiches and a hodgepodge of invasion talk. You'd think this was Iwo Jima. I hope we're Goliath this time."

Below, enough rumors had leaked out that I was besieged by questions about who would be in the force. My reply, "only volunteers," elicited everyone's volunteering. How could I lay on enough punches to satisfy these warriors?

After lunch Higgins, our navigational quartermaster, laid out the biggest chart he had of our target, Kaihyo To. Swish came in with Chief Electrician Gordon Wade, Chief Radioman Gierhart, and motor machinist chiefs Williams, Whitt, and Noll. Following them came leading men Newland, Maher, Todd Parker, and Shoard.

I explained the meager intelligence information, then went on. "The eastern side of the island has reefs, rocks, and shoals, plus a stockade, which I presume is for the rookery. We can't land there, nor do we want to kill any seals. The south coast is much the same. On the west coast are a few barracks and warehouses for the summer government people. From these buildings on the southwest plain the terrain rises quickly to a 50-foot cliff with the beach beneath. After a successful bombardment to eliminate all opposition, we will land our assault force of nine people on the northwest beach.

"Now to the business of selecting volunteers for the elite force that Ensign Masek will lead. Swish, do I need to ask for volunteers?"

"Well, captain, I can assure you that everyone has heard about capturing Kaihyo To and they've all volunteered. Soon as they see you checking charts—especially an island—they zero in on what's up."

"Okay. Now, some with specialist capabilities. Singer, our photo whiz, must go along. That leaves seven more. We need a radioman to destroy their communications."

Gierhart's hand flew up. "I'm the only one who can do that quickly."

"Good, you're in. Six to go. A signalman?" Higgins volunteered. "No, we need you navigating to keep us from grounding; name one that's not on other parties."

"Bluth."

"Good. Five to go. A gunner's mate? No, not you Swish. I need you on the 5-inch gun."

"Petrasunas."

"Good. Four to go. A machinist to put their trawlers or sampans out of commission."

Noll insisted no one knew small engines and auxiliaries better than he did, so he won this berth.

"Now, men, I've selected the last three from lists given to me. Tom Flaherty, Carl Spencer, and Owen Williams. Conference adjourned. Swish, Jim, Max, stay to plan the gunnery part of the invasion."

As to the gunnery, it was decided to bombard the western coast, opening fire at 1000 yards with the 40-mm gun alone to find the response, then lay on the 5-inch for destruction. The twin 20-mm guns would take care of small arms return fire.

Swish left to commence training the assault force in their use of our Tommy guns. Max gathered his .50-caliber machine gunners Durbin and James Epps, 20-mm gunners Travis Kirk and Lunde, and 40-mm gunners McKee and Mason Edwards. The lesson he imparted was economy of ammunition and shooting in short bursts instead of a wild stream of fire. Both these tactics were most important in light of the further plans that were jelling for use of our guns.

In spite of perfect visibility and high-periscope, radar, and sonar searching, we spied no lens, no ships, no planes all day.


*1 July.* Our first fog in the area blanketed us. Plans for taking Kaihyo To became more detailed as we honed away, sharpening our claws. After our initial bombardment from 1000 yards, the *Barb* would haul out to 4000 yards for two hours to await Japanese air or surface support. If none arrived, rubber boats would be inflated, the assault force readied, and the landing area closed to 500 yards. With the five lightest men in one boat and four heaviest in the other, the landing party would arrive on the sandy beach to the northwest, under the barren cliffs. Then they were to head south, the 40-mm gun laying a barrage 200 yards ahead.

One question that arose during lunch was the estimated speed when paddling a fully loaded boat to the beach. We decided on an actual test using four 200-pound officers in the enveloping fog. Jim, Max, Bill, and Chuck were the only eligibles. They donned their foul-weather gear. Two-and-a-half laps around her equaled 500 yards. Swish inflated a rubber life raft and smartly insisted on everyone using rubber Mae West life jackets.

After a trial run using four paddlers doing circles, we shifted to two paddlers, a steerer using a paddle, and a bow lookout with weapon ready.

With Singer filming, and without exhausting effort, Swish trained the officers down to less than five minutes, which suited our schedule admirably. After a final rehearsal with weapons and gunners, we were ready.

"Men, tomorrow *Barb* strikes at dawn. This fog must lift sufficiently so we can see the whole of Kaihyo To. Provided our bombardment totally eliminates all opposition, so that danger is minimal, 'Masek's Mavericks' will land. Their job is five-fold. First, destroy radio communications. Second, eliminate weapons and ammunition. Third, eliminate motorized sea transport. Fourth, round up the settlers and bring back one prisoner with charts. Fifth, film a flag raising for the submarine movie. I regret that we cannot wait two more days to accomplish this on the Fourth of July. But the area is quiet, the weather calm and favorable, visibility diminished. The time to strike is now!"

As I wandered through the boat I felt the level of excitement to be beyond the contagious stage—it was epidemic. Men were offering two hundred dollars to those assigned to the assault force to take their place.

*2 July.* Rounding Cape Kita-Shiretoko Misaki, 10 miles north of the island, visibility improved as we entered Patience Bay again. Kaihyo To was soon sighted at 10,000 yards. With the white-gray background, we trusted our camouflage paint to conceal us before we moved in for bombardment. The large stockades of the seal rookery were there on the southeastern side. Two lighted beacons, a radio antenna, radar antennae, and an observation post had been added on the high, flat top of the island.

*Time 0625.* As we headed in I observed, "Jim, this is a bit of a surprise. Business has expanded beyond our expectations."

"I'll say. I count 23 large barracks, warehouses, and buildings."

"Great, the more the merrier. All ahead full!"

The Japanese report sheds some light on those developments:

> For the last 12 years Assistant Police Inspector Shimura has worked as the Head Watchman of Kaihyo To. Before Pearl Harbor, it was his kingdom . . . stretching from north to south some 800 meters and west to east 200 meters. He always came to his island in early April when the solid ice in Taraika Wan [Patience Bay] broke up. Accompanying him were his Junior Officer and five helpers. Soon thereafter the workers of the Sea Animal Company would come to slaughter and harvest the seal skins until November.
>
> On their departure his group would lock up and return to their fertile lands

at Shikuka, before the waves became so high they would hide the island. Then came the deep winter freeze. . . .

Changes with the war irritated and shocked him. . . . More sealskins were needed for pilot suits. The seal slaughter was greater than the herds should support. Due to the increased importance, a company of soldiers arrived. This year a Navy Radar division had worked at setting up and manning the necessary equipment and building for a permanent radar station. It starts operating the first week of July.

Inspector Shimura had risen at 0530 to check the seals in stockades before the daily slaughter commenced. It was a clear day because of the rain the previous day. Someone yelled, "A ship! Coming from the south." Everyone awakened. At 0620 he saw it and was puzzled by the strange ship which had a black thin hull. The soldiers on the beach washing themselves and the workers on the ever-increasing buildings were all perplexed. The ship came close very fast.

*Time 0641.* "MAN BATTLE STATIONS GUNS!" Inspired gun crews sprang to life with gongs in their ears and wings on their feet, with victory in sight and no thought of defeat.

Crossing the 10-fathom curve, we could see men running pell-mell for shelter. The *Barb* knifed straight in to 1000 yards from the buildings; sounding six fathoms, sky overcast.

"Look at them run! Some have navy uniforms."

"Right, Tom, and still more have army uniforms. Shankles, keep steady on north to help gun accuracy. Max, take the building with the radio antennae first when you're ready."

*Time 0651.* "FORTY-MILLIMETER GUN! COMMENCE FIRING!" At our first shot, island personnel opened machine-gun fire on us that walked across us without hitting. Most shots fell astern. Max immediately returned their fire with our twin 20-mm guns and two .50- and two .30-caliber machine guns. Then he aimed the 40-mm gun to an exposed 37-mm gun mounted on top of the 50-foot cliff, blowing it up. Opposition ceased. "Max, chew them up with the 5-inch gun!" That did it.

We commenced firing systematically, destroying all buildings. Lookout Don Miller reported, "Sir, I've spotted a 75-mm cannon on the cliff. I think the soldiers manning it are lying flat." Max ordered the 40-mm gun on it. The fourth shot knocked a wheel off, leaving it laying on its side.

*Time 0704.* We reversed course to remain close in; sounding five fathoms. At 800 yards I ordered, "ALL STOP. GUNNERS TEAR THE PLACE TO PIECES! DON'T WASTE ONE SHOT!"

The 40-mm gun destroyed the observation post, three luggers, or sampans, close to the beach, and an oil dump of about 30 drums. As the oil poured down the beach, we concluded it had to be seal oil, for the automatic fire striking it set off large flashes, but no combustion. The 5-inch gun destroyed the new radar station and installation and the radio station, and set the large buildings on fire, which rapidly spread to the adjoining ones. With sections of buildings flying up in the air and huge fires burning, an immense black smoke mushroomed its way to the high overcast. Singer described the scene perfectly. "Through the movie camera it looks like Vesuvius in full eruption."

*Time 0724.* "CEASE FIRE! Right rudder to course 270. All ahead full. Max, secure your guns as soon as we're out of range of any machine guns or small arms they might still have. Jim, we're heading out to observe results and give the planes a few hours to arrive in case they sent a transmission before we destroyed their transmitters. Then, if things look good, we'll land. Tom, take the conn and lie to at 3500 yards. And don't forget our bet that *Barb* will sink 15 vessels this patrol. The guns just eliminated 3 more."

"Captain, that's a 'no-no.' The key word in our wager is 'sink.' Those had a line to the beach. *Barb* must *sink* 12 more for you to win."

"You're a hard man, Tom."

*Time 1030.* No planes arrived. Radiomen Wallace Lindberg and Wilson, guarding the Japanese frequencies, had heard nothing.

"AWAY THE ASSAULT FORCE! Swish, while we're lying to, inflate the two rubber boats. We'll flood bow buoyancy and safety tanks to flood down forward. This will make it easier to debark. We'll hold here until everything is ready." When Masek's Mavericks were ready, I handed Willy the flag to be raised on the island flagpole when they hauled down the Rising Sun.

I offered his group a few thoughts. "Men, your primary mission is to keep from being wounded while you're on Kaihyo To. Gierhart and Bluth have both radio and signal flag communication with *Barb*. At the first sign of any trouble, everyone take cover while we eliminate it with our guns. Take only one prisoner, preferably one wearing a navy uniform with chevrons. Don't kill any Japanese unnecessarily. Don't go near the fires, since ammunition contained in a building may explode. Don't enter any building for charts unless you can see that no one is hiding inside. If anything happens to your boats, I'll put *Barb*'s bow on the beach and haul you aboard. Any questions? No? Be careful. Good luck and Godspeed."

*Time 1045.* "MAN BATTLE STATIONS GUNS!" Now at the northwest corner of the island, the *Barb* worked in slowly toward the beach, 400 yards from the nearest buildings. Not trusting the enemy, we took an angle of approach such that all guns could bear if we encountered opposition.

The incineration was increasing as the flames leaped from damaged building to damaged building.

Both periscopes were raised for conning tower lookouts. On the bridge and periscope shears above, 10 pairs of binoculars were engaged in searching every square foot of the island. Since we needed only two engines, motor machinists and electricians augmented the lookouts. Whitt, Wearsch, Zamaria, Koester, Swearingen, Charles Johnson, Schmitt, Powell, and David Cole (our stewards' mate) were perched above.

A thousand yards. Sounding seven fathoms. "Swish, how about using a couple of your gun loaders—Hatfield and Turnage—to start heaving the lead. I'd like a check on our fathometer." A few minutes later, "Sounding six fathoms." Hatfield called out, "By the deep SIX."

"Checks, but keep them coming. All stop." We coasted closer. "Range?"

"Six hundred yards. Sounding five fathoms."

"By the mark FIVE." Just another hundred yards.

*Time 1102.* Our signalman, Stan Wells, hollered, "Captain! There's a pillbox on top of the cliff!"

Tom and I swung our binoculars along the flat and found four. "Max! Put the 40-mm gun on the one to the left with the antenna sticking out. It's the largest."

*Time 1103.* "FORTY-MILLIMETER GUN, COMMENCE FIRING! They returned fire with machine guns and rifles. Then all our automatic guns opened up and silenced the opposition. After 18 shots, the 40-mm put one projectile through the slit of the pillbox. This blew open most of the front side. Amazing accuracy, but obviously our landing was too risky.

Bluth shouted from forward, "Hey, Koester, you offered me two hundred dollars to give you my place. Now you can buy me for a nickel!"

*Time 1110.* "All back two-thirds. Secure the assault force, but stay clear of the guns. Hold your fire unless fired upon. Secure the lead-line soundings. Deflate and stow rubber boats. Blow bow buoyancy and safety tanks dry."

*Time 1125.* "Range 4200 yards."

"Secure from battle stations. Right full rudder. All ahead standard. Jim, set course for Shiritori."

As a face-saving measure the islanders again opened up with their machine guns, the splashes way short. I pressed the intercom. "Men, I realize that the rubber boats are not all that have been deflated, myself included. But I cannot praise the gun crews too much nor the alertness of our lookouts nor our volunteer assault force. Regretfully, there were too many military, too well dug in. Yet what has been definitely accomplished is the ideal submarine bombardment. The island is out of commission. No communications, no radar, no power, no buildings, no boats exist. The wreckage still burns. A hearty well done to all. I apologize for forgetting again to put four cases of beer in the cooler. Do so, now! On departure from Kaihyo To, I rechristen it little Iwo Jima!"

Back on the island, Inspector Shimura's junior officer labored, slowly writing out the report of the battle while sitting on a hibachi:

The ship came close very fast in 10 minutes and suddenly began firing cannon and machine guns. In a short while all the buildings were on fire.

All the power boats which were to be used for communication with the mainland were honeycombed. A first class Warrant Officer in the Radio Room, a Technician, and three workers were killed. Though the Navy Division returned fire with machine guns, the enemy burned up the buildings and then concentrated on the flagpole and objects in the center of the island without a stop until 11 A.M. Inspector Shimura barely escaped into the underground shelter on the southern tip of the beach. When the firing stopped, he went out of the shelter, but saw no ship, perhaps it went under water.

The dead bodies were cremated on the beach. All the rest of us have no place to sleep. Fortunately, we had some rice and miso, which has been stored under the sand.

On 3 July, the following day, the ferryman of the Sea Animal Company, Mr. Nakone, and another person took the only boat left, a rowboat, and rowed 15 kilometers to Cape Kita Shiretoko. On the morning of the 4th, the Rescue Division of the Army arrived. The Navy Radar Division, with everything destroyed, left immediately.

Only the seals came back to the island as if nothing had happened.

*Japanese Army News* reported: "It appears that the enemy ships are still on the Sea of Okhotsk. At Kaihyo To, they heard about 600 to 700 bomb explosions and recognized three enemy ships." And Radio Tokyo news proclaimed: "Kaihyo To bombarded by six Capital Warships. One submarine surfaced later and joined the engagement."

# Chapter 26 Oh, You Cutie!

With Shiritori in sight at supper time, the *Barb* dived and closed the beach until we had only three fathoms under the keel. Our plan was to reconnoiter for a future rocket attack. We were happy to find how much it had grown. The spine of mountains running from south to north undoubtedly provided vast sources of mineral and forestry products. The town was thriving with numerous large factories. The newness of many portended its great importance in the war effort. As usual, the Japanese were working late. At 2000, trains were still shuttling back and forth on the many local spurs— and up into the mountains. Impressed, we decided to reserve 36 rockets for a future attack, for that night we had other plans.

*Time 2126.* Clear of Shiritori as the long twilight dimmed, the *Barb* surfaced. Some 35 miles to the north lay our target area.

Our presence in Patience Bay would be a sore subject to the Japanese after the destruction of Kaihyo To. We decided to repay the compliment of harboring the pontoon planes that had given us such a fit in our battle with the Terutsuki by giving Shikuka a rocket massage after midnight. There were other reasons connected with the development of my harassing theory. Shikuka was a most difficult target. The *Barb* would have to be a good 15 miles inside the 10-fathom curve to be able to attack with rockets, an even deeper penetration than that at Namkwan Harbor on the China coast.

The Shikuka-Kaihyo To attacks certainly could alert or bring forces to Patience Bay. Consequently, this might be our last safe opportunity to attack Shikuka. In addition, while the Japanese searched this area, we planned to torment the La Perouse Strait–Aniwa Bay area, which should have cooled off, there being no subs in the Sea of Japan. Basically, for harassment to be effective, it must force the enemy to raise the level of his defenses in all areas, countering the lone *Barb* effort. Ultimately, an out-of-proportion diversion of enemy forces that should have been applied toward a more serious threat would pay dividends—a takeoff from the maxim "divide and conquer."

*3 July.* At midnight a happy crew spliced the main brace at Swish's suggestion. Everybody was up: no submariner was able to sleep when on the surface in enemy waters too shallow in which to dive. Yet we were safe since no one was searching here.

*Time 0100.* "Captain, I can make out the range lights of Shikuka Harbor. With the heavy overcast and light drizzle, we've a perfect night for rocketeering."

"Good work, Dave. Slow to one-third. Max, set up your rocket launcher."

On the chilly bridge, dressed in two sweaters and foul-weather clothing, with fouler deeds in mind, we enjoyed the miserable weather. Below on the fo'c'sle, red flashlights glowed as Max and his men checked out the launcher circuits.

*Time 0210.* "MAN BATTLE STATIONS ROCKETS! Max, use the maximum range. Our target aim will be 3° left of the center of the mass of high chimneys so that the right drift of the rockets will still fall in the factories. Pray that some fires will be started to greatly embellish the explosion damage."

"Aye. Swish, set range at 5250 yards. Load 12 rockets."

We eased in. "Jim, keep an eye on those soundings. I don't want to have less than four fathoms. Rig in the speed swordarm. Check with Jack at plot and radar to see if we have a definite shoreline-to-chimneys measurement."

"We do, captain. Coming in at slow speed, we're definitely picking up the docks and piers. The chimneys are 300 to 400 yards beyond. There are no ships."

"Perfect. We'll shoot at 4900 yards from the docks. Coach us in as we squirm into this girdle."

"Sounding four fathoms. Range 5000 yards."

"All stop."

"Sir, sounding now is less than four fathoms!"

"Okay, we'll coast in. Stand by, Max. Range 4900 yards."

*Time 0240.* I ordered "ROCKETS AWAY!" with an upward wave of my arm. (Like a kid, I enjoyed the novelty of these rocket-launching orders and felt theatrical at it.) We raised our polaroid-goggled heads to watch the blastoff. Only Singer's camera whirred. Nothing happened! Embarrassing? Like forgetting your lines in a play. I felt stupid as we just looked at each other.

After a few frantic moments, McKee, one of the loaders, found a loose wire. Someone had tripped over it in the dark.

"All back one-third." The *Barb* repositioned herself. My theatrical cloak fell off.

*Time 0247.* "ROCKETS AWAY!" This time the system worked, and the 12 5-inch rockets went swishing out in just over four seconds. Their flare eclipsed 20 feet above the deck. "CLEAR THE DECKS! ALL AHEAD FLANK. RIGHT FULL RUDDER."

*Time 0247.31.* The rockets exploded among a mass of buildings with the thunder of bombs. Lights blinked on and off in town for a few minutes, followed by complete darkness. No fires were started. Several different sources of aircraft search radar came on as we scampered for deep water. "Secure from battle stations; a well done to all hands! Jim, set course down the middle of Patience Bay for Otomari in Aniwa Wan. Regardless of the fog we'll stay at flank speed until we have 20 fathoms of water, then slow to full."

On my innerspring mattress, I fell asleep in three minutes, believing that I had removed at least one shoe.

*Time 0935.* "Captain, radar contact 13,000 yards. I'm heading for her."

"Coming up. Station the tracking party."

Within 10 minutes we had sight contact in the haze. A small coastal freighter of at least 1000 tons: a coal burner with slant bow, kingpost, mast just forward of her engines-aft bridge, funnel, kingpost, and a spoon stern. "Jim, she has a small pilothouse built on top of her bridge, fastened to the mast, and she's loaded. Angle on the bow is 120° starboard. I can't see any Soviet markings, but I need a periscope look close aboard to make sure. Let's end around on the edge of visibility."

"Plot shows she's zigzagging and headed for Shikuka."

The fog thickened. In half an hour, with the target making 8.1 knots, we were ahead and dived. Passing 200 yards off her starboard side, I could see about a dozen men on deck in black, woolen, high-collar merchant marine uniforms. They were all Japanese. The fog was too heavy for filming through the periscope. As soon as she disappeared, we surfaced and ended around again, satisfied that she wasn't Soviet.

"Max, we'll shoot one of your four small Mark 27 homers that you have aft. Let's hope it operates better than those Mark 28s three times its size."

*Time 1206.* The *Barb* submerged 4000 yards ahead on the target's base course. We knew his zigzag plan. Nicknamed a "cutic," our homing torpedo would swim out of the tube. It had to be fired at a close range with the submarine below 150 feet; the warhead was inactivated at 100 feet to protect the submarine.

*Time 1227.* A last look at 1800 yards, then the *Barb* went deep. I noted two lifeboats partially lowered from the davits of our target. Before the last look, I reminded myself to "think beer."

"Jim, I'm becoming superstitious. Every time I forget the beer, something comes a cropper. Have Cole put four cases in the cooler and see if it doesn't change our luck." Closing the target on sound bearings and single-ping ranges, we came in nicely, but Dave had a diving problem. "Open the outer door on tube 10. Final ping range?"

"A hundred yards."

"Dave! I'm ready to shoot. Our depth is only 135 feet. I said 150 feet."

"Can't help it. We've run into an 18° negative gradient. I'm barely holding our depth at dead slow speed with a 3° down angle. Do you want to abort the approach?"

"Hell no! Our position is good. Safe enough, I guess. Range?"

"Seventy-five yards."

"Set."

*Time 1233.* "FIRE 10!" The sensation was weird, unlike firing a normal torpedo by high-pressure air. Neither a whoosh nor a jolt as the fish swam out. Nor was there a venting of the torpedo tube that raised the pressure in the boat till our ears popped. With this torpedo all was quiet except the flap-flap-flap of the target's propeller passing overhead.

"Sonar?"

"Gierhart reports torpedo has been picked up—hot and normal— curving toward target's propeller. Now both noises are clear and melding togeth . . ."

WHAM! "Wow! That was close enough to knock your fillings loose." Wry smiles and cheers below. A torpedo hit! At long last, a torpedo that worked.

"Gierhart says the ship is breaking up and sounds like it's coming down on top of us."

"All ahead flank! Right full rudder!" The *Barb* zoomed clear. In 45 fathoms of water we could hear the loud thud through the hull as the ship smashed into the bottom with hissings and crunching noises as compartments collapsed. Another enemy ship had died.

*Time 1247.* On the surface again within yards of the flotsam, we were surrounded with coal dust and the whirlpool eddy where the ship upended and sank stern first, sucking crew members down with her. Singer started filming. "Max, Bill, Swish," I urged, "get some men down on deck. The top section of the pilothouse ripped off and is slowly sinking. Leave the bow planes rigged out. Bring up two grapnels and boat hooks to pick up the rolls of charts and signal flags."

Men scrambled down on deck to retrieve flotsam as I eased the *Barb* over alongside the pilothouse. A big Rising Sun was painted on its top. "Tom, that ship upended and sank so fast it snapped off the top of the mast with the pilothouse attached. Hey, Max! While you're on the bow plane, try to retrieve those large red and green running lights for me."

Max bravely walked across the top of the pilothouse, picked up the port light, passed it to Howard Brunton on the bow plane, and got the starboard one back just as the whole shebang sank. Singer filmed it all.

Seventeen charts were recovered. With a data base from 1936 (versus

Max Duncan picking up sidelights and charts from a 1000-ton cargo ship sunk by a homing torpedo. Taken from movie film of the salvage operation.

ours from 1894), they were invaluable to us. The soundings were in meters and far more numerous. Also, they had the mined areas lined off, which agreed with the information we had obtained from Kito. Each chart was stamped with the name of the ship we sank. Unfortunately, this stamp was in kanji instead of the kana that we could transliterate. So we didn't know the name of the ship.

One smaller-scale chart had the complete track of the ship laid out from the time she started her voyage in the Inland Sea. Through the Sea of Japan she stopped at Niigata, Tsugaru Strait, thence through La Perouse to her sinking point, which agreed with Jim's navigational position.

Clearing the deck, the *Barb* got under way for Aniwa Wan. After supper, we spliced the main brace. Phillips and Ted Wilby (a first patrol ship's cook) rigged the cake with a single ship, her bow perpendicular, and the inscription, "OUR FIRST TORPEDO HIT—BUT NOT THE LAST." Tom reminded me, "Four down, 11 to go."

*4 July.* Rounding Cape Naka-Shiretoko, we sighted a frigate and dived to attack. Our target did a figure S at 8000 yards, then returned to his patrol between the minefield lines. Disappointed that he didn't want to play in our ballpark, we surfaced as soon as his routine patrol took him out of sight.

Max and I were on the bridge discussing the frigates that patrolled the La Perouse minefields. "Max, do you realize that this last ship is the first merchant ship that has ever been sunk by a cutie? It has great disadvantages in its slower speed and having to be fired at such abominably close ranges. What submariner ever considered firing at a ship at 75 yards? They'd have called it suicide. I have an idea for you to work on, however."

"Oh, boy. Here we go again."

"Calm it, Max. At least the Mark 27 worked where the Mark 28s failed. Since we have large negative temperature gradients at the 150-foot cutie firing depth in deep water, which deflect a sonar beam and make it erratic, we could use a cutie to attack a lone frigate searching for us."

"All right. If we're sure we have an 18° gradient above us like our shot yesterday. Captain, why don't we put another cutie in tube 10 while we study the ramifications of such an attack. Let's go aft to do it now."

In the after torpedo room, Roark, Wells, and White were busy giving the torpedoes a freshening battery charge. Max told Roark to load another Mark 27 in tube 10. Roark walked to the torpedo skid containing one, caressed it, then leaned over and kissed it, saying, "Oh, you cutie! Sink them all!" I grinned; I had never seen a torpedo kissed before.

In the after engine room we were stopped by a knot of motor machinists—

Johnson, Schmitt, and Alden Hutchinson. "Captain, the whole crew knows about your bet with Mr. King to sink 15 vessels. You seem to win all your bets. We'd like to bet with you, but he says that this time you're sunk. With 11 vessels more to go, bum torpedoes, and only two more weeks left on this patrol, is it possible?"

Max said, "Is anything impossible for the *Barb?*"

*Time 1846.* "Captain! Radar contact 14,000 yards! I can see his masts. Looks like a destroyer, nothing else."

"Men, bet with me on the *Barb*. This may cut it to 10." On the bridge Willy pointed out the two-stack destroyer with a small port angle. When we were submerged again, the tracking party clocked her speed at 12 knots, not hurrying anywhere. "Dave, make your depth 150 feet. Check the gradient. Jim, battle stations torpedoes and make ready all bow tubes."

"Negative gradient 21°."

As I looked at the TDC, Max whispered, "How about tube 10, the cutie?"

"I've got to identify."

Jim added, "But this is a warship."

"Bow tubes, Jim. If those fail, we'll end around for a deep 'cutie' shot. We're coming in nicely."

At 3500 yards, with the outer doors opened, I made out the markings on her side: a Soviet destroyer.

It was extremely important that we be notified of Soviet warship movements. I had not heard of any, been warned of any, nor read of any in other reports. It was fortunate that we were not planning on using special weapons, for bow tube doors had been opened before we identified her merchant ship markings.

*Time 2023.* Surfaced. "Radar contact!" Again? At flank speed we got ahead and dived. Another Soviet ship. The Fourth of July passed with no hits, no runs, no errors, and no fireworks, but, oh, what a cutie we had had the day before!

*5 July.* Karafuto resembles a bowlegged cowboy whose left arm is hooked out to enclose Patience Bay. His legs encircle Aniwa Wan, 30 miles long and 30 miles wide. At his crotch lies Otomari, where train ferries and main commerce land, having come through La Perouse Strait, then up the inside of the western leg. The coastline is mountainous. Shipping passes as close as 200 yards offshore in clear weather. In fog, they move off a little. The minefields stretch north from Hokkaido to an area 4 to 10 miles off the western ankle of Aniwa Wan.

*Time 0450.* We submerged two miles north of Cape Chishiya on the western leg, five miles north of the minefields. After studying our Japanese charts, we realized this was the only spot on the whole coast up to Otomari where we could come within 3000 yards of the coast and still have 15 fathoms of water. We would wait in this two-mile stretch for prey, our only exodus to the east. Also, a close watch would have to be kept on the current that swept down into the minefields.

*Time 0910.* Swinging the scope around I spotted a frigate coming down the coast outboard of us. "Chapman, right full rudder to east. MAN BATTLE STATIONS TORPEDOES!" Twelve minutes later, stopping for a look, I sighted two train ferries swinging down the coast, almost invisible against the mountainous background. One was a beauty: two stacks athwartships, three decks of superstructure under a large bridge. Hugging the land, she passed less than 200 yards from the rocks we used as navigational beacons. The *Barb* reversed course at full speed. The approach ended with us in the middle, 3400 yards from inboard and outboard targets. I kicked myself for being sucked out by the escort. Lesson learned.

*Time 0953.* Smoke coming north through La Perouse Strait was from two ships: a small coastal freighter clinging to the coast and a medium freighter that eased out to let the two ferries pass in between. I had a bad moment noting that the large rust patch on the medium freighter's side resembled Soviet markings. I could hear the howls if we fired fish across the bows of a neutral ship to hit a small Japanese freighter. It was only rust.

"Okay, Singer, place your camera on the scope every other time that it's raised. We'll film this entire approach." Singer, pleased at long last, took several movie scenes.

*Time 1038.* "Up scope. Wow! Bearing—mark. Range—mark. Down. I nearly clipped off the pontoons of a twin floater seaplane doing figure eights across the freighter's bow. Make ready all tubes. Open the outer doors, tubes 4, 5, 6. Sounding?"

"Thirteen fathoms."

"We'll turn left after we shoot."

*Time 1047.* "Up scope. Angle on the bow 80° starboard. Final bearing— mark. Range—mark."

"Fifteen hundred twenty yards. Set."

"FIRE 4! FIRE 5! FIRE 6! Left full rudder."

"Hitting time, 20 more seconds."

"Up scope. Singer, put your camera on to catch the explosion." He peered through the scope and filmed away. Now we would get what Lockwood wanted.

Nothing happened! "Damn it, there's something wrong with their depth mechanism and mine. Put four cases of beer in the cooler. Jim, open the outer doors on tubes 7 and 9 for a stern shot. Max, quick setup. She's come right 20° rounding the rocks. That helps. Angle . . ."

BOOM! A torpedo hit the beach. "Angle on the bow 90° starboard. Final bearing—mark. Range 1670 yards."

"Set."

*Time 1052.* "FIRE 7! FIRE 9!" BOOM! Another torpedo hit the beach. Gierhart reported, "Torpedoes hot, straight, and normal."

*Time 1054.* WHAM! "Up scope. Heavy clouds of smoke under her stack. She's listing to starboard. The tops of her screws are coming out of the water. The small freighter is stopped on her port beam—not worth a torpedo in ballast with the bow part of her keel showing. Singer, start filming." Success this time.

"Sonar reports breaking-up noises have commenced."

"Down scope. All ahead standard. Head for 20 fathoms."

*Time 1106.* "All stop. Final check. Up scope. Three lifeboats are in the water. She's listing to port 10°, anchored probably by the explosion."

"Gierhart has breaking-up noises louder, and he hears pinging coming in with an up doppler."

"Well, she's not settling. Much as I dislike this, we must go back in and deliver the *coup de grâce*. Reverse course! Make ready tube 1. All ahead standard."

*Time 1123.* "Open the outer doors on tube 1. Sounding?"

"Fourteen fathoms."

"All stop. Final setup. Up scope. Range—HOLD IT! She's sagging amidships. She's breaking in two like a vee for victory. Singer, roll it. Reverse course. Jim, take a last peek as she sinks. All ahead two-thirds."

Jim swung the scope around. "Down scope. Captain, there are pairs of escorts coming in from north and south and the plane is returning."

"Dave, take her down to 80 feet. I may have to slow to be quiet. Let me know if you need speed to hold her so close to the bottom. Rig ship for silent running. Rig ship for depth charge."

As the pinging became louder, we slowed; as it diminished, we speeded up. "Dave, anything on the bathythermograph?"

"Nothing. Straight isotherm. We're too shallow, just burying our heads hopefully."

BOOM! "Can't tell whether that distant thunder was a bomb or depth charge. In an hour we'll reach 30 fathoms. Then, with a good gradient, we may be able to use a cutie. Max, let's load another cutie in tube 9 now."

The concentrated sweep of the area continued all day.

*Time 2116.* The *Barb* surfaced in much-longed-for air. The sky was still light, but the escorts seemed leery of hanging around a sub after dark. We set course for Hokkaido to cool this area off and to give the crew a chance to relax. At dark we opened the forward torpedo room deck hatch and let the engines aft dehydrate the boat. (The cold, clammy sweat from long submergence in ice water covered everything.) That job completed, this hatch was closed and the process repeated with the after torpedo room hatch.

Lying in my bunk reviewing the day, I realized that Aniwa Wan affected the personnel more than any other area. Even with a sinking, without close bombs or depth charges, it seemed to permeate most of us with an electric, nerve-racking claustrophobia. I put this down to shallow water, proximity of minefields, and the obvious fact that our fish were not hitting. We had had the best of setups with targets on constant courses at constant speeds and the sub undetected. With the same team having a batting average of 22 hits out of 24 fish on our last patrol, we now stood at 2 hits out of 15. Having no desire to shoot fish to run on the surface, I resolved that the remainder would have their depth set at three or four feet. My motto: we don't have problems, just solutions and other options.

*6 July.* In variable fog and rain, the *Barb* swept the northern coast of Hokkaido from the minefields east. Courses were adjusted so that breakers on the beach were visible without the surfaced *Barb* being sighted. The area was empty.

Swish's welcome call to splice the main brace brought everyone up with a round turn at 1100 sharp. The men were happy again. A sinking, some suds, and a sublime cake by Bentley and Phillips were all it took in our restrictive, all-male world. The ship sinking in a "V" supplied the motif. Tom announced that we now stood at "Five down, 10 to go!"

Radioman Edwin Schilke waved a message in the air. "Listen to this intercept of Japanese Army News from Radio Tokyo: 'At 0145 Tokyo Time'—that's 0247 *Barb* time—'on 3 July, Shikuka was attacked by five enemy ships. It appears that enemy ships are still on the Sea of Okhotsk.'"

This brought down the house. Bill Walker called over, "Hey, Higgins, is your chronometer wound? Somebody's two minutes off."

*Time 1152.* "Captain, we're in position to dive and close the city of Shimoyubetsu. Weather is still sloppy."

"Jack, I'll give you the word from the conning tower. Dave, we're reconnoitering to see her docks and factories. The underwater slope is all sand and very gradual. We'll go in to 12 fathoms at slow speed, no closer."

Two hours later we had plotted in the numerous small factories and railroads, not the best rocket target. Surfaced, fog thickening in the empty area, the *Barb* turned back toward Aniwa Wan to take advantage of the weather and the ships at anchor there. The following day en route to Aniwa Wan the fog gave way to cloudless, 10-mile visibility, which was not the object of this journey. Twice we dived near the minefields and made approaches on individual frigates, then gave up as they reversed course and stayed among their mines.

Not having had a navigational fix for 16 hours, further approach on these targets was too risky in the unknown currents. We made a landfall to obtain a fix. A correction of our track showed the *Barb* had been submerged within 500 yards of the eastern mine line, a no-no because entry into the minefield kills.

By supper time the *Barb* was squatted off the point where we sank the *Sapporo Maru* on 5 July. During the day the only merchant ship sighted was Soviet. The area was devoid of normal traffic, including small craft. We waited, submerged, in a water depth of 30 fathoms where we could use our cuties against the frigates. Soon two came slowly down the coast, pinging. We dived, hoping they would extend their search seaward, yet they never ventured outside the 20-fathom curve. With the heat still on in this area, we set course for Patience Bay.

*8 July.* Surfaced, ambling along Karafuto's east coast, the *Barb* turned into the southern reaches of Patience Bay, searching the coast there for anything. Visibility dropped to six miles.

*Time 1833.* "Captain, we've sighted a lugger on opposite course, hugging the coast, range 10,000 yards, light fog. I'm bringing *Barb* around to trail."

On the way to the bridge I looked over Jim's shoulder at our position on the chart. "We're 15 miles north of the town of Sakayehama, almost the southwestern corner of the bay. It's pretty shallow." With sunset in an hour and a half, and twilight as long, we followed on the edge of visibility. Once the lugger saw us, she would probably run herself aground before we got a shot off. We had to catch her in reduced visibility. The *Barb*

The celebration cake following the sinking of the *Sapporo Maru* on 5 July 1945.

eased inside the 10-fathom curve at sundown. Now in water too shallow in which to dive to protect herself, the *Barb* closed in the shadows. The sound heads were rigged in as the water shallowed. We approached Sakayehama—which was evidently the lugger's destination. Only two miles to go. "Jim, Max, MAN BATTLE STATIONS GUNS, but have the 5-inch gun crew standing by below decks. Use the 40-mm gun. We have just two fathoms beneath the keel if we have to leave the harbor."

*Time 2055.* At a range of 1100 yards, she started to turn into the harbor. "COMMENCE FIRING!" Shells flew and the town came to life as the lugger returned fire with a 25-mm machine gun. Visibility was rapidly decreasing. After 32 rounds and 20 hits, the lugger was destroyed and sinking by the stern. "CEASE FIRE!"

*Time 2058.* Robert Tague, a lookout, cried out, "Two-engine plane taking off!"

"Clear the decks! All ahead emergency! Course 090! Constant helm zigzag 20°!" A crush of people came tumbling down into the conning tower. Pasting my eye against the high periscope, I shouted, "Shut the hatch!" In three minutes we crossed the 10-fathom curve. The plane disappeared to the south. "All stop. Open the hatch."

*Time 2103.* Back on the bridge, Tom, Bluth, and I watched the Saka-yehama Defense Force roll out two medium-caliber cannons on wheels and pump out shots every five seconds. They fell well astern in the vicinity of the lugger.

On the high periscope, Jim called up, "They're shooting at the lugger. Some of our misses must have landed ashore, and they believe they came from the lugger. I doubt if they can see us without binoculars."

The lugger sank. Grudgingly, Tom said, "Six down, nine to go, though their hits may have gone right through her. We're lucky. That plane must have been on a routine flight." Radioman Delameter stuck his head up the hatch and handed Tom a message. "Captain, it's a lifeguard assignment for air strikes, 12 July, off the east coast of Hokkaido by Admiral Halsey's Third Fleet Carrier Task Force 38. The date may be advanced."

I sighed. We were over 500 miles away from our downed aviator rescue station. Though we might be needed, this would occupy about a week of our 12 days of patrol time remaining. We had lots of plans jelling, but no choice. "Jim, set course for Nemuro Strait, the northeastern tip of Hok-kaido."

*10 July. Time 0918.* Two freighters were coming up through the strait, along Hokkaido. The *Barb* dashed across the strait for a close-inshore attack position. Willy, now qualified for the officer of the deck watch, was amazed that we stayed on the surface, so close that we could see the fishing shacks on shore with the naked eye. "Don't you think they can see us?"

"Sure. Odds are they have no telephones or communication with passing ships. The *Barb* has to be close-in."

*Time 1015.* We dived for a torpedo attack. The approach was simple: steady course, steady speed, using periscope radar for ranging, and no escorts, though air bases were close by. One target was 2000 yards astern of the other. "Make ready tubes 1, 2, and 3. Set depth at 3 feet."

*Time 1117.* "FIRE 1! FIRE 2! FIRE 3!" Sonar reported all torpedoes hot, straight, and normal, but we heard no explosive hits with torpedoes set at three feet. Only one vague rumble mixed with wave noises against the rocks sounded like two objects colliding.

"Up scope. No change in course. We're dead ahead of the next ship. Dave, take her down to 150 feet. Max, we'll use a cutie on her. Make ready tube 9. All sonars pick up the screws." We heard one torpedo explosion on the beach.

*Time 1125.* "Come on, sonars, we need bearings!"

Gierhart reported, "Sir, nothing on sonar except wave noises on the beach."

"You must pick up the ship, Dave, 130 feet, smartly."

*Time 1130.* "JP sonar has picked up target well astern of us. She's speeding up and has already passed overhead." Back at periscope depth, extremely disappointed, I watched the targets go over the hill and round Cape Shiretoko Misaki. A battle surface action was considered, but targets were alerted and armed with 37-mm guns, plus I expected the Hokkaido Coastal Air Patrol to arrive momentarily. Two did. Finally, in the late afternoon, the area was clear, so we surfaced. Meanwhile, the torpedomen practically took our last five Mark 18 electric torpedoes apart searching for our troubles. The whole fire control system was checked and rechecked in vain. We set course for Kunashiri Strait.

In the wee hours of the morning the *Barb* transited the strait. Our cable station was still in a fog. I followed the advice of the station manager at Midway and passed it up.

After breakfast we encountered a large diesel sampan 15 miles from the airbases on Shikotan Island to the east of Kunashiri. The 40-mm gun stopped her but she wouldn't sink. To save ammunition I had Tom try a high-speed sweep five yards off her beam to wash water through the holes. That failed, so two shots from the 5-inch gun at the sampan's waterline sank her. Tom grumbled, "Seven down, eight to go."

Toward midnight we made the first submarine transit of Shikotan Strait, south of the island. Since our Japanese charts showed that some miles of its length were only 80 to 90 feet deep, there would be no point in placing mines against submarines there. I was right, and we saved about 50 miles running around the northern end to get to our station.

*12 July.* On station we waited for word of an early strike by Halsey's Task Force 38 while maintaining our own radio silence. Nothing came. Noting the position of our lifeguard station, I wondered whether anyone had heeded the report I had sent in a year ago about a new minefield off Akkeshi Bay. For this strike, some planner had put our station only 10 miles from this minefield—not too comforting if we needed to go inshore to rescue some aviator.

Since we had nothing to do, the *Barb* patrolled the coast on the surface from Akkeshi Bay northeastward to Suisho Shoto. In a light fog, the beach visible, we found nothing. The next day visibility was low, and the strike was canceled. Two days lost; only six left on patrol.

*14 July. Time 0524.* The strike was on. Our fighter cap of four planes arrived from the carrier. The strike commenced with wave after wave of planes. What a beautiful sight; what support for a change. During the day large fires were started at Kushiro and Akkeshi that burned all night. At 1720, our comforting air cap gone, the *Barb* was forced to dive from an enemy plane.

*15 July.* The day's strike had been moved up the coast 30 miles to Bokkuriso Bay. This delighted us. At 0450 our fighter cap of four planes arrived. With no mines around, we closed the beach to two miles to watch the fun. The first wave quickly set Nemuro on fire and knocked out all important targets. Then a search began for something else to strike. From our vantage point we pointed out radio and radar stations, which the planes immediately set ablaze. Fifty fighters scoured the area. I thought we might join them with a shore bombardment, but neither we nor our cover could find anything worth using our precious ammunition against.

We noticed horses grazing in a meadow, the last of the opposition. Over VHF radio we heard a section leader caution, "Leave those poor horses alone!" The strike subsiding, our presence known, I broke radio silence to request ComSubPac to grant us a week's extension on patrol.

*Time 1600.* The strikes completed, our cap bade farewell and good hunting. The *Barb* headed for Aniwa Wan, 32 hours away. Entering there, we crossed to the western side where we had sunk the *Sapporo Maru* north of the minefields. In the morning we found a frigate in deeper water and dived to attack her with a cutie. At 6000 yards she returned to the mine lines—tough luck. Other frigates did the same all day. In the evening ComSubPac granted our week's extension.

*Postscript:* Japanese records have the *Toyu Maru* (1256 tons) sunk by a submarine in the Nemuro Strait area on 15 July, where the *Barb* was the only lifeguard. No aircraft attacked any ships there during the 14–15 July strikes; no ships ventured forth; and no other submarine was within 100 miles. The *Toyu* was similar to the ship the *Barb* fired at on 10 July in that area. Perhaps our torpedo failed to explode but still caused sufficient damage to sink her later. The Joint Army-Navy Assessment Committee listed her, erroneously, as sunk by naval aircraft, though no attacks on ships were made within 100 miles on either day. Possibly the date is in error. Although this is yet another unsolved mystery, the *Barb* is the only possible cause of her sinking.

# Chapter 27   Hear That Train Blow!

At his headquarters in Guam, Fleet Admiral Nimitz finished his daily briefing to the gaggle of war reporters. It included Admiral Halsey's attacks on Japan's east coast. When asked about Halsey's attacks on the north coast of Hokkaido and Karafuto, Nimitz replied that no surface ships or planes had attacked those areas. Skeptical, the reporters countered with the *Japanese Army News,* Radio Tokyo, and Tokyo Rose's assertions that the north coast of Hokkaido was now preparing for invasions due to the continuing attacks by ships operating in the lower part of the Sea of Okhotsk. Nimitz reiterated his denial and laid it down to Japanese war jitters.

Afterward Nimitz asked Admiral Lockwood privately about his submarine operations in the disputed area.

"Well, Chester, there's only the *Barb* there, and probably no word until the patrol is finished. You remember Gene Fluckey?"

"Of course. I recommended him for the Medal of Honor. You surely pulled him from command after he received it?"

"No. Before his eleventh patrol he wormed a promise out of me that if *Barb* did well on that patrol, he could have a fifth patrol, a 'graduation' patrol."

"What is that?"

"He's worked up a theory of harassing when there are no ships around to sink. He loaded rockets."

"Rockets in a sub?"

"Yes, and one of the landing-ship-type launchers. I put a photographer in *Barb* to get some good footage for Project Three, the submarine film that we lack."

"Ask him for a rundown of *Barb*'s activities when it's safe." Thereupon, Admiral Lockwood sent this message:

> COMSUBPAC TALKS TO BARB X AT YOUR CONVENIENCE GIVE US A BRIEF REPORT YOUR ACTIVITIES AND RESULTS YOUR ROCKET BOMBARDMENT WITH REGARD TO TACTICS EMPLOYED COMMA PROCEDURE COMMA FACILITY OF USE COMMA ACCURACY OBSERVED AND RECOMMENDATIONS AS TO FUTURE USE IN SUBMARINES X

We received this while investigating the patrol activity of the minefields, trying to lure them out into deep water. I decided to send an immediate answer on our activities so far. (Breaking radio silence was immaterial, for we were on the eastern side of the minefield, away from the traffic and surfaced, allowing the submarine to be seen. No single frigate would accept our invitation to join us in the deep water.)

> COMSUBPAC BREEZY NIGHTLY NEWS X BARBAROUS BARB REPORTS SINKING ONE LARGE AND ONE SMALL FREIGHTER AND THREE LUGGERS ONE TRAWLER FOUR SAMPANS IN ADDITION TO ONE ISLAND SMASHING BOMBARDMENT AND TWO ROCKET MASSAGES GIVEN IN EXCHANGE FOR BOMBS AND DEPTH CHARGES RECEIVED X DUTCH SUBMARINE O-ONE NINE REPORTS ITSELF AGROUND ON LADD REEF WITH COD SPEEDING TO AID X

***18 July. Time 0425.*** After twice rousting the tracking party to approach ships ultimately identified as Soviet, the *Barb* dived off Vennochi Point— our customary stamping grounds—to take a crack at the daily traffic. A lone frigate patrolled the tip of the minefield 17,000 yards southeast of us.

I noted that the Japanese had placed a spar buoy at the spot where we sank the *Sapporo Maru*. Some passerby must have damaged her hull on the underwater wreckage. This menace to navigation forced traffic to pass outside the spar buoy in 13 fathoms of water. Thus, it allowed the *Barb* to hover 1000 yards farther out in 120 feet of water, an excellent torpedo range.

The huge train ferries we had aimed toward may have disappeared, since the news broadcast said three were eliminated by the air strikes at Hakodate in southwest Hokkaido. Our two may have been sent south to replace them.

Yawning the morning away, I laid out a chart on the wardroom table to see what mischief we could conjure up. My wandering eyes kept returning to the single train track along and near the western shore of Patience Bay. There was a fair stretch between two main junctures. Each juncture forked out into three tracks going to principal areas. How could we harass the enemy here?

I searched without luck for a bridge. Other officers became enthusiastic; our thinking caps were on. With no solution, this disturbed my eternal philosophy: we don't have problems, only solutions. I called for Swish to start some fresh blood circulating.

Swish sat down. As a chief gunner's mate, his eyes lighted up like firecrackers the minute explosives were mentioned. After a livening discussion, the problem was narrowed down. Our objective was neither to blow up the track nor to blow up the train. Both must be accomplished simultaneously to handicap Japan's war effort and transport for several days or a week or more.

Obviously, to do this, the scuttling charge had to be placed under the track. The problem was that the charge was equipped to be set off by a timer. No good; it was impossible to predict a train schedule. Of course, we could replace the timer with a remote control switch, but who was going to hang around waiting for a train and still be able to return to the *Barb?* Too risky!

*Time 1127.* "Captain, I've sighted a large freighter with a frigate escort coming down the coast. Permission to go to battle stations and ready the tubes?"

"Great! Do so, Max."

As the officers scampered off, I held on to Swish, who would relieve the diving manifold. "Swish, spread the word on this train target throughout the boat. Everyone's ideas are welcome, serious or silly. We have no 'not invented here' bias. Believe me, that train is going to blow!"

"MAN BATTLE STATIONS TORPEDOES! Make ready tubes 2, 3, 4, 5, 6, and 10." As the gongs bonged away, I landed in the conning tower. Max had the *Barb* moving in nicely. Jim checked our navigational position. "No problem. That spar buoy is a godsend, captain."

"Here's the pitch. Because of this flat calm I'll use the narrow attack scope only so we won't have the ST radar. We'll take our ranges on the frigate—since we know his masthead height exactly—and use the division marks on the scope. She's 1500 yards broad on the port bow of the freighter by Max's radar measurements. The frigate will be our first target for tubes 4 and 6. I'll let her ease by and shoot her from abaft her port beam. Then we shift immediately to the freighter for a new setup, shooting tubes 2, 3, and 5, the last of our torpedoes. Let's try a depth set at four feet."

"Sonar reports pinging."

"Jim, parallel his course so we'll present as small a target as possible. Dave, what's the bathythermograph show?"

"Isotherm—no help."

"Max, speed?"

"Targets' speed 12 knots on the nose. What a loverly, simple approach." A few periscope observations confirmed their courses and speeds as they approached. Sonars checked.

Soon we could hear the whining swish of the frigate's screws through the hull, her pinging still on long scale. "Up scope. Down. Gad! I'm staring into half a dozen binoculars sweeping. Their officer of the deck is berating one port lookout. A group of men are fiddling around or greasing the depth-charge racks. No range. We're so close, just part of the ship fills the whole periscope. We've got him. Put four cases of beer in the cooler."

"Captain! What's that you have in your hand?"

I was thunderstruck! I was holding the elevation handle of the periscope, which had taken this inopportune moment to come off. It tilts the view lens up or down. The imp in me came out: "Look, no hands!" Then I returned to the serious business of consummating an unheard of, untried, one-armed approach.

*Time 1158.25.* "Up scope. Range, 1200 yards. Final bearing—mark. Down. FIRE 4! FIRE 6! Sonar? Max, shift to the freighter for a quick final setup."

"Sonar has both hot, straight, and normal, right at her."

*Time 1159.18.* "Up scope. Freighter's tight in near the spar buoy. Angle on the bow 90° port. Range 1400 yards. Final bearing—mark. FIRE 2! FIRE 3! FIRE 5! Sonar?"

"Gierhart says all hot, straight, and normal."

*Time 1200.* "Hitting time—10 seconds." Riding the scope up with one hand on the handle and one on the eyepiece for balance, I swung it back to the frigate just in time. WHAM! The *Barb* shook! "My God! We hit him under his depth-charge racks! The whole stern has blown off! Men, bodies, the Rising Sun flag are flying through the air!"

Singer, who had been filming the early part of the approach, grabbed my shoulder, "May I, sir?"

"Get it!" I backed away. The frigate quickly assumed a perpendicular angle and sank until she hit the muddy bottom, leaving 20 to 30 feet of the bow sticking up vertically. Unaccountably, we missed the freighter. She headed home. The opposition temporarily vanished, we raised the larger periscope to give the conning tower personnel a look. Tom said on taking his turn, "Her fo'c'sle sure makes a better spar buoy. I guess we can count her. Eight down, seven to go, and we're fresh out of regular torpedoes. Captain, ready to throw in the sponge?"

"Not until I've squeezed you out."

Machinists Noll and Cullen Simpkins secured the arm back onto the attack periscope. We secured from battle stations and headed out from the coast. Retaliation would soon arrive and our only submerged arms were the three cuties. After lunch we surfaced, feeling very aggressive, with 40 fathoms of water under us and the three pills now loaded in the tubes aft.

Two frigates searching for us were sighted at 11,000 yards. Their angle on the bow was 20° port. We remained surfaced, hoping to draw them out from their shallow, 20-fathom position. Shortly thereafter, apparently on sighting us, they changed their angle on the bow to 90° port and made tracks for the minefield. A Soviet ship followed, which we dismissed when we identified her.

*Time 1807.* "DIVE! DIVE! Plane closing fast, nine miles."

At last the long-expected plane arrived, our first such contact since the air strikes on eastern Hokkaido. If we had known it would take this long, we could have exchanged gunfire with the freighter. No bombs were dropped, but he swept the area for an hour.

We took advantage of the lull to work on the train attack. Swish had some ideas to talk over, so he joined Max, Dave, Willy, and me in the wardroom with Hatfield, an electrician who had worked for a railroad in West Virginia before joining the Navy. "Captain, Swish here has told me about your problem of blowing up the track and the train at the same spot. I can tell you what I think you need to substitute for that timer to accomplish this feat. Mind you, I've never done this myself, but it's one idea."

"Let's have it."

"Well, I would remove that timer switch. We all know that 55-pound super-torpex high-explosive charge has to be buried under the track to catch both track and train. It measures in inches about 14 by 14 by 16, so we dig a hole and put that between two ties. Then we dig a hole between the next two ties and bury the batteries that are wired to and actuate the charge. Now, to complete the circuit we hook in a microswitch. We place it on the surface between the next two ties. That's all there is to it."

"Sounds simple enough, but please explain to us simpletons just how do *we* set off the microswitch?"

"You don't, sir. The train does."

"Okay. How?"

"The rail sags underneath the weight of the engine. So we mount the switch on two wedges, slip it under the rail, the engine comes along, the rail sags, closing the switch, and she blows!"

I clapped Hatfield so hard on the back in my exultation that I nearly knocked him off the bench. Max added, "Fine, do you have a microswitch?"

"No, but Teeters must have one with his radars."

Dave replied, "No prob."

Hatfield's ideas were brilliant. Were there any flaws? "Tell us, exactly how much does the rail sag?"

"Depends on the size of the engine."

"Well, on average?"

"Let me think. When I was a kid we used to crack nuts that way. Oh, I'd say the rail sags enough to crack a good-sized black walnut laid on a hunk of wood."

"How much is that?"

He held up his thumb and forefinger: "About yea big." "Yea big" appeared to be about an inch. His solution was feasible. It did need a check, however; we needed to improve on "yea big."

"Hatfield, is your family connected with the feuding Hatfields and McCoys of Kentucky and West Virginia that I've read so much about?"

"Sir, that was a long, long time ago."

"If they are all like you, I'd hate to be a McCoy. Sure glad you're on our side and in the Navy. Are you a volunteer for our saboteur squad to blow this train?"

"I'd be mighty privileged, captain."

Jim appeared. "Captain, we're still submerged and twilight is ending."

"Sorry, Jim, I've been submerged in our train project. Take her up. Swish, now that Hatfield has shown us a way, I want everyone to think the whole saboteur attack over from start to finish. We'll use both rubber boats, and I'll name the other members of the squad tomorrow."

*Time 2121.* Jim had the *Barb* at radar depth to take a sweep around when I arrived. "Believe we have a contact, captain, at 17,000 yards. Offhand, it looks like three frigates: two together and one 3000 yards astern; radar interference from each. I'll bet we're their quarry."

"Jim, we can't get ahead of them unless they leave that coast. Nor can we surface in this bright moonlight without being gunned."

Through the night periscope I watched them pass at 2000 yards. We couldn't get closer for a cutie shot. When they left, we surfaced and set course for Patience Bay and a backlog of unfinished business.

The train had become our top priority. Since early June it had kept me pondering. Now we had a way to accomplish exactly what I had wished, but the amount of rail sag bothered me. This was crucial. An overestimate could cause the risky operation to fail, and an underestimate was of no consequence. As a postgraduate of design engineering, it behooved me to determine the sag reasonably and mathematically. I broke out my thick bible of engineering—*Mark's Handbook*—my slide rule, and a sheaf of paper and started to work. Three equations later, I had fallen sound asleep over my desk.

*Time 2337.* "Captain, contact. Showing Soviet lights."
"Keep clear, Bill."

In the morning the *Barb* cut across Patience Bay to the western shore. There would be no harassment until after the train job unless it was too important to resist. To minimize risk we were careful to remain undetected while seeking the best landing spot for the sabotage. Studying the charts, we selected several locations to reconnoiter submerged.

Afterwards I isolated myself in my cabin with drawn curtain and service records to list the men in the landing party, for they might have to land this very night. With everyone eager to go, there had been hints that favoritism had influenced previous selections to the boarding party and the assault force for Kaihyo To. This time no one would know how the party was selected. I needed the strong, the smart, and the dependable, as well as widespread representation from the various ratings.

Lieutenant Bill Walker, probably our biggest and strongest officer, would be the leader. Chief of the Boat Saunders, a gunner, would be second in command. Auxiliary man John Markuson was strong and could repair anything. Newland, lead cook and a smart cookie, would find some way to feed them if they got stranded ashore. Jim Richard, a crack motor machinist and born pirate, represented the engine rooms. Signalman Sever would take care of blinker-gun communication and navigation. Torpedo-man Klinglesmith was big, strong, and capable. Electrician Hatfield, with his knowledge of railroads, was a must. The list was perfect. All types were represented.

My secret criteria were: first, no married men, except Hatfield; second, members of all departments; third, half regular Navy, half Naval Reserve; fourth, at least half had to have been Boy Scouts, where they learned how to handle themselves in medical emergencies and in the woods. Should the group have to flee through the mountains of northern Karafuto to the Siberian part of Sakhalin Island, such qualities would be indispensable.

I called Jim in. "I'll lead the saboteurs. This isn't a combat situation, yet it's too risky to turn over to someone else. I'm sure I could pull it off."

"Sir, I swear I'll send a message to ComSubPac if you attempt this. It's not in the best interests of the men."

"I suppose you're right, but I hate to miss this once-in-a-lifetime opportunity. Take a look at this list." Jim concurred with the names.

Next, Bill and Swish had a look. They agreed. The officers and chiefs gathered, perused, and agreed. I talked to each man individually; they beamed when informed.

Training began right away. Clothing, the inflatable Mae West life jackets, and weapons were all laid out and tested.

Hatfield said a pick and shovel were essential to dig the two pits in the railroad crushed-stone bed under the ties and to bury the charge and batteries out of sight. We had none. The engineers responded by ripping up steel deck plates in the lower flats of an engine room. A cutting torch soon made the appropriate shapes, which were bent and welded to pipe handles. Voilà! In two hours the magic pick and shovel appeared.

A waterproof firing system was needed for inclement weather, so a number-10 can was adapted to seal the three dry-cell batteries that supplied the current for the charge detonator. Two wooden wedges were made; the microswitch was screwed to one. Red flashlights and spare batteries were readied.

Before I again turned to the equations to determine the sag of the rail, I took a quick trip through the boat to feel her pulse.

Absolute unanimity. No one hinted at any favoritism shown to the first team. If we had had 20 rubber boats, the whole crew would have gone, leaving me alone on the *Barb*. I talked to engineers Thomas Ryan, Mason Edwards, Arthur Wood, Leonard Hill, George Simek, James McCarthy and to electricians Forrest Sherman, Raymond Mulry, Norman Larsen, and Frederick McDole.

Everyone was happy, helpful, and suggested angles that might not have been considered. One of these brought up the danger of dogs. Leaving them hurriedly, I called for our stitched-up prisoner, renamed "Kamikaze," to be brought to the wardroom. Then I broke out the English-Japanese dictionary Marjorie had obtained for me while I was home on leave after hearing about my plight with the one-page vocabulary.

Kamikaze was smart, had a high school education, and was picking up English fast through working with the torpedomen. He was smart enough to claim he was Korean, knowing full well none of us could tell the difference—and he knew that I knew his gamble. I briefed him on our overall plan. He was enthusiastic.

"Kamikaze, Japanese patrol coast men on foot, men in automobiles?"

"No, big part."

"What big part?"

"Cliffs, rocks."

"Patrol beaches, sand?"

"Hai. Yes. Beaches, sand, towns."

"Use automobile?"

"No, walk."

"Regular patrol?"

"Yes."

"When?"

"Every day."

"Night?" He nodded his head. "How often pass same point?"

"Two hours," fluttering his hand—more or less.

"How many men same patrol?"

"Sometime two, most one."

"Armed?"

"Rifle, pistol."

"Dog?"

"Always take dog."

"Why patrol?"

"Soviet spies."

"*Arrigato,* Kamikaze. Thank you information."

He stood still. Then, swept away by the excitement, he gushed, "Me, Kamikaze, want go party, Master, explode train. Can help dogs, patrol. Meet, stop other Japanese."

"*Arrigato.* Better Kamikaze stay with me."

"Jim, spread the word about the anticipated armed patrol with dogs passing every two hours. They must be ready for every eventuality, including dogs from neighboring houses. Perhaps tossing any dogs a few steaks would be better than fighting them. I must get back to work on that track sag."

A few hours of intense calculation, assuming the smaller Japanese engines weighed about one-third of the American ones, provided ballpark results. The rail would definitely sag about seven-tenths of an inch. I called for Bill, Swish, Hatfield, and Markuson to talk over my results.

"We don't need that much clearance. It is important that there is enough to ensure that the hair-trigger microswitch does not set the charge off when Hatfield makes the final hookup. So, we will halve that—roughly. Markuson, I want you to manufacture a metal feeler gauge three-eighths of an inch thick. Hatfield will use it to adjust the wedges holding the microswitch.

Once he has set that clearance accurately, he and he alone makes the final live connection. Bill, when he does that, all others must be at least 20 yards away, flat on the ground, with exposed body parts turned away and eyes closed. If Hatfield makes a mistake, I don't want any other person killed or blinded by the flash."

*Time 1837.* "Captain, we're ready to dive to close the coast for observation."

"Take her down, Dave. The rest of you skedaddle and get some shuteye. It may be tonight if we have the right place and the right weather."

Reconnoitering a fairly good position, we sighted and logged one train. Then we started compiling a timetable of location, number and types of cars, and direction of journey, north or south.

*Time 2201.* We surfaced in bright moonlight and a cloudless sky, which prohibited any operation that night, so we headed across the bay to look for a lone sampan that we might capture to assist in the landing. No joy.

*20 July.* After breakfast all the officers studied the charts we had recovered. Their hydrographic accuracy and contour lines on land showed elevations and configurations that our charts lacked. By noon we had the optimum landing spot.

*Time 1517.* The *Barb* dived to reconnoiter our choice. Coasting with two fathoms beneath the keel, we sighted trains on schedule. The beach was sandy; no houses were within 700 yards. We were a mile offshore at 60 feet depth in 12 fathoms. The Japanese chart matched perfectly. Studying the local trains, we recorded that the number of cars varied from 7 to 32. The average train consisted of 12 freight cars, 3 passenger cars, and 1 mail or baggage car.

"Bill, more good news. I have the scope spot on your landing site. Beyond the beach a grassy meadow stretches for about 200 yards up to the military highway. No animals. The land from the highway to the railroad about 100 yards farther in appears to be small scrub. Take a look."

"Beautiful!"

"Can you see those twin peaks on the background mountains?" He nodded. "Those two peaks will be your guide while paddling in. I'll bring *Barb* in to less than 1000 yards from the beach by radar before you disembark. Use your magnetic compass, course 274. I doubt if tonight's the night. We need an overcast to veil that moon."

*Time 2132.* Surfaced in a calm sea; The bright moon fouled up any landing ideas. A night fit for lovers.

*21 July.* Patience Bay is well named. It tried mine. While awaiting cloud cover, we did something useful, however. We took soundings of uncharted areas for the hydrographers. During the day I held a conference of all the officers and the saboteur squad.

"*Barb* will approach the beach flooded down, ready to launch the rubber boats. She will come in on the batteries as silently as possible. The approach will be at slack water—timed to arrive at the debarkation point by radar at 2310. The whole coast observes blackout afloat and ashore. By that hour the inhabitants should be asleep. We'll be less than 1000 yards offshore, so speak in whispers. That includes Tague heaving the lead for soundings. Questions?"

"We'll lose sight of the black boats paddling in," commented Dave. "Why don't we keep track of them and where they land by making up a tin radar corner for each boat?"

"Great idea. Do it."

I went on. "Upon beaching, Sever and Newland will guard the boats. The other six proceed up the meadow and cross the highway to the track. At a suitable position for planting the charge, the party will divide. Markuson will go 50 yards south on the track near the road as a guard. Klinglesmith goes 50 yards north near the road, and Richard 20 yards inland if he's not needed to dig. Walker, Swish, and Hatfield dig under the tracks and plant the battery can and charge. Then adjust the microswitch clearance and recall the guards. All get well clear, flat on the ground, head turned away with eyes closed while Hatfield makes the final hookup of the firing circuit to the charge. Understood?"

Max piped up. "Shouldn't we build in a test circuit in case a wire has come loose in the waterproofed battery can?" Hatfield concurred.

"Good idea; do so. Communications are simple. Those of you who have been Boy Scouts will remember the two bird calls to be used. First, when you approach a group always whistle, 'bob WHITE!' This is the alert signal and will save you from being strangled by your shipmates." I had each man practice the call, and everyone passed save one. "Sever, you sound like a sick quail. If you can't whistle this, I'll have to substitute a quartermaster for you." With that threat he became a bluejay and could mimic any bird.

"Your other whistle signal is for assembling. Use the whippoorwill's, 'ti-la-ti.' Try it." All passed.

Willy shook his head. "Captain, these are American birds which may not exist in Japan."

"True, but I'm banking on my belief that the Japanese won't know any more about their bird calls than the Americans do. Now let's move on to other signals. A blast on the mechanical whistle signals for an emergency dash to the boats. Two Very stars indicate, 'We are in trouble, lay a barrage in the direction indicated.' One Very star from *Barb* signals, 'We are in trouble, will return every night.' 'W' on the blinker gun means the landing party is returning to *Barb*. Last, one Very star at 15-minute intervals informs us that you are unable to locate *Barb* after 30 minutes' paddling from the beach. You must stay together. Plan on the party debarking at 2320 and landing on the beach at 2335. Return no later than 0230, for twilight commences at 0245. No further questions? Okay."

Another lovers' night. Only one sampan was sighted moving along the beach while we hooted at the moon.

*22 July.* Continuing the sounding of uncharted areas, the *Barb* bided her time. The scant five days left on our patrol weighed like a 100-pound sack of potatoes on my shoulders. My fingerprints were all over the barometer from shaking it to eliminate this high-pressure area. After lunch, as I caressed it, there was a slight drop. Silently praying, I hopped up to the bridge.

"Chuck, how's the weather?"

"No change, sir. Calm, one sweater." I scanned the horizon in the hope that there just might be a wisp of stratiform in the white smudge on top of the mountains to the southwest. For half an hour I watched it develop as southerly breezes sprang up, flinging out plumes of cirrus clouds. Following these came a white stratus capping the peaks. I could feel my heart pounding.

"Jim, what's the moon for tonight?"

"Three-quarters on the wane; rises at 2100 and sets in the morning. Why?"

"Come take a look."

"What d'you know, clouds coming. Tonight's the night?"

"You can bet on it." I pressed the intercom. "Men, we have a stratus moving in that will cover the sky, a moon tonight that will glow above, and a soft breeze. At long last we have the ideal weather for which we've been waiting *four days*. The saboteur squad will land tonight!"

Cheers resounded below. "Permission to come up, sir?" It was Swish. "Captain, what you just said hit the boys below like a short circuit. It brought everybody to their feet. Can we splice the main brace when we return?"

"This event is as important as a ship sinking, so why not? Actually, you could take beer ashore as a picnic ration, but you would be leaving a clue behind. Certainly no official could object to each man having one beer after we sink a ship, the first submarine ballistic missile rocket assault in history, or the smashing of Little Iwo Jima. If I had followed the doctrine of providing each man two ounces of booze every time we were depth charged or bombed, we'd be under the weather at times. So far, in 40 days each of us has had only five beers; I'm sure a sixth would do us good. We need some yeast. Also, I've become superstitious. Every time I've forgotten to put the beer in the cooler before firing we've had trouble. Swish, put four cases in now."

Swish hurried away to do my bidding and returned in less than a minute. "Captain, there are already four cases in the cooler. Everyone's been so busy with the train attack that the cooks forgot to bake the cake and we forgot to splice the main brace after we sank the frigate."

"Blimey, Swish, forget the cake and pass the word to splice the main brace now. We shall not dishonor one of the noble traditions we inherited from the British."

As we sipped our cherished single libation after days of patiently waiting and observing, the undercurrent of expected action sweeping through the boat made each man's spine tingle. It struck Kamikaze. Again he pleaded with me to let him join the party, promising not to escape. Again I told him to stay with me.

*Time 2200.* Heading in to the drop point, we had a final briefing, and I took Walker aside for a private chat. "Bill, avoid trouble. If the spot isn't good or the odds poor, bring the squad back. Watch your men for signs of freezing at their task, not ordinary fear or apprehension. If you find one, escort him back to relieve Newland as boat guard. Keep your men together and quiet going and returning—no unnecessary conversation. All going well, don't get cocky and try something else. I want neither prisoners nor injuries to Japanese except those positively required. Slip in and slip out without being detected. Make sure your group and the boat guards have both meat and liver for any dogs or strays. You have one mission only: booby-trap the train and bring the men back safely. The latter is more important."

Flooded down, the deck almost awash, the *Barb* had a silhouette somewhat similar to that of a schooner or patrol boat. If we were seen from the shore, the rubber boats would not be. No one would suspect a submarine so close to shore or in such shallow water. It was very exciting. In the dim moonlight the rubber boats were being inflated, equipment was gathered together, and last-minute quiet joshing was well in progress.

*Time 2230.* "Radar contact. Two small vessels coming down the coast."

"All stop." We lay to, tracking them at five knots.

*Time 2255.* "Sir, they're zigging toward us."

"Ahead two-thirds. Left full rudder. All hands go quietly to your gun stations, all guns. Do not sound the gongs."

We were losing valuable time till they passed clear.

*23 July. Time 0000.* In position: two fathoms under the keel; twin peaks capped with clouds; shoreline 950 yards. "Launch the boats." I had planned to say something apropos of such an operation, like "Synchronize your watches." Instead I whispered, "Boys, if you get stuck, head for Siberia 130 miles to the north. Follow the mountain ranges. Good luck and God bless. Shove off."

We watched the boats all the way in. Radar easily tracked them by their radar corners. They made a couple of circles and zigs, taking almost half an hour. Momentarily we watched for shots, flares, and a general clamor, but the blackness of the night engulfed everything in a challenging silence.

*Time 0025.* The boats had been dragged up to the beach. Certainly the *Barb* could be seen but not identified. Not a light was visible anywhere. I felt positive that once the initial landing was made unopposed, the rest would go off smoothly.

*Time 0047.* "Captain, a train is coming up the track!"

"My God, Max, there are no lights except a rare glimmer from the firebox. White smoke is swirling back. The boys ashore must be in the middle of their digging job. I pray that train doesn't get derailed due to the holes under the track." Crossing my fingers, I held my breath, my imagination running rampant. Not having noticed any trains at night, it had not occurred to us that they could be so perfectly blacked out. Compared to us, the Japanese were past masters at blackout.

Trouble for the saboteur squad began as the rubber boats shoved off from the *Barb* with Bill in the lead boat, Swish following in the second. The twin peaks they expected to use as a reference point vanished in the clouds. The compasses were erratic because of all the metal objects around. They tried to keep on a line from the *Barb*, which they could see, perpendicular to the shoreline. Consequently, they took over twice the time for the transit and landed in the backyard of the first house to the north of the meadow.

Fortunately, though the party was only about 50 yards from the house, no dogs put in an appearance. Dog tracks accompanied by barefoot-human prints were noticed on the beach. After a short period of huddled reconnaissance, the main party left the boat guards. They proceeded cautiously inland, skirting the house, and arrived at the expected meadow.

Only—what appeared to be grass when seen through the periscope turned out to be waist-high bullrushes that crunched and crackled with their every move. All shapes took on human forms and scared a few of the party. Arriving at the highway, they held another huddled reconnaissance. All was clear, so Bill dashed across the highway, calling "Follow me." He fell headfirst into a four-foot drainage ditch full of branches. Pulling himself out, he whispered cautions to the men, then ran across the highway and tumbled into yet another ditch. One hundred yards farther up they arrived at the railroad track, reconnoitered, and selected their spot for the charge.

The three guards were posted, with Markuson being told to examine a water-tower-like structure down the track. Digging commenced. The pick and shovel striking against the large crushed stone of the rail bed clinked and clanked, broadcasting their presence in the eerie hush. They stopped. After a whispered conference they started digging again with their bare hands, pressing with the implements when needed.

Hatfield's ears pricked up. "What's that noise? Someone is running up the track." Hearing no "bob WHITE" alert signal, they grabbed their weapons to dispatch the intruder.

Swish cautioned, "Take it easy. At this time of night there's no one running up this track except a scared American."

Markuson appeared—breathless. "Why in the devil didn't you give the alert signal? We could have killed you."

"You know—that funny structure—I was to check? I climbed the ladder—and poked my head over the ledge. There's a guy in there asleep. I tiptoed down and ran to tell you it's a lookout tower! My mouth got so dry when I tried to whistle, all that came out was whooh—WHOOH." Markuson was sent out again to stay well clear. Work recommenced in earnest, progressing nicely except for an occasional whimper—"ow"—as a fingernail was torn off by the sharp stones.

Bill stopped. "I saw a light flash down the track." At Hatfield's direction everyone pressed an ear to the rail in good frontier fashion. They heard nothing. "Must have been someone getting out of bed to have a pee call. Back to work."

With the breeze from north to south, the surprise of hearing and seeing an express train roaring down on top of them less than 80 yards away brought panic. Bill jumped clear and into a briar clump, adding more scratches. Swish slid into hiding behind a scrub tree 15 inches high. Hatfield leaped into a shallow foxhole and felt two shots hit him as the train roared by with the engineer leaning far out of his cab, looking at them.

Thinking this must be it, Hatfield felt his body slowly rising out of the foxhole. The carbon dioxide cartridges on his Mae West life jacket had gone off and inflated the vest.

*Time 0052.* The passing of the train left some men crossing themselves as others, too, stepped up their faith. Though a bit jumpy, they increased the work pace with more muffled "ows."

At the end of the first hour of the operation, Sever and Newland (guarding the boats) were becoming edgy. If the beach patrolman came by every two hours as Kamikaze had said, the odds of his appearance were far greater than 50-50 and increasing. Both had brought welder's gloves, the heaviest on board. Neither having volunteered to approach the dog first, if the handler let him loose, a compromise was necessary. Each had his pistol in his right hand and the steak or liver in the left, gloved hand. They would challenge the patrol together.

On the *Barb* there was a sigh of relief as the train went by. Now it was like awaiting the birth of one's offspring; everyone was pacing the deck.

Twenty minutes after the train passed, the holes were completed. The scuttling charge and battery were carefully buried and disguised. Hatfield checked the test circuit by depressing the microswitch. "Mr. Walker, I'm all set to wedge up the microswitch and make the final hookup to the charge." The night became melodious with the soft whistles of the whip-poorwills and the guards creeping in. Each of the group had been trained for the other's job in case something happened. Now came the moment, with all assembled, to lie down clear and leave Hatfield to make the final connection alone.

Now came the moment when everyone except Hatfield mutinied. All five, including Bill, decided that they wanted to make sure Hatfield connected the circuit correctly. As they watched over his shoulder, he performed the task perfectly. The charge was now alive—but a discussion started.

"Let's see that distance gauge with which you set the microswitch under the rail." Hatfield drew it from his pocket and held it up for all to see.

"It sure looks like an awful lot for a rail to sag. Suppose the Old Man's calculation was wrong, and trains go over it and it doesn't explode? He'll never let us come ashore again to move it closer." All nodded in agreement.

Klinglesmith straddled the rail and patted the ends of the two wedges, moving the live switch closer to the rail. Markuson kneeled down with his red flashlight and called out the distance remaining. "About a quarter of an inch."

At that Hatfield interrupted. "For God's sake, stop it before you blow us all to kingdom come!"

Bill and Swish agreed: "Head for the beach together." En route through the bullrushes two men got the bright idea that the squad should grab a couple of prisoners or toss a couple of hand grenades into the two Toyota trucks back of the house. Bill was ready for that one, simply saying, "Nix. The skipper warned me about you two."

Swish whistled the "bob WHITE" alert call about 20 yards from the boats. Newland and Sever replied, relieved.

*Time 0132.* On the *Barb* we muffled our cheers as we saw the blinker signal—dot dash dash—that indicated that the boats were leaving the beach. Shifting my horseshoe to another pocket, all I could think was, what good luck. I had already eased the *Barb* in to 600 yards with the sound heads and speed swordarm raised, in case trouble arose. Less than a fathom beneath the keel was sufficient.

*Time 0145.* On the .50-caliber machine gun Epps yelped, "CAPTAIN! ANOTHER TRAIN COMING UP THE TRACKS!" The boats were only halfway back. Grabbing a megaphone, I broke the silence. "Paddle like the devil!" Wasted. The boats had already spotted the train; their paddles churned like eggbeaters.

The train, streaming white smoke, was getting closer and closer. Any second now. Even the boats stopped to look.

What a moment! Our world stopped. Everyone was awestruck with the expectancy of imminent destruction.

*Time 0147.* BOOM! WHAM! What a thrill! The flash of the charge exploding changed into a spreading ball of sparkling flame. The boilers of the engine blew. Engine wreckage flying, flying, flying up some 200 feet, racing ahead of a mushroom of smoke, now white, now black. Cars piling up, into and over the wall of wreckage in front, rolling off the track in a writhing, twisting maelstrom of Gordian knots. Fires sprinkled among them.

Then a gap of seconds before the sounds of the explosions were hurled across the water, sounds of the grinding, snapping, crushing, tortured steel and wood. Stunned speechless, I began to breathe. Then I grabbed the megaphone and hollered, "Paddle! Paddle! We're leaving! Starboard ahead one-third, port back one-third, left full rudder." The *Barb* twisted around to head out.

Members of the *Barb*'s saboteur squad with the battle flag, the only Americans to land
on Japanese soil in World War II. From left to right: Paul Saunders, William Hatfield,
Francis Sever, Lawrence Newland, Edward Klinglesmith, James Richard, John Markuson,
and William Walker.

*Time 0151.* Boats alongside, 100 proud hands hoisted the victorious
saboteurs on board and hauled the boats up on deck. The *Barb* slowly
slunk away at two knots from her victory over unforeseen odds. The gods
were good to us.

I could have secured from battle stations, since we would be in water
too shallow to dive for a while. Instead, I said on the intercom, "All hands
below deck not absolutely needed to maneuver the ship have permission
to come topside through forward and aft deck hatches."

The hatches sprang open and men poured forth to gawk. Torpedomen,
radiomen, stewards, engineers, cooks—the *Barb* disgorged her flock. Maneuvering on batteries, only the man on the controllers, the helmsman,
and the radar watch in the conning tower remained below.

A full-circle binocular sweep showed the *Barb* to be alone. Ashore, the
chaos continued, with lights being turned on in some houses. The lights
of an auto coming down the road joined the fires flickering alongside the
tracks. Shortly thereafter, military vehicles came racing to the scene with
sirens screaming. From macro to micro our stage disappeared as we pulled
away, but never to be forgotten without a flush of pride.

"Quietly, secure from battle stations, clear the decks, and close the hatches when everything is stowed. Set the regular sea detail."

*Time 0230.* "Splice the main brace!"

Jim looked at me. "I thought we would have it with the cake at 1100 this morning."

"Jim, do you think these tigers are going to bed now? They've got sea stories to tell until their great great grandchildren drift off to sleep. Listen to them. The questions will never stop."

*Time 0400.* By now a parody had developed, a takeoff on the song "Down in the Valley." This was being sung by the full *Barb* chorus of saboteurs over the intercom. Naturally, some inspiration came from Pharmacist Mate Laymon's snakebite cure for stress.

> Down off Kashi-Ho,
> Kashiho so shallow o-o o-o
> hear that train blow love
> hear the Barb blow o-o o-o o
>
> Late in the evening
> hear that train blow (so dear to me)
> the train did stop love
> it did not go through hoo-ahoo ahoo-oo
> and now it's gone love
> and so are you hoo-ahoo ahoo o
>
> The train did blow love
> it blew and blew hu-uhu uhu-u
> thank heavens it's gone love
> and so are we hee-ahee ahee-ee
>
> That train is gone love
> and out of sight a-ight a-ight
> good night my darling
> darling good night a-ight a-i-ight
> good night!"

With one last shout—"HEAR THE BARB BLOW!"—they split the main brace and turned in, exhausted, to dream of their loves.

# Chapter 28  Countdown to Graduation

*23 July. Time 1115.* Brunch over, I gathered the officers together for a conference. "Now that the train job is history, we only have the rest of today, the 24th, and the 25th; then *Barb* must be through Kunashiri Strait by midnight the 26th. Our priority now is Shiritori, where we'll reconnoiter for a triple rocket massage. Following that we lay a single massage on Kashiho to finish off the last of our rockets. That accomplished, there are the large fish canneries at Chiri on the northeast leg of Patience Bay to bombard. Then we must head for Kunashiri to depart at the time granted by our extension. Chiri will finish off the rest of our 5-inch ammunition. Questions?"

Max expressed some minor doubts. "Quite a full schedule. We'll have to clobber one a day to meet it, in addition to the normal *Barb* philosophy to expect the unexpected. We really ought to hold back a bit on some of the smaller ammunition to have a modicum of self-defense with no torpedoes."

"Max is right. Meeting's over."

The *Barb* submerged off Shiritori in the late afternoon. Our reconnaissance was postponed due to haze; no rockets this night. At least that's what I thought.

Late that afternoon at Admiral Nimitz's headquarters in Guam, a courier plane had just arrived from the States. His orderly met the plane, signed for his mail, and begged the latest copy of a San Francisco newspaper from the pilot. This went immediately to the admiral. Nimitz took a hard look at the front page, then sent for Admiral Lockwood.

386

"Charlie, I've had enough trouble with the reporters hinting that I am lying about the attacks and movements of the Third Fleet. Look at this."

Front page center lay a map of Japan from Kamchatka to Okinawa. An inch-wide arrow swung out of the Pacific. It split at Kunashiri Strait, with nearly half its width swinging off into the Sea of Okhotsk toward Karafuto. The other part passed down eastern Japan. He pointed to the lead column: "The deepest penetration of Japanese waters yet by a section of Halsey's Fleet." Nimitz smirked. Lockwood frowned. The article went on to say Tokyo Radio had previously reported that the north coast of Hokkaido was preparing for an invasion. Meanwhile, the attacks continued. On 18 July the frigate *112,* under the command of Captain Ishiwata, was sunk by torpedo about noon while escorting the ferryboat *Soya Maru.* The fortress of Nishinotoro fired cannons and three navy warships went to mop up the sea, but the effects were unknown. Fleet Admiral Nimitz had no comment.

"It must be *Barb,* Chester."

"Charlie, find out what she's doing."

> FROM COMSUBPAC TO BARB X EYES ONLY FLUCKEY X WHAT ARE YOU DOING X OPEN UP AND GIVE US THE DETAILS X

"Jim, take a gander at this rocket that I must answer."

"What's up, captain?"

"The problem is not what we've done since our last message. What scares me is the logistic report required at the end of all our messages. When Uncle Charlie sees 'Torpedoes zero,' it'll set off his alarm bell. He'll think that *Barb* is defenseless."

*24 July. Time 1219.* We submerged and worked in close to Shiritori, happy with the cloud cover far above. A close look, with only a smidgen of water beneath the keel, proved not only that Dave had become an outstanding diving officer but also that Shiritori was an ideal target, with huge factories. One chimney disappeared in the clouds. Using the ST radar in the periscope to obtain its exact range, and the linear divisions of the exit eyepiece, I calculated the height to be better than 400 feet. Radar Technicians Lehman and Maher checked radar's pip to use it in conjunction with the data computer for the attack. Now we could get an accurate fix that would ensure our smacking the largest factories.

*Time 1934.* Well out of sight and overjoyed with our key find of that chimney, the *Barb* surfaced. Everyone was eager to get on with the night's work. We were no sooner up and starting supper than Gierhart handed me two messages. I read them aloud.

FROM COMSUBPAC ACTION BARB INFO BARNSTORMERS X COMSUBPACS HEARTY CONGRATULATIONS FLUCKEY UPON SINKING FRIGATE AND DE-STROYING TRACK AND TRAIN X DEPART AREA AND PROCEED MIDWAY X

"Damn it, I knew this would happen. He's already given *Barb*'s area away to the Barnstormers wolfpack on the polar circuit."

"Captain, he's ordered us out. Do we secure now and set course for Kunashiri Strait, or will you go ahead with the attack planned for tonight?"

"You bet your boots we will, Jim."

"Are you going to disobey the admiral's orders?"

"No. It's not disobedience. He didn't specify the route we were to take to depart. So we depart down this western coast. To expedite, we'll rocket Shiritori just after dark, then proceed at flank speed down the coast and rocket Kashiho before dawn. From there we skid across the bay to bombard Chiri. After that I'd like one more crack at luring a single minefield frigate into deep water. Then we'll head for Kunashiri to leave the Sea of Okhotsk. Jim, did I ever tell you that the Russian word 'okhotsk' means 'hunter'? Though they probably were referring to seals, that's still our job, and we're going to hunt as long as we can. Oh, I guess one can call some of this 'sea-lawyering,' but previously Lockwood did give us an extension to 2400 on the 26th of July."

"How about the other message?"

"It's just the nightly news. Here, read it."

DAILY SCUTTLEBUTT X BARB REPORTS SINKING ONE FRIGATE AND BLOW-ING UP ONE CHOO CHOO TRAIN WITH LANDING PARTY AND OWN DEMO-LITION TEAM PUT ASHORE X SEAROBIN REPORTS TWO AMMUNITION LOADED SEA TRUCKS AND TWO LUGGERS JOINED THEIR SISTER SHIPS ON THE BOTTOM OF THE SEA X RUNNER AND GUNNEL ARRIVED GUAM X

*Time 2117.* "MAN BATTLE STATIONS ROCKETS! Tom, while we are easing in to Shiritori, I'll go below to check the chart. The Japanese charts we recovered are invaluable. The one for this area has an enlarged insert, including the factory district plan."

Below, Jim and Higgins were all set. The navigational plot aligned exactly with the chimney, and Jack's attack plot was also exact. With visibility decreasing, the *Barb* snuggled silently into position at slow speed on her batteries.

On my way to the bridge, Max gave me the thumbs-up sign that the rockets were ready. "All stop. Tom?"

"Sir, I can barely make out that chimney."

"Forget it, we've a beautiful picture on radar. Jim, let me know when we're 5100 yards from the chimney. Max, put us on the launching course."

"Captain, don't forget that we're launching at maximum range, so we have to steer 3° left of target to allow for the right drift of the rockets."

"Understood. We only have four fathoms of water now. Any closer than that 5100-yard range on the chimney and we touch bottom. Yell when you're ready!"

"Set."

*Time 2236.* "ROCKETS AWAY!" The first batch of 12 went swishing out on their way to the biggest factory. "RELOAD!" Thirty seconds later they landed, sounding like bombs. Three minutes later the second batch zoomed off, landing with heavy explosions. We reloaded once more and twisted the ship slightly to get the town hall and large factory. The third batch went in a flash to turn the town upside down. Secured from battle stations, we left the launcher set up and set course for Kashiho where, offshore, the *Barb* had had her ears pinned back by that terrible *Terutsuki*-class destroyer.

*Time 2310.* Lookout Parker cried out, "Captain! Two large fires just broke out back at Shiritori!" Muffled explosions were ripping the factories apart.

"Broocks, come right and reverse course. We can't miss this. Fire is our most destructive agent ashore."

More fires started. Many more explosions shot flames up into the heavy overcast. "Willy, pass the word to open the forward and afterdeck hatches. Permission is granted for all to come up on deck to enjoy and witness the effect of the rockets. Shift propulsion to the batteries."

Now back only 4000 yards from the shore, Willy circled the *Barb* at slow speed while fires spread amid continuing explosions. We gawked, feeling a bit heroic and Neronian at the same time.

*Time 2350.* "Captain, we must get moving for Kashiho to attack before dawn."

"Thanks for waking me up, Jim. Fire is the answer. Look at it spread."

"I know. I've measured its breadth parallel to the coast: nearly three miles of buildings aflame."

"Wow! Men, clear the decks quietly and close the hatches. As soon as you're clear, flank speed coming up."

*25 July. Time 0224.* "MAN BATTLE STATIONS ROCKETS!" Visibility being reduced to about 6000 yards and water shoaling fast caused a bit of a delay as dawn approached. Finally, the *Barb* stood still long enough to hiccup her rockets at 0310. Through my polaroid goggles I watched the flash as each rocket slid down the pipe rack onto the needle-sharp firing pin, igniting the powder to blast off. The flare died about 10 feet above me in a diminutive shower of sparks. Each took an explosive charge ashore more than double that of a 5-inch shell.

The *Barb* and her crew stood still, breathlessly waiting for the 30-second voyage in space to terminate in the rolling thunder of falling bombs. When it came, buildings blew, but there were no fires. "Jim, set course for Shiritori to review our damage." Securing from battle stations, we felt a bit sad. That ugly, pipe-rack rocket launcher had touched our lives like a stray, hungry mutt that one adopts. Max and I had often watched Swish with his emery cloth rubbing and honing the firing needlepin to remove rust and to sharpen it. As Max said, "The way Swish caresses it, you'd think it was alive." I regretted knowing I had given my last order forever of "ROCKETS AWAY!"

Nara Katsumasa, a Provost Corporal who worked for the Shikuka Division of the Police Force, filed this report:

Military strategy in Karafuto is changing from anti–U.S.S.R. to anti–U.S.A. Several incidents taking place have indicated the coming of a U.S.A. attack, though only a few were informed of this. The bombardment of Kaihyo To Island for five hours without a stop, Shikuka attacked by five enemy ships, and the blasting of the freight train are cause for alarm.

I was working at the Shikuka Air Base. When the ice in the Sea of Okhotsk began melting, enemy submarines frequently approached along the coastline. The torpedo planes continued searching along the coastal line, but could not find them. One day I joined the searching flight, but did not find any.

About that time, the nearby Shiritori Police Center sent information that there was a possible spy on land. So I took one of my men with me to help them. Judging from the fact that the enemy submarines never missed their target from the first firing, possibly a spy communicated to the submarines.

Last night, a civilian reported that he witnessed an exchange of light signals between the mountains and the sea. The two signalers said there was going to be an attack soon.

Tonight we returned to our inn. After we finished our dinner, we heard a sound like thunder over our heads. We immediately went out. They bombed the gasoline storage of the Oji Paper Factory [the largest in Japan] in Shiritori. Then the gasoline drums started exploding. The firemen could not stop the fire from spreading.

I thought, "I will catch the spy." I watched the people on the spot carefully.

Then thinking that the spy would escape to the sea after the second bombing, I ran toward the wharf along the railroad. From the tracks we moved toward the beach. Crouched and looking around, we saw a small glowing ember in the darkness thirty meters ahead. It looked as if someone was smoking a cigarette. We thought, "That must be him" and pulled out our pistols.

We went close—within 10 meters from the ember which suddenly became brighter and larger, and the next moment very weak as if he were exhaling a cigarette. We heard nothing but the sound of the waves. Holding the pistols on our eye level, and aiming at the bright ember, I yelled, "Who is there?" There was no response. The flame did not move. Next, I tried a body crash, but instead I landed in the sand. Just an ember fanned by the sea breeze. We were totally dumbfounded.

*Time 0445.* The Barb sat 6000 yards off Shiritori again. For the last 45 minutes we had been in a heavy smog. Five miles south of the town we began hearing continual light and heavy explosions similar to those we heard seven hours previously. The smoke was now thicker, heavy with the odor of burning wood and paper. Unfortunately, visibility was about 50 yards, so we couldn't assess the damage. A sudden loud blast and rumbling explosion shook us, driving the crew out of their bunks. I explained the circumstances over the intercom to allay any misgivings. Some thought we were being bombed or depth charged on the surface. Cheers resulted, though men were coughing due to the smoke.

"Jim, no use staying here. Set course across to Chiri."

*Time 0946.* "MAN BATTLE STATIONS GUNS!" Visibility increasing, Singer had sighted a large sampan.

*Time 0959.* "Forty-millimeter gun, commence firing." In jig time the sampan crew abandoned ship. "Away the boarding party!" Tom and his group scrambled aboard as the *Barb* drew alongside. One Japanese volunteered to come aboard as a prisoner. The others took to their rowboat, as we were only 4000 yards offshore.

The boarding party snooped into everything. Swish took their steering wheel. Klinglesmith and Petrasunas passed three anchors over to the *Barb*. Tom picked up some snow crabs for supper. Another brought signal flags and the company flag.

In half an hour I welcomed Tom back aboard: "Nine down, six to go." Backing clear, we sank her with one 5-inch shell. These single shots at the waterline from 100 yards saved a lot of ammunition.

"Swish, as soon as you have the 5-inch gun secured, have Doc Laymon bathe, disinfect, and check the prisoner for any diseases. When he's finished,

*Top:* Boarding an armed sampan. Taken from movie film. *Bottom:* Prisoner on board.

bring him to the wardroom for questioning. I want to ask him about Chiri. We'll be bombarding in about two hours."

When the prisoner arrived, our linguistic tussle commenced. Evidently, he wasn't married. He claimed that he was a slave to the cannery company to which he had originally been indentured for 10 years, but they wouldn't let him go home. This could be false, yet he was older and happy to be on board.

Breaking out a Japanese chart of Patience Bay, I asked him if he had heard of any train wrecks lately. He nodded and put his forefinger on the exact spot. "Two nights ago, enemy aircraft bomb train, kill 150 men. Right there, say local newssheet." Our aircraft got the credit again. But my primary interest lay in the type of defenses around Chiri. We started to develop this when . . .

*Time 1208.* "Captain, we've sighted the masts of a sampan toward Chiri and another farther down the coast. I'm heading for her."

"Good, Dave. I'll be up in a minute."

We finally elicited from our prisoner that there were no guns larger than 37-mm, but there were some 15 seaplanes. These pontoon planes, he said, were based at the lake close inland from Chiri. I had my doubts, but to play it safe, we would remain outside of 15 fathoms for the bombardment.

*Time 1226.* "MAN BATTLE STATIONS GUNS!" Tom took over the watch from Dave as we approached the sampan and Chiri. "Sir, I concede the sampan before it sinks. Ten down, five to go." Ten minutes later we opened fire with the 40-mm gun, which stopped her. The 5-inch gun then fired a single shot at the waterline and she went under.

"Well done, Max. Now eliminate the cannery."

"Aye, sir. Swish, shift the pointer and trainer right to the powerhouse. It's the smaller building with the chimney. When I check fire, move them left to the adjacent largest building. After each is demolished, move up the row. Resume firing!"

With that the 5-inch gun poured 43 rounds into the cannery. Sections of buildings blew up and large holes were created, but no fires. Incendiary ammunition is a must. We ceased fire when our 5-inch, high-capacity ammunition was down to five shells. At the bottom of the magazine we found three star shells. Someone's bright idea. They have no explosive, and the last thing a submarine needs is to illuminate herself at night.

*Time 1250.* "Gun stations secure and stand by below. There may be some planes nearby. We're going after a sampan returning up the coast."

*Time 1314.* The sampan sank as Tom croaked, "Eleven down, four to go."

"Tom, I'm going below for a bite to eat. It's your regular watch now. Find me four more sampans to secure our bet and I'll send you up a sandwich and some ice cream. Otherwise, you can starve."

"Aye, Captain Bligh, sir."

In the wardroom, Willy was chuckling over the dispatch I had sent to ComSubPac that morning at 0214 after our successful attack on Shiritori, providing the how, where, why, what, and when they had requested to outfit other subs with rockets. It also detailed the mischief we had done. "Captain, you recommended a big order—1000 rockets with a range of 8000 yards to displace most of the torpedoes. Golly! We could rip up the coasts with that quantity. Your last sentence is a beaut":

DON'T BELIEVE THE ROCKET WILL REPLACE THE TORPEDO ANY MORE THAN THE AUTO REPLACED THE HORSE BUT THEY ARE DEFINITELY HERE TO STAY AND FUN BESIDES X

"Will that get you in trouble?"

"Willy, remember this: The truth can't be broken, but it can be warped. Perhaps it will wake up someone who can make decisions. We shall see."

*Time 1334.* "Captain, we've sighted a lone sampan and two more returning to the cannery. One of the latter has the other in tow."

"Coming up, Tom. You've practically done yourself in. Your reward is on its way. Ragland, please bring a couple of tuna sandwiches and two soup bowls of ice cream to the bridge before you man your gun station."

*Time 1350.* "MAN BATTLE STATIONS GUNS! AHEAD FLANK!" Between munches on my sandwich, I apologized for giving orders with food in my mouth, but we had to hurry before the sampan and her tow beached. "Jim, Max, Tom, forget the lone sampan. She's outboard of us. Use the 40-mm gun to clear the deck of any opposition from the towing sampan. Then put in a waterline shot to sink her. With luck, if the towline doesn't part, she might drag her tow down with her. This'll save us ammunition. If not, use the same system to finish her off.

*Time 1355.* "COMMENCE FIRING." Within a minute of the waterline *coup de grâce,* the target sank and snapped the towline. I taunted Tom, "twelve down, three to go."

Max shifted fire to the towed sampan, which stubbornly refused to sink. The 5-inch star shell passed right through, leaving a clean, round hole six inches above the waterline. Our last three high-capacity shells finally put her under. "Thirteen down, two to go. Let's go to flank speed and catch the last sampan."

***Time 1413.*** The last sampan submerged with our last two star shells drilling waterline holes in her hull. We secured from battle stations with nothing in sight. Swish and Petrasunas, our 5-inch gun captain and pointer, patted the gun fondly as they retired it until Midway, all ammunition having been expended. Tom whimpered, "Fourteen down, one to go."

"Where to now, boss?"

"Jim, Kaihyo To is only 30 miles southeast of here. Let's take a look at the reconstruction."

Fleet Admiral Nimitz called Admiral Lockwood in again. "Charlie, the reporters are giving me a fit. Radio Tokyo reported the bombing of *Barb*'s train and the bombing of Shiritori, and a smaller town to the south. I thought *Barb* left the area. She's upstaging Halsey. Just between us, that's hard to do."

"Chester, I ordered him to depart his area two days ago. He did send a message yesterday giving the details we had previously requested concerning their rocket attacks. We believe he's on to a good thing. Following up on his advice, 15 subs are being fitted out with rockets."

"Splendid. Still, he may not have received your orders. With no torpedoes, he should leave. Give him a prod."

> FROM COMSUBPAC TO BARB X EYES ONLY AND ACTION FLUCKEY X YOU COME HOME X ACKNOWLEDGE X

Being in the same time zone as Guam, the Barb received this before we arrived off Kaihyo To. I replied:

> BARB TO COMSUBPAC X ACKNOWLEDGING X BUSY DEPARTURE DAY X 25 JULY AT 0310 GAVE FINAL ROCKET MASSAGE TO FACTORIES IN KASHIHO X AT 0500 ATTEMPTED REVIEW DAMAGE SHIRITORI MUCH FOG AND SMOKE ASSORTED LIGHT AND HEAVY EXPLOSIONS SPORADICALLY CONTINUING PERHAPS HIT JACKPOT X AT 1235 BOMBARDED CANNERIES AT CHIRI X SANK SEVEN MORE DIESEL SAMPANS X NEW PRISONER SAYS LOCAL JAP NEWSPAPER STATED OUR TRAIN DESTROYED BY PLANE BOMB ONE HUNDRED AND FIFTY MEN WOMEN AND CHILDREN KILLED X LATTER PROBABLY PROPAGANDA ANTI-USA X HE SAYS TROOPS ONLY TRAVEL AT NIGHT X

Observation of Kaihyo To on the surface revealed that only two buildings had been patched. The place was still in ruins, with over half the buildings wiped out by the fires.

Headed south to leave our area via the minefields and the north coast of Hokkaido, we hoped to entice a frigate out.

After supper I was giving advice to a lovelorn officer, Jack Sheffield.

Married to a lovely gal, he was leaving the Navy after the war. The separations of Navy life were not for them. They would move to California and set up his business, but first he wanted to take his wife on a second honeymoon. He knew that I had married there.

"Jack, prewar times were a bit tougher. I did borrow enough money, however, to take my bride honeymooning at Lake Arrowhead Lodge in the San Bernardino mountains. It's a beautiful vacation spot—well laced with movie stars and starlets.

"Phoning to reserve, I asked for a room with a double bed. The only cancelation available had twin beds. I complained to the clerk. He gave in and said they would substitute a double. It was a lot of fun there in the rustic wilds, but with one oddity. When you went down to breakfast, you had to leave the doors open for the rooms to be made up. All the stars passing by always peered into our room.

"The last night there the ballroom was opened for dancing. Marjorie and I were dancing when the music stopped for an announcement. 'The orchestra wishes to dedicate the Anniversary Waltz to our newlyweds, Ensign and Mrs. Fluckey. Will the other guests please leave the dance floor while we honor them.' We were embarrassed, because we thought we had been acting like an old married couple. After ages the music ceased. We crept sheepishly back to our table with a starlet couple. Marjorie asked, 'How did they know?'

"The starlets replied, 'Know! You have the only room in the lodge with a double bed. Everyone here has seen it and has been badgering the management. Their answer is that they had to truck that double bed all the way up the mountain from San Bernardino at your insistence, and they don't intend to redo every room in the Lodge.' So, Jack, don't expect a double bed."

*Time 2040.* "Captain, we've run into a heavy fog. This will scotch your plans for another frigate. Recommend we head for Kunashiri, though we will arrive there in the morning and won't be able to transit the strait."

"Concur, Jim. Do so."

Gierhart entered with the nightly news from ComSubPac.

DIURNAL CHATTER X WHALE AND SCABBARDFISH EACH RETRIEVED ONE AVIATOR WHILE TORO PICKED UP THREE BRITISHERS X BARB CONTINUED COASTAL DEPREDATIONS WITH ROCKET TREATMENT TO FACTORIES AT KA-SHIHO AND BOMBARDMENT OF CANNERIES AT CHIRI RPT CHIRI X BELIEVES HIT JACKPOT AT SHIRITORI BECAUSE OVER THIRTY HOURS AFTER ROCK-ETING LIGHT AND HEAVY EXPLOSIONS ARE CONTINUING WITH MUCH

SMOKE X JAP PRISONER REPORTS NIPS BELIEVE PLANE BLEW UP TRAIN X
ALSO SANK SEVEN SAMPANS FOR DIVERSION X CERO ARRIVED MIDWAY X

Having read it aloud to the officers present, I yawned. "It's too bad the wolfpacks can't find something. Fellows, let's hit the hay and rest up. If this fog breaks, tomorrow will be a busy day. We'll sweep the west coast of Kunashiri Island for canneries and sampans. It's probably undefended."

*26 July.* Habitually, I was on the bridge a half hour before sunrise in what I know is the most beautiful part of the day. Yet to me it was also the most dangerous time for a sub on the surface. The watch was sleepy, their blood pressure low. The kaleidoscope of colored shadows was befuddling, and the white-hot intensity of the light at sunrise blinding. An enemy periscope or a plane skipping from cloud to cloud could go unnoticed. I, on the other hand, was bright-eyed and bushy-tailed from my rest.

Our last dawn in the area had us mummified in a cold, dripping shroud of fog. We kept our binoculars tucked inside our foul-weather jackets and depended on our radar and sonar to keep the unseen enemy at arm's length.

*Time 1130.* "Captain, we've rounded Shiretoko Misaki. The fog is behind us. I can see the mountains on Hokkaido and on Kunashiri."

"Great, Dave. Close the coast to 2000 yards."

All officers studied the chart on the wardroom table. "Sure we'll be seen, but we've got 15 fathoms. Let's go!" Sweeping close enough in that we could wave at a few inhabitants on Kunashiri, we passed several small villages and pipsqueak fishing stations. Our Japanese charts gave promise of something at Shibetoro in 20-mile-wide Nemuro Strait. I counted on Admiral Halsey having eliminated nearby planes.

*Time 1435.* We positioned lots of lookouts topside, but Bluth on the high periscope won the prize. "Cannery dead ahead headed to be dead."

His boss, Higgins, cracked back, "What a deadhead."

"Looks good, Max. Not a cannery, though; lumbermill with a single, tall, black stack. There are three huge buildings and many smaller houses and shacks. To the right and toward the beach is a sampan building yard with 16 rectangular, box-like cradles on tracks that run into the water. To the left of the cradles are neat rows of brand-new sampans."

"Sir, I can count 26 there, probably more behind them. Then look to the right: 17 more in three rows, three buildings behind them, and lumber piles to the right. Some are hard to see. Those large fir-tree branches are scattered over everything."

"Keep counting, Willy. All cradles are full."

Dave added, "Those branches are for camouflage from aircraft. Despite all the strikes, this place is untouched. It fooled Halsey."

"Captain, battle stations?"

"Not yet, Jim. Ahead flank. Shankles, come right 10°."

"We're leaving?"

"No, Jim, just a high-speed sweep by to draw any opposing fire. All we have is the 40-mm gun." None came. "Reverse course."

*Time 1502.* "MAN BATTLE STATIONS GUNS!"

"Tom, have Ragland and Cole put the last four cases of beer in the cooler. This is a valuable target. Though we lack big ammunition, I feel lucky today."

"Aye, sir. Don't forget that these sampans are destructible, but non-sinkable. They don't count for that last one you lack to win our bet."

Max lowered his binoculars. "I've found their assembly-line system. They build the hull in the cradle, slide the cradle into the water for launching, then bring the sampan back up on rollers to install the engines and topside."

"When we shoot, Max, put the 40-mm on the powerhouse with the smokestack. After the initial bursts with the twin 20-mm guns, pepper the general area to keep individuals from responding with any small-caliber weapons. At 2000 yards you can start."

"Ready now, sir."

"All stop. COMMENCE FIRING!"

Within two minutes, a fortunate 40-mm shell missing the powerhouse lobbed over and landed in a fuel tank behind. It caught fire and exploded, spreading the fire everywhere. Fire-fighting squads appeared with hoses. Bursts of 20-mm fire forced them to drop everything and flee into the forest above. Meanwhile, the 40-mm kept pounding away at machine shops, cradles, and any structure of importance. The mill was now a blazing inferno. Time and again firefighters tried to stop the rapidly spreading blaze that consumed everything. Our gunners—admiring their courage—peppered close enough to drive them back without killing them. I never saw a Japanese even wounded. Now out of 40-mm shells, we waited.

In the face of our peppering bursts an older man appeared, stripped down to the waist, walking down to the beach with two buckets. Max called out, "CHECK FIRE ALL GUNS!" Everyone watched him. He must have been the owner. At the water's edge he stopped, shook his clenched fist at us, then filled his buckets. Turning around, he slowly walked back toward the mill, stooped over from the weight of the buckets. Suddenly he stopped, dropped both buckets, looked at the holocaust, then turned and faced us again. Looking at us, he threw up his arms in hopeless grief, his lifetime's work shattered in 30 stinking minutes. Turning once more, his shoulders stooped, his spirit broken, he wobbled up the path and into the forest. I could almost feel his tears running down my cheeks, or were they mine? War is such hell.

*Time 1630.* "Captain! There's a ship on the horizon coming up Nemuro Strait. She's turning toward us. Looks like a trawler. Must have seen the smoke and is coming to investigate. She was headed for Hokkaido."

"Good work, Don. How come you're a lookout again?"

"No torpedoes, no battle station. I'm enjoying it."

"Welcome back. Max, secure the 40-mm gun. Tom, head for the trawler. Swish, what else are you hiding in your locker besides hand grenades? We need to sink this trawler."

"There's a case of rifle grenades, captain."

"Break them out. She's our 15th and must not escape."

*Time 1656.* "COMMENCE FIRING!" The automatic weapons opened up at 400 yards as we swept by. The trawler had a 25-mm gun, larger than our 20-mm, but quickly abandoned it. As we turned to come back, Swish was ready with his rifle grenades. He fired 18, yet only 3 penetrated the hull, making small holes about an inch in diameter. Each one exploded against the tough hull with little effect. This amazed me. When our boarding party was trained by the Marines at Midway, I saw this same type grenade rip through large timbers and even through eight inches of reinforced concrete. Baffling. It appeared the only way to sink this trawler would be to board and break some seacock. Tom's men could be ready in five minutes. In the interval, a fire broke out on board, and her crew abandoned ship. We picked up two volunteer prisoners, ceased fire, and let her burn.

Certain that she would sink, the *Barb* left this column of black smoke and trekked back four miles to the smoking inferno ashore.

*Time 1815.* The fires had done our work for us. All buildings were nearly leveled, leaving the tall stack standing alone, red hot. The sampans to the left were completely destroyed. A fire had broken out to the right of the cradles in the lumber piles. The dry camouflage branches and the rising wind helped enormously, spreading the flames toward the remaining 17 sampans and cradles. A bucket brigade on the beach fled our final burst of 20-mm fire.

*Time 1845.* Fire-fighting attempts having ceased ashore, the *Barb* returned to her trawler. En route we saw and heard several oil explosions on board. These had no other effect. Tom was quiet, not offering to grant me my inevitable win on our bet. Swish finished off all his rifle grenades with no results. Our remaining machine-gun ammunition would be wasted on this rugged hull, so we waited, unable to board. An hour later Max remarked, "I do believe this trawler is so tough she'll burn all night." I agreed. Tom smiled.

*Time 1950.* "Send Singer to the bridge with his movie camera. Men, this trawler appears capable of burning till tomorrow, and *Barb* is required to transit the strait at midnight. Our only recourse is to ram. Do not be alarmed. When the collision alarm is sounded, lock the forward torpedo room watertight door. Leave all other doors on the latch."

Tom responded, "Captain, you can't ram! You'll damage the superstructure shutters of the torpedo tubes."

Singer arrived and climbed up on the shears. "Camera is ready, sir."

"Tom, I relieve you of the conn just in case you're right and the bow is damaged. Besides, you might collide too softly and have that trawler stuck on our bow for days. To rub your nose in it, however, I'll give you the honor of participating. Sound the collision alarm!"

Tom grinned and did so. The sirens below wailed. The bridge hatch slammed shut. In seconds Jim reported, "Ship rigged for collision."

"Shankles, come left to course 242°. All ahead one-third. Maneuvering room, make turns for seven knots." Without a waver the *Barb* turned in to ram. It seemed like driving a car into a burning garage. The radiating heat put a glow on our faces.

"Come right a hair. Make it 243°. We're coming in to ram from upwind. Make it 244°, 20 yards to go. Hold her tight, Shankles!" Pressing the intercom, I shouted, "Stand by! Hold on! Here we go!" BAM! The *Barb* shuddered. "The bow hit—we're riding up on her—her starboard side is caving in as she rolls toward us! All back full! She's sinking stern first. Watch out topside!" Her foremast snapped across the bow, sending things flying. "She's . . . gone! Sunk! We are clear. All stop. Secure from collision. Let's have Bill, Swish, Parker, Miller, and Arthur on deck to check the shutters."

Tom was a good sport. He walked over, raised my hand on high, and shouted, "The winner and we're still alive! Fifteen down and zero to go! Captain, the bet was worth it just for this unforgettable moment. Permit me to say 'well done.' "

We flooded down aft to lift the bow high. Max joined in the shutters check. When he returned to the bridge he reported, "No apparent damage to bow or shutters other than some paint scraped off. They checked mechanically on opening and closing, so no warps or bends."

"Thanks, Max. One fire extinguished. Let's return to the other."

*Time 2030.* Lying to offshore. The moon rose like a moldy lemon through the heat waves.

As darkness fell, the flames licked hungrily throughout the area, finally reaching the last group of sampans and the 16 building cradles. So far, 35 sampans had burned. The fire was licking its chops over the last eight, spanking new, fresh out of their cradles. In another hour the cradles too, with their 16 unlaunched sampans, would be ashes. The singsong crackle of flaming wood and fir branches strummed a lullaby. The *Barb* would claim only the 35 she had witnessed burning, but the true total would be an astounding 59.

*Time 2200.* Jim shattered my reverie. "Time to leave, captain, if we're going to get through the strait tonight."

"Lullaby of the Leaves. I was just thinking what a perfect swan song this is for the *Barb* and for harassment. Shibetoro is a prize at my graduation." I thought, farewell dear Okhotsk. You've been an adventure and a revelation of what can be accomplished with new concepts and beefed-up weapons, which are certain to come some day with our ever-changing world."

"Now, captain?"

"Now, Jim. Kick her ahead and let's get moving. Secure the few gun stations. Set the regular sea detail. Then splice the main brace for a job well done."

*27 July. Time 0047.* We transited Kunashiri Strait. As we passed the cable station, it was still shrouded in that fog pocket. Searching for the patrol boat, I spotted her beached in the same spot where we had chased her aground. Could it be that she holed herself and was unsalvageable? Some day I'd like to visit the cable station to see what we passed up.

Clearing Kunashiri Strait, the *Barb* entered the rough Pacific Ocean, which doused the bridge. I went below and sent out our departure message to ComSubPac.

> BARB SIXTH TO COMSUBPAC X 26 JULY AT 1500 WHILE WAITING TRANSIT
> KUNASHIRI BOMBARDED LUMBER MILL AND SAMPAN BUILDING YARD WITH
> 20 MM AND REMAINING 40 MM AT SHIBETORO LATITUDE 44-20 LONGITUDE

146 X LUCKY HIT IN OIL TANK STARTED FIRE WHICH BURNED MILL TO THE
GROUND X ALSO BURNED 35 BRAND NEW SAMPANS X BUILDING-BOATS-
CRADLES AIR CAMOUFLAGED X AT 2200 DEPARTED WITH FIRES AT 16 CRA-
DLES HOLDING SAMPAN HULLS AND REMAINING EIGHT SAMPANS X DURING
INTERLUDE SANK TRAWLER BY RAMMING AFTER UNSUCCESSFUL ATTEMPT
WITH MACHINE GUNS AND RIFLE GRENADES X NO DAMAGE BOW X TWO
MORE PRISONERS X

Then I turned in, thankful for my innerspring mattress, which had lulled
me to sleep within two minutes whenever I had had the chance to lay
down during the last two busy days.

Awakening, I joined the officers in the wardroom, where they were
occupied writing up their specific sections of the technical addenda to the
*Barb*'s Twelfth War Patrol Report. Radioman Lindberg handed me a
message. "Just picked up the Nightly News from Guam, sir." I read it
aloud.

DAILY NEWS X THREADFIN PICKED UP 3 AVIATORS NORTH OF SAIPAN X
STERLET RECOVERED ONE BRITISHER AFTER HE HAD BEEN IN WATER 2½
DAYS AND DRAGONET PICKED UP ONE OF OURS X BASHFUL BARB BORED
BY NECESSITY OF WAITING TO TRANSIT KUNASHIRI BOMBARDED LUMBER
MILL AND SAMPAN BUILDING YARD AT SHIBETORU WITH 20 MM AND 40
MM X HIT IN OIL TANK STARTED FIRE WHICH BURNED MILL TO GROUND
AND 35 BRAND NEW SAMPANS X DEPARTED HOURS LATER WITH FIRES AT
16 BUILDING CRADLES AND LAST 8 SAMPANS X ALSO SANK TRAWLER BY
RAMMING WHEN FIRE FAILED SINK IT X PARGO AT GUAM X

"Fellows, I know these messages sound good to us, compared to the
dearth of ships sunk by submarines throughout the Japanese Empire since
we left Pearl on 8 June. The two different wolfpacks in the polar circuit
above our area have only sunk two ships total. Yet, I've made an awful
boo-boo."

"Boo-boo?"

"Definitely. That last 150-ton trawler was a mighty fine ship. After our
initial firing pass, which cleared away her gun crew, we should have checked
fire and boarded her. It was stupid of me to continue shooting. This set
her ablaze. Since we were less than two hours from Kunashiri Strait, we
should have taken her as a prize. Tom would be given command to bring
her back with us to Midway."

"Me?"

"You'd achieve a command early and would have won our bet."

"But that's a long trip. Suppose another craft had come to investigate."

"Don't worry. We'd have been close-by to protect you."

"With what?"

"Scare them off—or use machine guns. We had enough diesel oil to refuel the trawler, or we could have towed you, if necessary. You'd probably end up at Pearl. By Prize Law she would probably be auctioned off as a museum. This could have netted the Dolphin Welfare Fund a cool half-million dollars, opening up college to a lot of submariners' kids. Can't you understand the opportunity we missed?"

"Captain, I'm glad that you won our bet."

The evening featured two heartening bits of information.

> FROM COMSUBPAC X SEA CAT CARRY ADDITIONAL 5-INCH AMMUNITION X INSTALL FIVE ROCKET LAUNCHERS SIDE BY SIDE MAIN DECK BETWEEN FORWARD ESCAPE LOCK—TORPEDO LOADING HATCH X ATTACK SUITABLE TARGETS HOKKAIDO—KARAFUTO AT DISCRETION COMMANDING OFFICER X EXPECT THEM TO BE SIMILAR IN NATURE TO THOSE CHOSEN BY BARB X

The second: "Radio Tokyo: On 26 July the Okhotsk side of Kunashiri Island suffered 240 bombs. Two submarines also attacked." Admiral Lockwood replied to Admiral Nimitz that the *Barb* had departed 27 July.

As the week of transit to Midway passed, I added the following to my technical remarks on rocketry:

> The torpedo has fulfilled its purpose. Its day in this war is passing. I believe that in the near future, with the increased tempo of air strikes and lack of air opposition, lifeguard duties will be taken over more capably and more efficiently by PT boats. Submarines not equipped with mine-detecting gear to enter the Sea of Japan must stagnate or slowly sink into oblivion. That is, unless they look to a new main battery—rockets. The rocket is not a toy. Its possibilities are tremendous, strategically and tactically, but not beyond comprehension.
>
> They are self-selling, simple, and easy to operate. They do not require gun sights. Most of our opposition does.
>
> What tremendous advantage we possess, each submarine a submersible task force. Let's make the rocket our final devastating blow against the Japanese with one idea in mind—DESTROY AND PULVERIZE!

Then I closed with remarks on my beloved personnel:

> How difficult it is to close this chapter in the *Barb*. What wordy praise can one give such men as these. Men who, without the information available to the commanding officer, follow unhesitatingly when in the vicinity of minefields, so long as there is the possibility of targets. Men who offer half-a-year's pay for the opportunity to land on Japanese territory to blow up a train with a self-trained demolition team. Men who flinch not with the fathometer ticking off two fathoms beneath the keel.

Men who shout that the destroyer is running away after we've thrown every punch we possess and are getting our ears flattened back. Men who will fight to the last bullet and then want to start throwing the empty shell cases. These men are submariners!

*2 August. Time 0800.* With Midway two hours away, I completed another project. This, the solution of the mystery as to where the Midway Island gooney birds went for seven years before returning to that island. The riddle was a riddle no longer. I broke out a sheet of stationery and wrote the *National Geographic* in Washington, D.C.:

> Dear Mr. Grosvenor:
>    Once a Boy Scout, always a Boy Scout—I was a member of Troop 15 in Washington. Knowing of your keen interest in wildlife, particularly birds, I have solved a longstanding puzzle. Your magazine should be the first to know.
>    This is the answer to the mysterious disappearance of the gooney birds of Midway Island for seven years before they return. They go to Etorofu Island in the Kurile chain just northeast of Kunashiri Island, the island closest to Japan. The following is not classified information because the Japanese have publicly announced that submarines are operating in the Sea of Okhotsk. Routes to and from combat areas are varied at the discretion of the Commanding Officer.
>    As Captain of the submarine Barb the first time we went from Midway to the Kuriles, I noted some goonies resting on the water en route. Some were tagged. On our second visit, our courses followed the great circle route to and from Etorofu Island. Day after day, all the way, we came across the goonies, some were going, some returning. En route in other areas we have never seen any of these very tame birds. Why they do this, I leave to your experts.
>    With great admiration for your splendid magazine, and
> With kindest personal regards,<br>Eugene B. Fluckey, Cmdr. USN

Knowing that I would be relieved soon, I packed up my belongings for the last time. Cleaning out my drawers, I laid a design I made of a new submarine weapon on the wardroom table. Bill (rechristened Choo-choo) Walker remarked, "Just exactly what is that?"

"It may be nothing or it may be a 'W' gun. Somewhat like the 'Y' gun on escorts for firing depth charges to make a bracketing pattern. This would be mounted on the fantail of a sub for use against attacking destroyers. The right, left, and center tubes will hold a spring-loaded bundle of buoyant mousetrap projectiles. These will be connected by wires to form a line perpendicular to our beam. When an escort chases us at night while we are on the surface or possibly submerged, we fire it. The escort coming at us wraps the wire around her sides. The mousetraps, now armed, explode on contact and cut her hull in two, horizontally. Will it work?"

"Perfect for that Terutsuki."

"Maybe it'll sell."

***Time 1000.*** Arrived Midway with our large battle flag flying, one Rising Sun pennant, and two red-ball pennants decorating our guywires, along with 50 smaller red-ball pennants—to the greatest welcome ever. The pennants indicated the number of ships and vessels downed, afloat and ashore, on this patrol. As the band played, the crazy, tame gooney birds, oblivious of the fact that we had discovered their hideaway, danced and strutted for the *Barb*.

At Midway, the Marines take *Barb*'s four prisoners into custody.

# Chapter 29 Swan Song

Lieutenant Commander Cornelius Patrick Callahan stood quietly on the pier, scrutinizing the *Barb* and her crew at quarters during our uproarious welcome to Midway. Spotting him in the background as the high brass poured on board, I hoped for a moment that time would stand still.

Word had arrived that he was destined to relieve me as commanding officer. He would take my *Barb* away from me. Now I realized what a grasping mistress she had become, not only for myself but for all on board. No one wanted to leave her. Yet, reason and the surety of change pressed such thoughts aside. I thought, be happy that you were permitted to have the forbidden fifth patrol in command. Be happy that no one under your command ever received a Purple Heart Medal for being wounded. Be happy that in spite of having erratic- and circular-run torpedoes and more shells, bombs, and depth charges directed at the *Barb* than at any other submarine, she came home unscathed, whereas others were lost to the enemy or their own error. Be happy that the *Barb* would receive more medals for factual achievement than any other submarine. Finally, realize that "when one door shuts, another door opens"—even if you have to nudge it. Our pride showed—and I intended that it should.

As usual, Commodore Kenneth Hurd, the squadron commander, and Captain John Waterman, the division commander, gathered in the wardroom for briefing on the past patrol. We also went over the recommendations that I should forward for a given number of personnel to receive medals. They agreed that a separate recommendation should be forwarded for medals for the saboteur squad that went ashore to blow up the train.

406

The departure of the brass gave me an opportunity to introduce Pat Callahan to the officers, chiefs, and leading men. Pat had just come from a patrol on the *Segundo* with skipper J. D. Fulp in the Yellow Sea. The patrol had been successful, with one ship sunk, even though traffic had vanished. Affable and sharp, he was readily accepted. It made me feel good to know the *Barb* would be in top-notch hands.

Clutching our mail and a ditty bag, we trickled off to Gooneyville Lodge as the relief crew took over. Only Jim Webster and I were to be transferred. Finally, Max Duncan would become the executive officer under Pat.

First I sorted out the love letters from Marjorie and Barbara by date, then read the latest ones, thanking God that both were in good health. My mind relieved, I opened an official letter from the Commander Submarine Squadron 24. It was a most thoughtful surprise from Mike Fenno. As skipper of the *Trout,* he had slipped through the Japanese gauntlet and supplied Corregidor with antiaircraft ammunition. As ballast

Returning from the Sea of Okhotsk to Midway Island, the whole crew poses with the final battle flag.

on his return he carried the gold and silver bullion from the Bank of Manila.

Mike sent me the front page from the San Francisco paper that had caused so much trouble for Fleet Admiral Nimitz, the one showing the broad arrow representing Halsey's Third Fleet sweeping the east coast of Japan. The headline said it was the deepest penetration of Japanese waters yet by a section of Halsey's Fleet.

Mike's note added that if the *Barb* kept going for the rest of her patrol at the pace to date seen in dispatches, Admiral Halsey would end up with the smaller part of the arrow. Such a true-blue friend deserved my prompt reply, making him president of the "*Barb* Booster Club."

A personal word also came from Admiral Nimitz. He had himself studied the mining plan that we had submitted on 10 February. This envisioned simultaneous mining by 10 wolfpacks of inshore locations at 29 points on the China coast, from the Yellow Sea to Amoy. He concurred with the concept. It should have been carried out in March. Since Okinawa was secured on 21 June, such a mining was no longer necessary. He commended the study, however. Mining the China coast had also been recommended by Captain Walter Ebert of ComNavGrp China in his message of 31 January.

The *Barb*'s Twelfth War Patrol Report was widely disseminated before the normal endorsements of the division commander, squadron commander, and force commander. This was due to the rocket attacks and our recommendations, the erratic deep running of the batch of electric Mark 18 torpedoes, and the complete failure of the homing Mark 28s.

Later I was informed that the Mark 28 torpedoes had not been tested in shallow water. Tests showed that ships' propellers emitted both a direct sound path and a strong sound path that reflected or bounced off the bottom at lesser depths. Thus, those that we fired headed for the bottom to bury themselves in the mud or sand, or they failed to start. Our little "cutie" hit was the only one ever to sink a merchant ship, but such shooting at a range of 75 yards was disturbing. Weeks later Captain Herb Andrews, an outstanding skipper on ComSubPac's staff, sent me the private, "off the record" letters sent after each patrol report had been digested. These contained the reaction and comments of Admiral Lockwood and his chief of staff. After the *Barb*'s twelfth patrol, they were as follows:

> My hat is off to you, your fighting ship, and your fighting crew.
>
> Poor torpedo performance is to be regretted. Something which has been with us throughout the war.
>
> The performance of your gun crews was superb and demonstrates in a

thorough manner what can be accomplished with a well drilled and well disciplined crew.

This being the first rocket attack launched by a sub, the success of this new weapon clearly demonstrates its possibilities.

This is an amazing story packed with thrills and very well written. I hope you will write this one up for the Saturday Evening Post. It would make Jasper Holmes [head of decryption and an author] look to his laurels.

Your gun attacks and frigate attacks were daring in the extreme and your saboteur expedition was considerably too risky. However, fortune favors the bold and you make a splendid success of it.

The qualities of leadership you possess as evidenced by the spirit of your ship are inspiring to those of us who warm our swivel chairs, and if you don't kill yourself off in some bit of recklessness, I predict high altitudes for you in your Navy career.

This is a fitting final chapter to Barb's and your war time career.

May you not find the piping times of peace too dull.

Charles A. Lockwood, Jr.

The submarine base at Pearl Harbor, now under Captain Edwin Swinburne, sent me a package. It contained Medal of Honor color placards for the entire crew who had taken part in our foray into Namkwan Harbor. Having been in the *Barb,* there was nothing that Ed would not do for us. These kept me busy over in the crew's barracks during Happy Hour, getting writer's cramp amidst the revelry. What a happy gang!

Word arrived that Admiral Nimitz had verbally approved the award to me of a fourth Navy Cross for this patrol. So, my recommendations for the allotted medals stemming from my award were awaited by the Board of Awards.

I did have a problem with my recommendations for the eight men ashore to blow the train. The rules restricted them individually to a Bronze Star, with a Silver Star for their leader, because they encountered no opposition. They deserved a higher medal. As far as I was concerned, each of them ought to have been awarded a Navy Cross, our next-to-highest combat award. This problem was resolved by a personal letter to Swinburne, a member of the Board of Awards. He agreed to upgrade my recommendations as a specific exception.

Our days of leave were full of fun, laughter, volleyball, swimming, barbecues, and pranks. Entertainment arrived. A USO troupe of three lovely young ladies dared to enter our all-male cage. With a daytime and a nighttime performance they did boost our morale—then they moved on to the next outpost. Speaking on behalf of the entire crew, I offered heartfelt thanks for the lift they had given us with their fascinating shows.

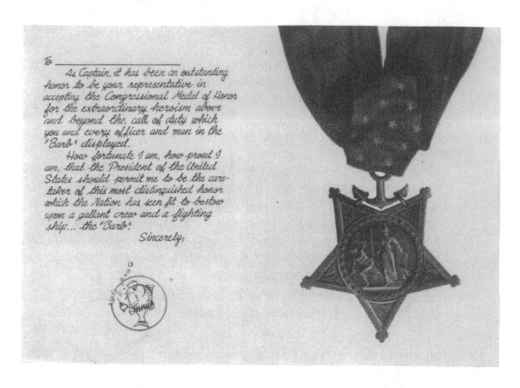

To _____

    As Captain, it has been an outstanding honor to be your representative in accepting the Congressional Medal of Honor for the extraordinary heroism above and beyond the call of duty which you and every officer and man in the "Barb" displayed.

    How fortunate I am, how proud I am, that the President of the United States should permit me to be the caretaker of this most distinguished honor which the Nation has seen fit to bestow upon a gallant crew and a fighting ship... the "Barb".

              Sincerely,

*Top:* Congressional Medal of Honor placard given to each shipmate. *Bottom:* The *Barb* alongside at Midway, dressed to show all the vessels she eliminated on her twelfth war patrol.

It was the first time in two months that we had seen a female, and they were more exciting, more talented, and lovelier than ever. On behalf of the *Barb* I received a kiss from each one—and withstood the resultant caterwauling—in the line of duty.

Pat became familiar with the *Barb* while I swept up the cobwebs and backlog of paperwork.

*6 August 1945.* Hiroshima lay flattened by the atomic bomb at 0815 Tokyo time. The bomb dropped by the B-29 "Enola Gay" exploded at 2000 feet, killing some 75,000 men, women, and children over a four-square-mile area. At 1015 Washington time, 6 August, President Truman announced it to the world. The radioed devastation seemed incredible to all military personnel. All day long we stayed by radio sets awaiting the latest reports.

There was no indication of a Japanese capitulation. The war must continue. The *Barb* was to be based at Guam for the next six months and to assume lifeguard duties on her next patrol.

*7 August.* On awakening, I immediately joined Pat. "Pat, no country can withstand the terrible toll of this atomic bomb. If you want to have a submarine command in this war, we must have the change of command now. Frankly, to have the traditional ceremony done with a speech of farewell to this gallant crew at quarters, with the high brass as guests, is too much of an emotional hurdle for me to ride. I fear I'll break down. If you have no objection, I'll gather the officers together and have Swish, the chief of the boat, get them in uniform with cap and tie, and we'll have the change-of-command ceremony at high noon."

"No objection, Gene. Whereabouts?"

"Well, the bar probably has the most space where we can line up our witnesses. Let's have it there, then I'll sign for the yeasty champagne to celebrate this auspicious occasion."

"Done! This must be the first time in naval history that a change of command has taken place in a bar. I do understand and appreciate your consideration for me instead of waiting the normal 10 days."

"Pat, I'll need your yeoman, Lunde, to pull some files for me, but I'll move up to Captain Waterman's office until I leave for the staff at Pearl in the training command."

"Gene, please feel free to use whatever services we can offer until *Barb* shoves off for Guam."

*Time 1200.* The officers and Chief Saunders lined up, caps and ties on, appearing more polished than I had seen them since we left Pearl a year ago. Solemnly, I read my orders. Pat read his. We saluted as he said, "I relieve you, sir." We shook hands, and Pat gave his first order. "Splice the main brace!"

*8 August.* The Soviets declared war and entered Mongolia, intent on getting in on the gravy. Our gravely ill President at Yalta had laid the groundwork for the great robbery of the Okhotsk Sea without a fight or any aid whatsoever against Japan. In the peace treaty to come, the Soviet Union received Karafuto and all the Kurile Islands. This sea became a *mare clausum,* their private lake, in which they forbade the Japanese to even fish. I was happy the *Barb* had destroyed a part of it.

*9 August.* At 1015 the second atomic bomb was unleashed on Nagasaki, killing 40,000. Still the Japanese resisted.

Portrait of the skipper by Al Murray, the famous combat artist of World War II.

A week later the *Barb* was under way for training, followed by loading for the next patrol. No rockets were available.

*21 August. Time 1000.* Though Admiral Nimitz had declared all offensive operations ceased on 15 August, the *Barb* was still being sent to Guam as a precaution. Pat gave the order, "Take in all lines." I grabbed the spring line, the last to be let go. Lifting it off the bollard, I tossed it on board with a farewell wave of my arm. "GOOD LUCK, PAT! AND GOD BLESS ALL YOU BARBARIANS!" Then I sat down on the bollard and kept waving until they were passing out of the channel, all the while the crew waving back.

The pier and dock areas were empty. I sat there until my *Barb* was out of sight. Finally, standing up slowly, I brushed the tears from my cheeks, ready to face the future. But how I loved that *Barb* girl!

Then it dawned on me that the men in the *Barb* who gave her life had taught me the most valuable philosophy for my life. Regardless of all the dangers they accepted at my command, and without all the knowledge that was available to me, a reciprocal trust glowed. I find it applies totally for success in life, love, marriage, and business. Simply put, "I believe in you."

# Epilogue

## Part I: Namkwan Harbor Revisited — Truth Revealed

All convoy records taken from Japan by Commander Submarine Squadron 20 after the surrender were finally returned to Japan by our National Archives. For reasons unknown, the Japanese records of the *Barb's* attack at Namkwan Harbor on 23 January 1945 are missing. The Japanese do know that there were two convoys anchored there totaling 27 ships, but none of the merchant ships are identified. Names are essential to finding out the factual sinkings and damage.

Within minutes after the *Barb's* strike, someone sent a message from the same position that the *Taikyo Maru* blew up and disappeared from this earth. This was intercepted in Tokyo, and Commander Sogawa, on operations watch, noted it in his personal diary. Other than the *Barb's* hurried visual assessment of three ships sunk, one probably sunk, and three damaged, and the coastwatcher's report of four ships sunk and three damaged, no records have been located.

Postwar, the Joint Army-Navy Assessment Committee credited the *Barb* with sinking only the *Taikyo Maru*. We had reported a radar count of 30 ships there and requested an aerial reconnaissance to ascertain full damage. Since no U.S. planes had ever flown over this Japanese-held area, the coastwatcher's report was considered conclusive by Admiral Miles. Marine Sergeant Stewart, who sent the report, was awarded a Legion of Merit and sent back to the States for taking such dangerous risks.

For years I searched for Stewart in vain, yet could not break through the bureaucratic "right of privacy" for those separated from the service. Even the Marine Commandant, General Lou Wilson, was blocked in his attempt.

Finally, Stewart's immediate boss in China, Lieutenant Karl Divelbiss, located him after 10 years of searching. Stewart had died two years earlier and never talked about the war to his father, wife, or family: a true intelligence agent, who according to another coastwatcher (Navy Radioman Bob Sinks) was the best and most daring, even riding in the junks of the Chinese pirates who were working with him.

Baffled, and determined to set the record straight after studying all the intelligence sources in the United States and Tokyo, I realized that my only window of hope lay in a personal visit to Namkwan Harbor. There must be village records or some elderly fishermen who had witnessed the event.

As the *Barb* had searched the Chinese coast for the convoy that night in 1945, I had studied the *Coast Pilot Manual,* seeking our most dangerous escape route to discourage any destroyer who dared follow us after our attack. One fact lodged in my memory as unusual: a small, walled fishing village lay at the southern part of Namkwan Harbor. This would be our objective.

In 1984, failing to obtain visas from the Communist Chinese Embassy, we also failed through the American Embassy in Beijing, the Governor General of Hong Kong, officials in Taiwan, and even Lindblad Travel Agency.

In June 1989, a stroke of luck brought some hope. After one of my speeches at the Nimitz Museum in Fredericksburg, Texas, during an International Submarine Symposium, the first questioner from the audience was Robert Sinks, the U.S. naval coastwatcher radioman who relayed all traffic from and to the *Barb,* the coastwatchers, the Chinese pirates, and the headquarters at Chungking. Now the chairman of a successful, world-renowned agronomic company, Applied Technologies, he had visited China on business and thought he could set up a revisit to Namkwan. Finally, in May 1991 the Chinese bureaucracy gave us a visa for Fujian Province. Namkwan Harbor (now Shacheng Gang) is just inside the northern border of this province.

On 1 June our group of agricultural experts, my wife Margaret, and I flew into Fuzhou from Hong Kong. We had a special invitation for a four-day official visit before leaving for Beijing on 5 June.

Arrangements had been made for us to hire a van, driver, and cashier. The cashier was needed to carry the necessary Chinese money into the remote region where our Chinese Foreign Exchange currency could not be used. A government interpreter would be provided, and we would pick up two government officials at a half-way point, Xiapu. This would be the only locale that had a Guest House for travelers, for they doubted we could make the entire journey in one day.

The straight-line distance from Fuzhou to Namkwan Harbor was 100 kilometers. We anticipated the round trip would be 250 to 300 kilometers. The officials estimated much farther.

That evening the group was served a 25-course banquet. At dawning, Margaret was too ill to get out of bed and travel, yet she insisted that I carry on—our one chance after seven years of failure. Murray Sinks, Bob's son and president of the company, jumped at the opportunity to fill in the vacant space in the van. Tales of his father (an aerographer's mate first class and radioman) sneaking through so much of this World War II, Japanese-held territory fascinated him. Off we went winding around the high mountains on questionable dusty roads with nary a guardrail.

Late that afternoon we arrived at Xiapu, met our officials, and stayed the night. This was our first encounter with straw mats, no sheets, and thundermugs (night potties). The helpful officials had telephoned as far as the lines went, discovering an old fisherman at the village of Sansha who knew all about the *Barb*'s attack. Though nearly all place names had changed in 1949 when the communists took over China, I knew Sansha was on Sansha Inlet about 30 miles south of Namkwan.

At 0800 we were off again in the van. The fisherman witness awaited us on the quay. He spoke of the material he had scavenged: food, tools, clothing, medical supplies, rafts, life jackets, guns, ammunition, etc. I asked him to describe the attack. His reply was that he had not seen it. The locale was farther north, and a lot of ships had been sunk. All he picked up was flotsam. We then agreed to backtrack to the north coastal road.

After we passed Funing Bay, our objective lay at the end of the next large promontory extending eastward for some 15 kilometers to the southern boundary of Namkwan Harbor. In the late afternoon we came to a fishing village, Jianguo, where the officials questioned many of the elderly.

Li Nendi, a witness, said he was 9 or 10 years old at the time of the attack. He happened to be awake in his father's junk at 0300, late January 1945, and saw the great explosions to the east about 12 to 15 kilometers along the coast. The next day his father went there after the Japanese ships fled south. Three or four ships had been sunk as well as the one that exploded. The driving room/cabin (bridge) of one of the sunken ships was above the surface with the main hull below. He saw five or six dead bodies. He said there were no roads to the old walled fishing village at the end of the peninsula.

Two other elderly gentlemen reported seeing the explosions and one reported hearing them. One fisherman said he picked up material (mostly food, medical supplies, guns and bullets) from a ship sunk near Seven Stars Isles. The Chinese Nationalist troops confiscated the other materials that he found at sea.

Now the government officials said our mission was accomplished. I disagreed and wanted to hire a motorized junk. We found one, but the 20-foot spring tide was out, and all junks were aground. My captain, Lu Xiao Da, said we must get under way at 0700 the next (our last) day. Any later would require us to leave around noon. Time was running out. Having no place to sleep that night, we headed back to the Guest House at Xiapu, a two-hour drive.

Now that I was taking charge of our group, reveille was set at 0430. The officials would arrange for a town eatery to serve us a 10-minute breakfast at 0500, then we'd leave.

At 0430 Murray and I were ready at the door of the Guest House, but our Chinese party were all asleep. After clapping our hands and yelling "Hubba, Hubba" a few times, we aroused them, and they began to appear. In the dark we found that all the outside doors were padlocked from the inside. After rousing a few employees, the doors were opened and off we went, only to find that the main gate was also locked. Back we went for another skirmish.

At 0500 we arrived at the eatery. The officials were pounding on the door which was answered by a yawning woman. Again I warned, "Serve what you can in 10 minutes." She did: baked pumpkin seeds, jasmine tea, and cold, boiled chicken feet!

On the dot of 0700 we arrived at my flagship, the *Min Fu Ding,* with my flag captain dressed in his best dress "Browns" and shouting for us to run, for the tide was lowering faster than he anticipated. As our group jumped aboard, his crew used poles to get the junk away from the dock and into the dwindling estuary where the motor took over. At full speed (six knots) we slithered past the numerous junks settling in the mud. Breathing a sigh of relief, Captain Lu estimated our arrival at 0930, despite the fog, which he said would clear.

The sea was calm, with little wind. Gathered on the deck forward, our group babbled with excitement about the unknown. Even the government officials knew not what to expect, having never ventured this deep into their own territory. Our only information was my memory that a small, walled fishing village lay close to the southeast end of the peninsula. Captain Lu verified this and said that about 100 people lived there.

Shortly before 0900, Huang Qi village lay before us. A cheer burst forth

On the flag bridge, en route to Namkwan Harbor in the good junk *Min Fu Ding,* 4 June 1991.

as we saw the 20-foot-high walls supporting houses on top. In the low tide we eased between the off-shore fishing junks and nets and put our bow on a ramp. The whole village erupted to see us as we hopped out and climbed the stone steps to the upper quay.

Now the government officials went to work. A man about 45 stepped forward, saying he was the village chief. We explained that our mission was to find some elders who might remember a night attack on Japanese convoys anchored in the harbor in late January 1945, and that we would like to view any village records he might have. He explained that there were no village records. Everyone knew of the attack, however, for the elders passed down history to the young verbally, and were keen to keep history in order.

Quickly he had a room cleared in an adjacent building for our group and called the most knowledgeable witnesses, who were happy to recount their stories.

Mr. Lin You Yin was a boy when the attack took place in January 1945. He recalled: "The thunderous explosions woke everyone, and there was one bigger than all the others, and fires, then smoke. The ships were sunk west of the Li Tao Lighthouse [Incog Islands], two ships north of it, and two ships south of it. That same morning when our junk went out there were many dead Japanese soldiers in the water. Also three other ships were damaged, and the Japanese fleet left later that morning."

Mr. Lin Zai Jun (age 18 in 1945), and Mr. Lu Shen Kin (age 17), remembered waking from the explosions in late January 1945. Two of the explosions between 0300 and 0400 were very big. The ships were warehouse ships loaded mostly with military supplies. The afternoon before there had been long lines of ships anchored in the harbor side by side. No one they knew saw the explosions. They both reported that four ships were sunk between the lighthouse and the coast. Other ships were also damaged: seven ships total were either sunk or damaged. Later that morning while they were fishing, the Japanese fleet left Huang Qi village, sailing to the south.

Both men also reported that they saw Japanese Army warships steaming toward the Seven Stars Islands to the northeast, looking for something. (The *Barb*'s escape route was south of Seven Stars and north of Tae Islands in the hope of running the destroyers aground.)

They did not know if the Army ships fired on any Chinese fishing junks, but there were many junks to the north, and some of the old fishermen still wear Japanese uniforms.

The pair reported seeing many Japanese bodies in yellow uniforms in the water. The villagers left the bodies alone, fearing them. The Japanese fleet did not recover their dead.

The villagers collected food, guns, bullets, medical supplies, and radios from the sunken warehouse ships. For many years one could see the bridge of one ship, the masts of another, and the turned-over hull of another. This harbor was used all the time by the Japanese ships, but after the attack the fleet never anchored there again.

Shortly after the attack, the Nationalist government conducted salvage operations for three months on the sunken ships and marked their location. In 1984 the communist government initiated another salvage operation for four months and removed the markers from all ships except one.

Asked about the names of the sunken ships, none of the witnesses knew. Regarding aircraft carriers, no one knew what they were. Aircraft often flew over the village, but their origin was unknown.

About 25 villagers attended our inquiry in the small hut and about 75 others were at the unshuttered windows or outside. All of them knew of the 1945 event and evinced great and touching interest in the two white visitors, the first Caucasians to set foot in Huang Qi, so the interpreter said.

As the whole town smiled and waved good-bye, we got under way leisurely, photographing everything. Our departure from "Namkwan Revisited" was a dream come true compared to my departure on the *Barb*'s first visit.

Back in Fuzhou my cashier totalled up the war debt for our trip, which I paid off in Foreign Exchange Chinese currency. My Chinese compatriots were correct on their distance estimate: my estimate of 250 kilometers round trip turned out to be 756 kilometers, not counting our junk travel. Yet such an experience was priceless and still seems like a dream from which some day I will awaken. The *Barb*'s estimates and the coastwatcher's radio message were never mentioned. All I wanted was the truth.

## Part II: The *Barb* Postwar

What was I most proud of as commanding officer of the submarine *Barb* during World War II?

My answer is simply this. No one who ever served under my command was awarded the Purple Heart Medal for being wounded or killed, and all of us brought our *Barb* back safe and sound—ready, eager, and willing to fight again after unparalleled patrols, lauded by naval seniors and authors.

No submarine can or should claim to be the greatest, particularly if her crew is lost. There were many great patrols during the various phases of the war, some probably unknown.

Quotes from Fleet Admiral Nimitz and Admiral Lockwood such as "probably the best conducted and most unselfish patrol of the war" were echoed by foreign leaders. Clay Blair, the most comprehensive submarine author, spoke of "the most daring patrol of the entire war." Although appreciated, this praise was not our objective.

The *Barb* was never in competition with anybody but herself. We were determined on each patrol to do better than the last one. And we should have, since we had more experience as tactics, weapons, targets, and the war moved on. Not only our own experience, but that of others, motivated us.

Where did our *Barb* go after departure to Guam on 21 August? Late on the 22nd, word came to cease offensive operations. Orders were issued for her to reverse course and return to Midway. She arrived there on the 24th. Then she was ordered to return to New London, Connecticut, where she had been built. En route there she passed via Pearl Harbor and the Panama Canal. After she arrived home, Max Duncan took over as skipper and mothballed her in April 1946.

Her rest ended in 1951 when she was modernized, streamlined, and had

*Top:* Chief witnesses Lin Zai Jun and Lu Shen Kin, who witnessed the attack and results on 23 January 1945, with the author and Murray Sinks, the naval coastwatcher's son and president of Applied Technologies Company. *Bottom:* An offshore view of the walled village Huang Qi.

a snorkel added. Prettied up, the *Barb,* was given to the Italian Navy in 1953. They rechristened her *Enrique Tazzoli.* She served them faithfully until finally sold for a measly $100,000 for scrap in 1972 at the ripe old age of 30. Had we known, we would have bought her and brought her home to rest as a museum.

What happened to her crew? Most of them returned to civilian life. Those who stayed in the Navy did well and progressed in the ranks. We have had three main reunions. The first was at the launching of the nuclear submarine *Barb,* with my late wife as the proud sponsor. Her nuclear crew treated us all like royalty. The second reunion was at her commissioning, when I had the fortune of being the principal speaker. As Commander Submarine Force Pacific in 1964, I made her my flagship as soon as she put her shining nose into Pearl Harbor.

Our third reunion, in 1986, was inspiring. After a banquet, each shipmate gave an account of his life after he left the *Barb.* The first one to speak reminded us all that we had been close friends in the *Barb,* whether officers or enlisted men. Each man had been born again by the *Barb* and love. She enriched our lives and gave us our philosophy: We don't have problems, just solutions.

# Appendix A

## Sailing List in the USS *Barb,* Eighth through Twelfth War Patrols

| Name/Rank/Rate | Home State | Patrols | Citations |
|---|---|---|---|
| Allen, Fred W., EM1c, USN | Virginia | 11 | LCR |
| Arthur, John J., TM2c, USNR | New Jersey | 10-11 | LCR |
| Baughman, Daniel S., LCDR, USN | Washington, D.C. | 9 | |
| Bentley, Warren T., SC2c, USNR | Wisconsin | 8-12 | LCR |
| Bluth, Paul D., QM3c, USNR | Minnesota | 8-12 | 2 LCR |
| Bochenko, Henry S., RM2c, USN | Illinois | 8-12 | LCR |
| Bowden, Dallas G., EM1c, USN | Massachusetts | 9-11 | LCR |
| Brendle, Louis J., CFC, USN | Pennsylvania | 8 | SS |
| Broocks, William R., QM2c, USNR | Texas | 8-12 | LCR |
| Brunton, Howard A., F1c, USN | California | 11-12 | LCR |
| Burnett, Joseph H., S1c, USN | Kentucky | 12 | |
| Burnett, Michael, FCS3c, USNR | Iowa | 10-12 | LCR |
| Byers, Clarence J., TM3c, USN | Arizona | 8-9 | |
| Campbell, Fred I., MoMM1c, USN | Wyoming | 8 | |
| Chapman, Paul A., S1c, USNR | New York | 10-12 | LCR |
| Cole, David F., StM2c, USNR | Michigan | 11-12 | LCR |
| Custer, Russel A., MoMM3c, USN | Pennsylvania | 10-12 | LCR |
| Davis, Ezra A., EM1c, USN | Iowa | 8-9 | SS/LCR |
| Delameter, John T., RM2c, USNR | New York | 11-12 | LCR |
| Dittmeyer, Jack L., RM2c, USN | Florida | 8 | |
| Donnelly, William E., CPhM, USN | California | 8-11 | NMC/LCR |
| Dougherty, Charles, SC2c, USNR | New Jersey | 8-9 | |
| Duncan, Max C., LT, USN | Maryland | 9-12 | 2 SS/BS |

| | | | |
|---|---|---|---|
| Durbin, Troy M., MoMM2c, USNR | California | 8-12 | 2 LCR |
| Easton, Jay A., LT, USN | New York | 8 | SS |
| Edwards, Mason L., MoMM2c, USNR | Pennsylvania | 11-12 | LCR |
| Elliman, Russell R., Bkr3c, USN | Hawaii | 9-11 | LCR |
| Epps, James O., EM1c, USNR | Ohio | 8-12 | LCR |
| Fannin, William M., S1c, USNR | Utah | 9-11 | LCR |
| Flaherty, Thomas J., S1c, USNR | New York | 10-12 | LCR |
| Fluckey, Eugene B., Cmdr., USN | Maryland | 8-12 | CMH/4 NC |
| Foster, Alvin B., GM3c, USN | Texas | 8 | |
| Gibson, Richard H., LT(jg), USN | Washington, D.C. | 8-11 | BS |
| Gierhart, Frank W., CRM, USN | Ohio | 12 | BS |
| Greenhalgh, Irving, TM2c, USN | Connecticut | 8-9 | |
| Hansley, John C., FCS1c, USN | California | 9-11 | LCR |
| Hatfield, Billy R., EM3c, USNR | Ohio | 10-12 | SS |
| Hazelwood, Emmit, MoMM3c, USNR | Arkansas | 11-12 | LCR |
| Higgins, John H., QM1c, USNR | Ohio | 9-12 | SS |
| Hill, Charles W., LT(jg), USNR | Illinois | 12 | |
| Hill, Leonard L., F1c, USN | Iowa | 12 | |
| Hinson, Edward E., CRM, USN | Texas | 8-11 | BS |
| Hofferber, Edward H., QM2c, USN | Wyoming | 8 | |
| Hogan, John E., EM2c, USNR | California | 8 | |
| Houston, Traville, MoMM2c, USN | Florida | 8-11 | NMC |
| Hudgens, Thomas J., CMoMM, USN | Connecticut | 8-11 | BS |
| Hutchinson, Alden, MoMM1c, USN | California | 11-12 | LCR |
| Jackson, Elmer.J., Ck3c, USNR | Ohio | 8-9 | |
| Jernigan, Charles, MoMM2c, USNR | Alabama | 8 | |
| Johnson, Charles, MoMM1c, USNR | California | 8-12 | LCR |
| Jones, William L., Stm2c, USN | North Carolina | 10 | |
| Kerrigan, John C., RM2c, USN | Pennsylvania | 8-10 | LCR |
| King, Thomas M., LT(jg), USNR | Connecticut | 10-12 | BS |
| Kirk, Travis G., S1c, USNR | Texas | 9-12 | LCR |
| Klinglesmith, Edward, TM3c, USN | California | 9-12 | SS/2 LCR |
| Koester, Glenn L., EM2c, USNR | Kansas | 8-12 | LCR |
| Kosinski, Julian, MoMM3c, USN | California | 8-9 | |
| Lamuth, William C., EM2c, USNR | Minnesota | 8-9 | |
| Lander, Robert B., LCDR, USN | California | 10 | |
| Langston, Claude, MoMM3c, USN | California | 8-10 | |
| Lanier, James G., LT, USNR | Alabama | 8-10 | SS/NMC |
| Larsen, Norman, EM3c, USNR | Iowa | 9-12 | LCR |
| Laughter, Wade V., TM2c, USN | Florida | 8-11 | LCR |
| Laymon, Cecil M., PhM1c, USNR | Indiana | 12 | |
| Lego, Herman F., CY, USN | California | 8-11 | LCR |
| Lehman, John H., RT2c, USNR | Pennsylvania | 8-12 | LCR |
| Leier, Ralph G., MoMM2c, USN | Minnesota | 8 | |

| | | | |
|---|---|---|---|
| Libby, Frank W., CEM, USN | New York | 8 | |
| Lindberg, Wallace, RM2c, USNR | Pennsylvania | 8-12 | LCR |
| Lunde, Everard M., Y3c, USNR | California | 12 | |
| Maher, Timothy P., RT1c, USNR | California | 8-12 | 2 LCR |
| Malan, Stephen A., CCS, USN | Oregon | 8 | |
| Markuson, John, MoMM1c, USN | New Jersey | 8-12 | SS/2 LCR |
| Masek, William, ENS, USNR | Virginia | 12 | |
| Maxwell, Richard S., TM2c, USN | Pennsylvania | 9-12 | LCR |
| McCarthy, James A., F1c, USNR | Connecticut | 12 | |
| McCloud, H. Dean, TM3c, USN | California | 8 | |
| McDole, Frederick, EM1c, USN | Illinois | 9-12 | LCR |
| McKee, Edward L., MoMM3c, USN | Oklahoma | 9-12 | LCR |
| McNally, Raymond D., Y2c, USN | North Dakota | 9-10 | |
| McNitt, Robert W., LT, USN | Maryland | 8-9 | 2 SS/NMC |
| Miller, Donald L., TM2c, USNR | Pennsylvania | 8-12 | LCR |
| Monroe, Paul H., LT, USNR | California | 8-11 | SS/BS |
| Mulry, Raymond D., EM2c, USNR | Massachusetts | 8-12 | BS |
| Murphy, Buel M., GM2c, USN | Washington, D.C. | 8-10 | LCR |
| Newland, Lawrence, SC1c, USN | Ohio | 11-12 | SS |
| Noll, Thomas J., CMoMM, USNR | Wisconsin | 8-12 | 2 LCR |
| Novak, Emil L., F1c, USN | Wisconsin | 8-11 | LCR |
| Parker, Todd D., TM1c, USN | Oklahoma | 11-12 | LCR |
| Penna, Jesse L., MoMM2c, USN | New York | 8-12 | LCR |
| Peterson, Howard, MoMM1c, USNR | Minnesota | 8-11 | LCR |
| Petrasunas, Joseph, GM2c, USN | New Jersey | 8-12 | BS/2 LCR |
| Phillips, Robert C., SC3c, USN | Michigan | 10-12 | LCR |
| Pitts, Morris C., EM2c, USNR | Kentucky | 8-10 | |
| Post, John R., LT, USNR | Colorado | 8 | LCR |
| Powell, Henry J., TM3c, USN | California | 8-12 | LCR |
| Price, Walter W., EM3c, USNR | Pennsylvania | 9-12 | LCR |
| Ragland, Paul F., CK3c, USN | Maryland | 8-12 | LCR |
| Richard, James E., MoMM2c, USNR | Virginia | 8-12 | SS/LCR |
| Roark, Rufus W., TM2c, USNR | Texas | 8-12 | BS |
| Romaszewski, Anthony, MoMM2c | Pennsylvania | 8 | |
| Ryan, Thomas J., MoMM1c, USN | New York | 8-12 | LCR |
| Salantai, Joe A., TM2c, USNR | Illinois | 8-9 | |
| Saunders, Paul G., CGM, USN | Florida | 8-12 | 2 SS/BS/LCR |
| Savage, Russell, T., Y3c, USNR | Pennsylvania | 8 | |
| Schilke, Edwin E., RM3c, USN | Idaho | 12 | |
| Schmitt, Rudolph H., MoMM1c, USN | California | 10-12 | LCR |
| Sever, Francis N., SM2c, USNR | Iowa | 9-12 | SS |
| Shankles, Ellis P., Qm2c, USNR | Louisiana | 8-12 | LCR |
| Sheffield, Lawrence J., LT(jg) | New York | 10-12 | SS |
| Sherman, Forrest H., EM1c, USNR | California | 9-12 | LCR |

| | | | |
|---|---|---|---|
| Shoard, Sydney A., TM2c, USN | New York | 8-12 | BS |
| Simek, George L., F1c, USNR | Illinois | 10-12 | LCR |
| Simpkins, Cullen, MoMM2c, USNR | West Virginia | 8-12 | LCR |
| Singer, Robert L., PhoM1c, USN | New Jersey | 12 | |
| Smith, Gaines, TM1c, USN | Massachusetts | 8 | BS |
| Spencer, Clarence, MoMM2c, USN | Ohio | 8-12 | LCR |
| Starks, Frank E., CMoMM, USN | California | 8 | LCR |
| Stowe, Carroll D., EM1c, USN | Oregon | 8 | |
| Swearingen, Charles, EM3c, USN | California | 9-12 | LCR |
| Swinburne, Edwin R., CAPT, USN | Florida | 9 | NC |
| Tague, Robert H., S1c, USNR | Maine | 11-12 | LCR |
| Teeters, David R., LT(jg), USNR | New Jersey | 8-12 | SS/BS |
| Tomczyk, Charles A., TM1c, USNR | New Jersey | 8-9 | BS/LCR |
| Turnage, Sam M., MoMM1c, USNR | Florida | 8-10 & 12 | BS |
| Vogelei, James P., SC2c, USNR | Pennsylvania | 8 | |
| Wade, Gordon L., CEM, USN | California | 12 | LCR |
| Walker, William M., LT, USNR | South Carolina | 11-12 | NC |
| Wearsch, Norman, MoMM2c, USNR | Ohio | 8-12 | LCR |
| Weaver, Everett P., LT, USNR | Illinois | 8-10 | BS/LCRT |
| Webster, James T., LT, USNR | Tennessee | 11-12 | SS/BS |
| Wells, Robert W., TM3c, USNR | California | 8-12 | LCR |
| Wells, Stanfield, SM1c, USNR | Michigan | 8 | |
| Welsh, Wesley A., EM1c, USNR | California | 8 | |
| White, James A., TM3c, USNR | Pennsylvania | 10-12 | LCR |
| Whitehead, Gordon, MoMM2c, USN | Massachusetts | 8-11 | LCR |
| Whitt, William F., MoMM1c, USN | Virginia | 9-12 | BS |
| Wierski, Casmier C., TM2c, USN | New Jersey | 8 | |
| Wilby, Theodore S., SC3c, USNR | Pennsylvania | 12 | |
| Williams, Franklin, CMoMM, USN | Texas | 9-12 | BS |
| Williams, Owen, FCS3c, USNR | Wisconsin | 8 & 12 | LCR |
| Wilson, Randall, RM2c, USNR | Oklahoma | 8-12 | LCR |
| Wood, Arthur R., MoMM3c, USNR | Massachusetts | 12 | |
| Zamaria, Joseph, MoMM3c, USN | Ohio | 8-12 | LCR |

Abbreviations of Medals: CMH = Congressional Medal of Honor; NC = Navy Cross; SS = Silver Star; BS = Bronze Star; NMC = Navy and Marine Corps; LCR = Letter-of-Commendation Ribbon

The USS *Barb* earned one Battle Star on the European-African Middle Eastern Area Service Medal for participating in the Algeria-Morocco landings from 8 to 11 November 1942. She also earned seven battle stars on the Asiatic Pacific Area Service Medal for participating in the following operations: sixth, seventh, eighth, ninth, tenth, eleventh, and twelfth war patrols, plus participating in the assaults on the Philippine Islands (9–24 September 1944), the Luzon Operation (3–22

January 1945), and Third Fleet Operations against Japan (12–15 July 1945). The Philippine Liberation Medal was awarded for the eleventh war patrol.

The text of the Presidential Unit Citation follows:

For extraordinary heroism in action during the Eighth, Ninth, Tenth, and Eleventh War Patrols against enemy Japanese surface forces in restricted waters of the Pacific. Persistent in her search for vital targets, the USS BARB relentlessly tracked down the enemy and struck with indomitable fury despite unfavorable attack opportunity and severe countermeasures. Handled superbly, she held undeviatingly to her aggressive course and, on contacting a concentration of hostile ships in the lower reaches of a harbor, boldly penetrated the formidable screen. Riding dangerously, surfaced, in shallow water, the BARB launched her torpedoes into the enemy group to score devastating hits on the major targets, thereafter retiring at high speed on the surface in a full hour's run through uncharted, heavily mined and rock obstructed waters. Inexorable in combat, the BARB also braved the perils of a tropical typhoon to rescue fourteen British and Australian prisoners of war who had survived the torpedoing and sinking of a hostile transport ship en route from Singapore to the Japanese Empire. Determined in carrying the fight to the enemy, the BARB has achieved an illustrious record of gallantry in action, reflecting the highest credit upon her valiant officers and men and upon the United States Naval Service.

The text of the Navy Unit Commendation, requested by the commanding officer in lieu of another Presidential Unit Citation so his men had a different ribbon to wear, follows:

For outstanding heroism in action against enemy Japanese forces during her Twelfth War Patrol in the area north of Hokkaido and east of Karafuto from June 8 to August 2, 1945. Despite severe enemy countermeasures, including six aerial bombing attacks, gunfire from hostile ships and shore batteries, and relentless depth charging, the U.S.S. BARB fearlessly attacked the enemy at every opportunity. Striking with devastating force in a series of brilliantly executed torpedo and gun assaults, she sank three Japanese ships including a frigate, and sent to the bottom or destroyed in shipyards fifty small craft, totalling 11,225 tons. By skillful planning and ingenuity, she risked extremely shallow water to approach numerous enemy coastal towns and effected four highly successful rocket assaults, the first of their kind in submarine warfare, and three gun bombardments, inflicting extensive damage upon important installations. Climaxing her daring Twelfth Patrol by an audacious commando raid, she sent a party of saboteurs ashore in rubber boats to blow up an enemy rail train with one of her self-scuttling charges. This splendid record of achievement attests the BARB's readiness for combat and reflects the highest credit upon her gallant officers and men and upon the United States Naval Service.

# Appendix B

**Scoreboard of USS *Barb* attacks on Japanese shipping as determined by research, including Japanese and U.S. archives and interviewing witnesses at Namkwan Harbor**

| Patrol and Date | Ship Name | Type | Japanese Gross Tonnage |
|---|---|---|---|
| **Seventh War Patrol** | | | |
| 28 March 1944 | Fukusei Maru | Cargo | 2,219 |
| Shore bombardment Rasa Island with the *Steelhead* | | | |
| | | | |
| **Eighth War Patrol** | | | |
| 31 May 1944 | Madras Maru | Passenger-Cargo | 3,802 |
| 31 May 1944 | Koto Maru | Cargo | 1,053 |
| 11 June 1944 | Toten Maru | Cargo | 3,830 |
| 11 June 1944 | Chihaya Maru | Cargo | 2,738[a] |
| 13 June 1944 | Takashima Maru | Passenger-Cargo | 5,633 |
| Two trawlers by gunfire | | | |
| | | | |
| **Ninth War Patrol** | | | |
| 31 August 1944 | Okuni Maru | Cargo | 5,633 |
| 31 August 1944 | Rikko Maru-damaged[b] | Tanker | 9,181 |
| 31 August 1944 | Hinode Maru #20[c] | Naval Mine-sweeper XAM | 281 |
| 16 September 1944 | Azusa Maru | Tanker | 11,177 |
| 16 September 1944 | Unyo | Aircraft Carrier | 22,500[d] |
| One picket vessel by gunfire | | | |

### Tenth War Patrol

| | | | |
|---|---|---|---|
| 10 November 1944 | *Gokoku*ᵉ | Raider Cruiser XCL-11 | 10,438 |
| 12 November 1944 | *Naruo Maru*ᵉ | Cargo | 4,823 |
| 12 November 1944 | *Gyokuyo Maru*ᵉ | Cargo | 5,396 |

Three two-masted schooners by gunfire

### Eleventh War Patrol

| | | | |
|---|---|---|---|
| 1 January 1945 | *Naval Weather Ship*ᶠ | Picket | 300 |
| 8 January 1945 | *Anyo Maru* | Passenger-Cargo | 9,256 |
| 8 January 1945 | *Shinyo Maru* | Explosives-Cargo | 6,892ᵍ |
| 8 January 1945 | *Hisagawa Maru*ʰ | Cargo | 6,886 |
| 8 January 1945 | *Hikoshima Maru*ⁱ | Tanker | 2,854 |
| 8 January 1945 | *Meiho Maru*ʲ | Cargo | 2,857 |
| 8 January 1945 | *Sanyo Maru* | Cargo | 2,854 |

### Namkwan Harbor Attack

| | | | |
|---|---|---|---|
| 23 January 1945 | *Taikyo Maru* | Explosives-Cargo | 5,244ᵏ |
| 23 January 1945 | *Unknown Maru*ˡ | Cargo | 7,500 |
| 23 January 1945 | *Unknown Maru*ˡ | Cargo | 6,500 |
| 23 January 1945 | *Unknown Maru*ˡ | Cargo | 4,000 |
| 29 January 1945 | *Katsuura Maru*ᵐ | Passenger-Cargo | 1,735 |

### Twelfth War Patrol

| | | | |
|---|---|---|---|
| 3 July 1945 | *Unknown Maru*ⁿ | Cargo | 1,000 |
| 5 July 1945 | *Sapporo Maru #11* | Cargo | 2,820 |
| 10 July 1945 | *Toyu Maru*ᵒ | Cargo | 1,256 |
| 18 July 1945 | *Coast Def Vessel 112* | Frigateᵖ | 940 |

### Shore Gun Bombardments

2 July 1945: Kaihyo To Island; new naval radar station, radio station, and all buildings, boats, and supplies completely destroyed.

25 July 1945: Chiri cannery completely destroyed.

26 July 1945: Shibetoro lumbermill, sampan building yard, and all cradles—plus all new sampans—destroyed.

### Rocket Attacks

22 June 1945: Factories at Shari.

3 July 1945: Shikuka Air Base.

24 July 1945: Shiritori town and Oji paper factory.

25 July 1945: Kashiho factories.

*Vessels Destroyed by Gunfire*
3 luggers; 2 trawlers; 69 sampans

*Destroyed by Saboteur Party*
1 train with 16 cars

*Vessels Destroyed by Ramming*
1 trawler

Totals sunk in Japanese Empire waters by the *Barb* from all factual sources so far uncovered are as follows: 29⅓ ships sunk; 146,808 tons sunk.

---

*Notes*

a. Joint Army-Navy Assessment Committee (JANAC) listed cargo tonnage of 1,161 instead of stated gross tonnage.

b. Confirmed damage by ULTRA, beached, no attacks other than one hit by the *Sea Lion,* declared a marine casualty. Split.

c. Confirmed sunk by the *Barb,* no prior damage, witnessed ship break in two; a hunter-killer decoy. Commander Mine Squadron 45, Captain Suzuki, was killed.

d. ONI 208-J (used by JANAC) lists this tonnage, not 20,000.

e. Japanese confirm these sinkings at exactly the same time.

f. JANAC failed to credit wolfpack. The *Barb* boarded and sank.

g. JANAC clipped 1000 tons off in a typing mistake in their own records.

h. Damaged, on fire, *Hisagawa* beached to avoid sinking.

i. JANAC confirmed the *Barb* sank her, but denied that the *Queenfish* sank *Manju Maru,* which the Japanese now confirm. The *Barb* offered one of her sinkings to Captain Karl Hensel as a split for the wolfpack.

j. Damaged, beached to avoid sinking; abandoned. JANAC credited U.S. Army aircraft on 28 March 1945 at the exact spot.

k. Cargo tonnage (unknown gross tonnage is greater).

l. Japanese records of this attack were never returned from the United States and are missing. Together with Chinese officials, the author revisited villages adjacent to this harbor in June 1991. Three separate groups of witnesses from different areas confirmed that four ships—two of which were very large—sank and three were damaged, as originally reported by ComNavGrp China coastwatcher.

m. Japanese assessed this ship as probably struck by a mine inside the 10-fathom curve (date not given), beached, and abandoned on 2 February 1945. Her position was about five miles from the *Barb's* attack on a similar ship she hit and that disappeared off radar in shallow water during a rain squall.

n. This standard type "Sugar Able Sugar" was sunk by a homing Mark 27 torpedo. Movies were taken. JANAC omitted. The Japanese list four of these army ships missing, cause unknown. Author has her running lights. ComSubPac was given many charts stamped with the ship's name.

o. Japanese list her as sunk by submarine on 15 July. She is similar to a ship torpedoed wherein the *Barb* heard only a colliding noise, possibly a dud torpedo. There were no ships sunk by aircraft or other submarines that day.

p. JANAC records show this frigate as Type-D, 900-ton class, and credited her as Type C, 800 tons.

### Lessons Learned

1. We did not know our enemy—his history or his language—as well as he knew ours. Needed interpreters were not available. We should have had a Nisei-Japanese on each submarine and, depending on our location, an interpreter of each of the local languages.
2. Historically, when the Japanese deem war inevitable, they preempt it by destroying the enemy fleet. Note the Sino, Korean, Russian, and Allied wars. We ignored history.
3. Take a prisoner from every sinking, if possible.
4. The Japanese always sent a ship or escort back to pick up survivors after any sinking.
5. Historically, Japanese captains are trained to beach their ships when sinking appears probable.
6. The Japanese are disciplined, brave, professional warriors. As U.S. allies, they must be permitted to be an asset.

# Appendix C

## The USS *Barb*'s Final Battle Flag at the End of World War II

The final battle flag.

EXPLANATION OF SYMBOLS

*Top row:* left to right shows the number of combat decorations
1. Twenty-three Silver Star medals (the third highest combat medal).
2. Presidential Unit Citation (PUC) for war patrols 8, 9, 10, and 11. For the twelfth war patrol the commanding officer requested a Navy Unit Commendation in lieu of the second PUC offered, to give the men another ribbon to wear.
3. The Congressional Medal of Honor.
4. Six Navy Crosses (the second highest combat medal).
5. Twenty-three Bronze Star medals (the fourth highest combat medal).

Not shown are 4 Navy and Marine Corps medals, 82 Letters of Commendation with Ribbon, and a Navy Unit Commendation for the whole Crew for the twelfth patrol.

*Below these:*

Merchant ships sunk are shown on white flags with a solid red sun in the center. Those with hollow centers indicate the ships were damaged, some of which ran aground and were declared unsalvageable, then either credited to aircraft that bombed the derelicts or declared a marine casualty.

Japanese naval ships sunk are indicated by the Rising Sun flags. The large one in the center is the *Unyo* ("Falcon of the Clouds"), a 22,500-ton escort aircraft carrier carrying 48 planes in her hangars and 12 planes on the flight deck. She was larger than many of their fleet aircraft carriers.

The lower Rising Sun flag is for the Auxiliary raider cruiser *Gokoku* ("Guardian God"), XCL-11. The lower center Rising Sun flag is for the naval weather picket boarded and stripped by the *Barb,* then sunk by gunfire. The solid Rising Sun flag to the right is for the frigate *112.* The hollow Rising Sun flag above it is for a major aircraft carrier that was damaged.

The top white flag with 14 blue crosses is for 14 British and Australian prisoners of war rescued as a typhoon was breaking after their having been on flotsam for five days. Some 1,750 POWs were stranded when two prisoner ships in a convoy bound for Japan were unknowingly sunk by another wolfpack.

The Nazi flag is for a European tanker under German control sunk in the Atlantic.

The small merchant flags with numeral seven each represent seven vessels under 500 tons destroyed, such as trawlers, luggers, schooners, and oceangoing sampans, under our orders to sweep the seas clean.

The gun symbols are for significant shore bombardments on four factories, canneries, and building yards.

The rocket symbols are for ballistic rocket attacks on factories and a large air base. The *Barb* was the first and only submarine to fire rockets in wartime. One attack destroyed the largest paper mill in Japan. This pointed the way to the future of ballistic missiles. The war ended before follow-up.

The train symbol is for a 16-car train blown up by men sent ashore who placed one of our self-scuttling charges underneath the tracks. This was the sole landing of U.S. military on Japanese soil in World War II.

# Index

Citations for *Barb* personnel are first mention only; see appendix A for all patrols in which a crew member participated.

# A Note on the Author

EUGENE B. FLUCKEY graduated from the United States Naval Academy in 1935 and from Submarine School in 1938. In late 1942, after five war patrols in the Pacific, he was ordered to Annapolis for a postgraduate course in naval design engineering. He attended Prospective Commanding Officers' School in 1943 and joined the crew of the USS *Barb* for her Seventh War Patrol. In April 1944 he took command of the *Barb* and saw her through her next five war patrols, earning the Congressional Medal of Honor and four Navy Crosses, unequaled by any living American. He relinquished command of the *Barb* in August 1945.

After the war, Fluckey commanded various Navy vessels and served as Naval Attaché in Lisbon from 1950 to 1953, when he was decorated with the Medalha Militar by the Portuguese government, the only attaché to be so honored. In 1956 he was ordered to the Naval Academy, where he directed the fundraising drive to build the Navy Marine Corps Memorial Stadium. Serving in the National War College and later on the National Security Council, he was advanced to Rear Admiral in 1960. From 1966 to 1968 he directed Naval Intelligence and sat on the United States Intelligence Board. He returned to Lisbon in 1968 as the NATO Commander-in-Chief of the Iberian Atlantic Area and was bombed twice by communists while under contractor security and personally targeted.

Admiral Fluckey retired from the Navy in 1972 and settled in Portugal with his wife, Marjorie. In 1974 they took over an orphanage and ran it together until her death in 1979. In 1980 he married Margaret Wallace McAlpine, who assisted him with the orphanage until it closed in 1982. It was Margaret who insisted, after their relocation to Annapolis, that he write this book.

On 17 November 1989, the United States Navy honored Admiral Fluckey by naming the nuclear submarine Combat Systems Training Center at New London, Connecticut, after him. Fluckey Hall is the only building at the submarine base named after a living person.

UNIVERSITY OF ILLINOIS PRESS
1325 SOUTH OAK STREET
CHAMPAIGN, ILLINOIS 61820-6903
WWW.PRESS.UILLINOIS.EDU